T0201770

Cellular Processes in Segmentation

Evolutionary Cell Biology

Series Editors
Brian K. Hall
Dalhousie University, Halifax, Nova Scotia, Canada
Sally A. Moody
George Washington University, Washington DC, USA

Published Titles

Cells in Evolutionary Biology
Translating Genotypes into Phenotypes – Past, Present, Future
Edited by Brian K. Hall and Sally Moody

Deferred Development
Setting Aside Cells for Future Use in Development in Evolution
Edited by Cory Douglas Bishop and Brian K. Hall

Cellular Processes in Segmentation
Edited by Ariel D. Chipman

Cellular Processes in Segmentation

Edited by

Ariel D. Chipman

Associate Professor, Department of Ecology, Evolution, & Behavior
Silberman Institute of Life Sciences
The Hebrew University
Jerusalem, Israel

CRC Press
Taylor & Francis Group
Boca Raton London New York

CRC Press is an imprint of the
Taylor & Francis Group, an **informa** business

Cover caption:Cellular level processes are at the base of all segmentation mechanisms. Cells interact to generate repeated structures in the embryo, which over developmental time lead to the segmented adult structure. (Illustration by Netta Kasher)

CRC Press
Taylor & Francis Group
6000 Broken Sound Parkway NW, Suite 300
Boca Raton, FL 33487-2742

First issued in paperback 2021

ISBN 13: 978-1-03-224235-4 (pbk)
ISBN 13: 978-1-138-38991-5 (hbk)

DOI: 10.1201/9780429423604

Visit the Taylor & Francis Web site at
http://www.taylorandfrancis.com

and the CRC Press Web site at
http://www.crcpress.com

Contents

SECTION I   The Diversity of Segmentation

SECTION II   Cellular Mechanisms of Segmentation

SECTION III   Beyond Segmentation

Series Preface (Evolutionary Cell Biology)

In recent decades, the central and integrating role of evolution in all of biology was reinforced as the principles of evolutionary biology were integrated into other biological disciplines, such as developmental biology, ecology, and genetics. Major new fields emerged, chief among which are Evolutionary Developmental Biology (or Evo-Devo) and Ecological Developmental Biology (or Eco-Devo).

Evo-Devo, inspired by the integration of knowledge of change over single life spans (ontogenetic history) and change over evolutionary time (phylogenetic history) produced a unification of developmental and evolutionary biology that is generating many unanticipated synergies. Molecular biologists routinely employ computational and conceptual tools generated by developmental biologists (who study and compare the development of individuals) and by systematists (who study the evolution of life). Evolutionary biologists routinely use detailed analysis of molecules in experimental systems and in the systematic comparison of organisms. These integrations have shifted paradigms and answered many questions once thought intractable. Although slower to embrace evolution, physiology is increasingly being pursued in an evolutionary context. So too, is cell biology

Cell biology is a rich field in biology with a long history. Technology and instrumentation have provided cell biologists the opportunity to make ever more detailed observations of the structure of cells and the processes that occur within and between cells of similar and dissimilar types. In recent years, cell biologists have increasingly asked questions whose answers require insights from evolutionary history. As just one example: how many cell types are there and how did these different cell types evolve? Integrating evolutionary and cellular biology has the potential to generate new theories of cellular function and to create a new field, which we term *"Evolutionary Cell Biology."*

A major impetus in the development of modern Evo-Devo was a comparison of the evolutionary behavior of cells, evidenced in Stephen J. Gould's 1979 proposal of changes in the timing of the activity of cells in development (heterochrony) as a major force in evolutionary change and in Brian Hall's 1984 elaboration of the relatively small number of mechanisms used by cells in development and in evolution. Given this conceptual basis and the advances in genetic analysis and visualization of cells and their organelles, cell biology is poised to be transformed by embracing the approaches of Evo-Devo as a means of organizing and explaining diverse empirical observations and testing fundamental hypotheses about the cellular basis of life. Importantly, cells provide the link between the genotype and the phenotype, both during development and in evolution. No books that capture this cell focus exist.

Hence the proposal for a series of books under the general theme of "Evolutionary Cell Biology" (ECB) to document, demonstrate and establish a long-sought level in evolutionary biology, viz., the central role played by cellular mechanisms in translating genotypes into phenotypes in all forms of life.

Brian K. Hall
Sally A. Moody

Preface

The evolution of the process by which segmented body plans are formed during development has been a central theme in evolutionary developmental biology since the reemergence of the discipline in the late 1990s. Understanding of the *Drosophila* segmentation cascade, followed by the realization that many developmental genes are conserved among species, made the evolution of the segmentation process an obvious and accessible area of study. At about the same time as the segmentation process was being studied in additional arthropod species beyond *Drosophila*, it became clear that the different phyla with segmented body plans are not closely related, making the question of how segmentation evolved even more intriguing for evolutionary developmental biologists.

In the ensuing decades, we have gained a better understanding of how segments develop in embryos of different taxa, and how this segmentation process varies over phylogeny. In recent years, researchers in the field have come to appreciate the central role of cellular-level processes in segmentation. It is this appreciation that provides the underlying theme of the current volume. Segments are formed by the arrangement of cells into reiterated structures. What are the cells involved in the process, what drives their differentiation and arrangement into segmental structures, and what are the molecular players involved in the process? These are the questions this volume attempts to answer. More important, to what extent are the answers to these questions similar in different segmented taxa?

This volume is divided into three sections. The first section includes two general overview chapters: an historical overview of the changing ideas about segmentation and theories about its evolution (Chapter 1), and a survey of the different modes of segmentation found in animals (Chapter 2). The second section contains detailed reviews of the segmentation process in the three main segmented phyla (Arthropoda in Chapter 3, Annelida in Chapter 4, and Chordata in Chapter 5). Each of these phyla has a general chapter devoted to the cellular aspects of segmentation in that phylum. Two additional chapters discuss special cases of unusual segmentation processes in malacostracan crustaceans (Chapter 6) and in clitellate annelids (Chapter 7). The final chapter of this section discusses the advances in live imaging and its contribution to studying segmentation (Chapter 8). The third section goes beyond the traditional discussion of segmentation with one chapter devoted to repeated structures in taxa that are not considered to be segmented (Chapter 9), and one chapter devoted to regeneration and its relationship to segmentation processes (Chapter 10).

I want to end by thanking Brian Hall and Sally Moody for initiating this book series and inviting me to put together the current volume; Chuck Crumly from Taylor & Francis Group for editorial support; the authors of the chapters; and last but not least Netta Kasher, whose excellent illustrations enhance many of the chapters and grace the cover of this volume.

Ariel D. Chipman

Editor

Dr. Ariel D. Chipman is Associate Professor in the Department of Ecology, Evolution & Behavior of the Silberman Institute of Life Sciences at The Hebrew University, in Jerusalem. He is the author or co-author of dozens of peer-reviewed scientific journal articles. His research focuses upon: (1) The evolution of developmental processes. Using comparative embryology as a tool for understanding evolutionary processes. (2) Early stages in patterning the arthropod embryo and the evolution of the segmented body plan. (3) Evolution of the arthropod head and the processes differentiating head from trunk. (4) Genomics of novel model systems.

Contributors

Ariel D. Chipman
The Department of Ecology,
   Evolution & Behavior
The Hebrew University of Jerusalem
Jerusalem, Israel

Benjamin Martin
Department of Biochemistry and
   Cell Biology
Stony Brook University
Stony Brook, New York

Lisa M. Nagy
Molecular and Cellular Biology
   Department
University of Arizona
Tucson, Arizona

Andres F. Sarrazin
Instituto de Química, Facultad
   de Ciencias
Pontificia Universidad Católica de
   Valparaíso, Chile

Gerhard Scholtz
Institut für Biologie
Vergleichende Zoologie
Berlin, Germany

Bruno C. Vellutini
Max Planck Institute of Molecular Cell
   Biology and Genetics
Dresden, Germany

David A. Weisblat
Department of Molecular & Cell
   Biology
University of California
Berkeley, California

Christopher J. Winchell
Department of Molecular & Cell
   Biology
University of California
Berkeley, California

Terri A. Williams
Department of Biology
Trinity College
Hartford, Connecticut

Eduardo E. Zattara
Instituto de Investigaciones en
   Biodiversidad y Medio Ambiente
Consejo Nacional de Investigaciones
   Científicas y Técnicas
Universidad Nacional del Comahue
Bariloche, Rio Negro, Argentina

Section I

The Diversity of Segmentation

1 Segmentation
A Zoological Concept of Seriality

Gerhard Scholtz

CONTENTS

1.1   SEGMENTS AND SEGMENTATION

The physical organization of arthropods, annelids, and chordates is characterized by serially arranged body sections along the longitudinal axis that can be referred to as segments (Figure 1.1, top).* In zoology, segments are generally understood to be integral units that are characterized with respect to structure, function, ontogeny, and evolution. According to this view, segments are seen as fundamentally different from other repetitive structures along the body axis (see later). Segmentation is thus considered constitutive for arthropods, annelids, and chordates, distinguishing these three groups from the numerous other unsegmented animals such as cnidarians, mollusks, flatworms, roundworms, and echinoderms, although these too display some serially arranged external and internal structures (Figure 1.1, bottom; also see Chapter 8).

---

* Segments are often referred to in zoological literature as metameres. Some authors, however, use the term "metamere" to refer only to the subdivision of the coelomic cavities, distinguishing it from the term "segment" (Remane 1950). Accordingly, segmentation is considered a specific case of metamerism (see Schmidt-Rhaesa 2007, 48).

FIGURE 1.1 Examples of segmented (top row) and unsegmented (bottom row) bilaterians. A. Myriapod (Arthropoda), B. marine bristle worm (Annelida), C. fish (Chordata), D. polyclade flatworm (Platyhelminthes), E. chiton (Mollusca), F. kinorhynch (Cycloneuralia). Chiton and kinorhynch exhibit a serial body arrangement that, however, is not interpreted as segmentation.

Depending on the animal group, segments are structurally characterized by the joint occurrence of a set of serially aligned internal and external elements such as mesodermal body cavities (coeloms), sexual organs, nephridia, nerve ganglia, muscular systems, extremities, and external rings (Scholtz 2002). In particular the serially arranged coeloms play a crucial role for this concept of segmentation. The comprehensive presentation of an elaborated coelom theory by Oskar and Richard Hertwig (1881) had a tremendous influence on the zoological thinking of the time, and the presence of a coelom was interpreted and used as the essential morphological, systematic, and evolutionary property. The coelom was regarded as identifying a more complex body organization and a higher evolutionary development. Accordingly, the presence of a serial coelom was for a long time the decisive and essential criterion in distinguishing between "true" segments and other serial structures (called pseudo-metameres) (for a current view of the coelom, see Rieger and Purschke 2005).

Ontogenetically, segments are differentiated from non-segments through their sequential formation from a posterior budding zone (see Scholtz 2002, 2010). The serially arranged sections of a tapeworm (proglottids), which also depict a series of repeated structures, would not be segments according to this definition, since they are formed from an anterior growth zone behind the head (see Scholtz 2010). In today's age of molecular biology, the explanation of morphogenetic phenomena in development has shifted to the level of gene expression and gene regulation. Accordingly, the genes involved in forming the segments, especially the expression of segment polarity genes, are viewed as the key criterion for distinguishing between segments and other serial structures (e.g., Tautz 2004; Hannibal and Patel 2013).

The significance attributed to segmentation comes from a variety of sources. For one thing, aesthetic reasons play a role here. Serial structures are formed by figures translated onto themselves through shifting, thereby representing a special form of symmetry. This regularity, referred to as "translational symmetry," has since antiquity formed a key aesthetic criterion, evidently having an appeal of its own, as confirmed by numerous examples in art and architecture (Figure 1.2) (Schramke 2017). The frequent use of anatomical terms such as ribs, legs, wings, or bodies in architecture even allows one to speculate that in addition to mathematical proportions,

FIGURE 1.2 Aesthetics of seriality in nature and culture. Whale skeleton in a gothic church building (German Oceanographic Museum, Stralsund).

knowledge of human and animal anatomies can be seen as a point of departure for forms of symmetry in artifacts. This certainly also applies for seriality in plants and animals, at least when they are not too long and comprise too many repeated elements. Möbius (1908, 36) used the examples of earthworms and millipedes, and wrote in his *Ästhetik der Tierwelt* (*Aesthetics of the Animal World*): "Most (symmetrical animals) are comprised of sections, metameres, or segments in the direction of their longitudinal axis, from anterior to posterior. A large number of metameres is not pleasing. You see nothing new and become fatigued and bored." Nevertheless, the aesthetic perspective on seriality also directly refers to its function, since the repetition of structural elements as a construction principle includes technical and economic aspects in addition to all questions of proportion and symmetry (see Riedl 1975). Finally, the observed structural repetition in segmentation represents an intellectual challenge. This concerns in particular the questions as to the evolutionary emergence and significance of segments. According to a popular view, segmentation forms a key characteristic that enabled and advanced the great evolutionary success of arthropods, annelids, and chordates. Thus, the arthropods—with more than one million species—are by far the most species-rich and biodiverse animal group; and the vertebrates include the human being, a species capable of reflecting on such problems.

The scientific literature on segments and segmentation is correspondingly vast (e.g., see reviews by Budd 2001; Scholtz 2002; Minelli and Fusco 2004; Tautz 2004; Seaver 2003; Deutsch 2004; Chipman 2008; Hannibal and Patel 2013; Altenburger 2016). From the problem of defining a segment to the ontogenetic and evolutionary emergence of segments, and finally to the role of segmentation for the evolutionary success of individual groups of animals, all aspects are being discussed, sometimes in a very heated, controversial manner. There is a particularly intense debate whether the segments of annelids, arthropods, and chordates are homologous and, therefore, if their last common ancestor was already segmented (e.g., Balavoine and Adoutte 2003).

This chapter discusses whether segmentation and segments are biologically meaningful concepts in order to understand the ontogenetic and evolutionary genesis and the modifications of repeated structures along the body axis, or if the term "segment" should instead be deconstructed and replaced by an emphasis on the concept of series and seriality of individual structures and processes.

1.2 SEGMENTS AND SERIES

The term "series" is used in numerous contexts and with very different meanings (Toepfer 2017). If "series" is used to refer to the repetition of similar yet somewhat different structures or units along a temporal or spatial axis, then segments in animals are prime examples of series. Segmentation in this sense is a relational concept. Segments never appear in isolation; there is no such thing as a body comprised of just one segment. Thus segments only assume the quality of segments when occurring in serial repetition. This reveals a difference to the use of the term in other contexts. A circular segment or arc, for example, is simply cut out of a whole and can exist singularly, without being dependent on other segments.

Because animal segments are formed through a set of serially concordant internal and external substructures that occur together, they represent multiple series that are spatially and temporally interrelated. An implicit basic assumption of this concept is that originally, all segments of a body are organized in the same way. Over the course of evolution, this originally homonomous pattern undergoes a diversification, and heteronomous segmentation results. The body is subdivided in this way into functional and morphologically distinct units (tagmata) such as head, thorax, and abdomen (Figure 1.3). Segments thus form a theme with variations and are viewed as elements or modules that can be consolidated into larger units, which in turn can be interpreted as modular (Toepfer 2017).

Animal segments are therefore assigned a special quality that essentially distinguishes them from other serial structures. As always, in biology such a comparison

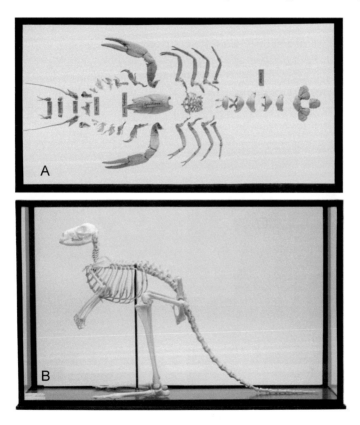

FIGURE 1.3 Diversification of segments and formation of spatial-functional units along the body axis. A. Exemplified by a crayfish as a representative of the arthropods (exploded exoskeleton, dorsal view). In the crayfish, the head and the thorax are fused to an anterior tagma called cephalothorax (dorsally, the segmentation is concealed due to a fusion; segmentation is only visible at the ventral side indicated by the series of appendages) and a posterior tagma, the pleon with obvious segmentation. B. A kangaroo as an example of a vertebrate (skeleton, lateral view) showing the head (skull), the neck, the thorax (indicated by the rib cage), the abdomen, and the tail. (Photographs: Eberle and Eisfeld, © Humboldt-Universität zu Berlin.)

and the resulting categorization concern two interrelated aspects. For one thing, this is a typological, classificatory problem: What properties define a segment and distinguish it from other serial structures? Is that which is defined as a segment always the same?

For another thing, segments do not only make up a class of structures; they are also a product of the historical, genealogical process of evolution. Consequently, the ensuing questions assume a different form: What precursor structures led to the formation of segments? How often and where in the genealogy of the animals did segmentation emerge? From this perspective, segments and other serial structures can be understood as steps in a transformation process. The boundary between segments and non-segments becomes fluid.

1.3   ONTOGENY OF SERIALITY AND SEGMENTATION

Generally, we can distinguish two different modes of ontogenetic segment formation, which can be described as segmentation by addition and by subdivision. In the first case, the segments are formed through a successive addition of segment anlagen by means of a posterior extension of the embryo or the larva (Figure 1.4). In this process, two morphogenetic processes interact closely. First, cells must be

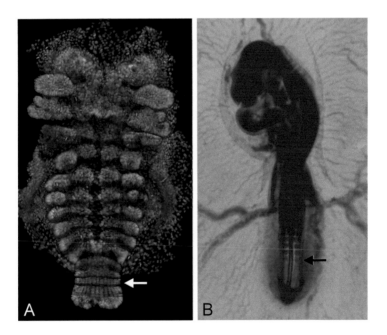

FIGURE 1.4   Sequential addition of segments during the embryogenesis of A. a crayfish and B. a chicken (modified after Scholtz 2017). The germ band of the crayfish can be seen in a ventral view; that of the chicken in a dorsal view. The anterior end with the head anlage is at the top of each picture. The yolk has been removed in both embryos and the cells have been dyed. The anterior segments are further developed than those more posterior. At the posterior end new segments are being added (arrows).

proliferated that supply the material for the segmentation process from a posterior budding zone and elongate the germ band (Scholtz and Wolff 2013; Auman et al. 2017; see Chapter 6). Second, the differentiation into segments takes place sequentially from anterior to posterior (Williams et al. 2012; Scholtz and Wolff 2013). This means that the anterior segments are formed earlier than the posterior ones, both molecularly and morphologically (Figure 1.4).

In the other mode the segments form by means of a subdivision of the already elongated embryo, and the actual segmentation process takes place almost simultaneously (Figure 1.5). This process has been examined in most detail in *Drosophila melanogaster,* the common fruit fly. In this case, the embryo is gradually divided into smaller and smaller compartments by means of a cascade of expressions of different gene classes (Gilbert 2003). After the maternal genes have determined the longitudinal axis and the terminal regions through gradients, the gap genes

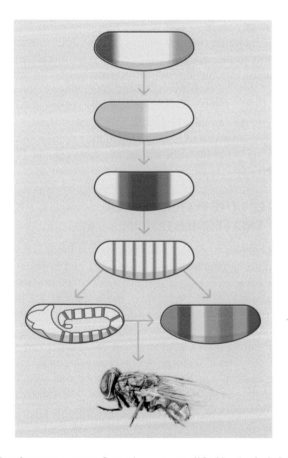

FIGURE 1.5 Synchronous segment formation as exemplified by the fruit fly embryo (modified after Scholtz 2017). The earliest stage of embryonic development is at the top. As the embryo develops, it becomes more and more subdivided by means of different gene classes (see text). The actual segmentation takes place in parallel to the diversification of the body regions. Here the head anlage is always facing left (Bottom drawing by Fabian Scholtz).

divide the body into an anterior, middle, and posterior region. These activate the pair-rule genes, which each mark every second of the future segments. Finally, the expression of the segment polarity genes leads to the determination and differentiation of the segmental boundaries. The last two gene groups are each expressed in transverse stripes in the germ bands. Parallel to this, the segments along the longitudinal body axis are specified differently through the Hox genes (Gilbert 2003).

Sequential segment formation can be found in most annelids, arthropods, and chordates. As compared with synchronous segmentation, this is viewed as the evolutionarily more ancestral mode. Some of the genes identified in the segmentation of *Drosophila* display similar functions and patterns of expression, also in organisms with sequential segmentation. This pertains in particular for segment polarity genes and the homeotic genes. In many cases, mixed forms occur in which some of the anterior segments appear synchronously and the more posterior ones are then formed sequentially (Scholtz and Wolff 2013). Regarding the details of these processes, the three major segmented animal groups differ substantially from one another. This pertains to the gene, cell, germ layer, and morphogenesis levels. For example, segmentation in annelids and arthropods begins on what will become the ventral side and then extends to the dorsal side, whereas the segmentation in chordates begins dorsally and later proceeds ventrally (Figure 1.4). Furthermore, in malacostracan crustaceans and in clitellate annelids specialized stem cells and stereotyped cell-lineages are involved in the segmentation process, which do not occur in other arthropods, in other annelids nor in chordates (Dohle 1999; Dohle et al. 2004; see Chapters 6 and 7).

1.4 THEORIES ON THE EVOLUTION OF SERIALITY AND SEGMENTATION

Since the late 19th century a series of hypotheses and theories have been put forward on the evolutionary development of segments and segmentation. It is possible to distinguish between concepts that are more function-based and those that are more structure-based. At the same time, these theories are influenced by the respective view of the phylogenetic relationships of the major groups in the animal kingdom.

Function-based theories have a teleological component and view the reason for a segmental subdivision of a wormlike body as lying in adaptive or construction-dependent necessities. For example, this can be the need for an improved blood supply in connection with an increase in size. The subdivision of an originally uniform coelom space is seen in connection with the emergence of serially arranged blood vessels (Westheide 1997). Alternatively, the possibility of improving the mobility of a worm-shaped body is interpreted as the driving force for segmentation to take place (Korschelt and Heider 1890). The external rings of a hard exoskeleton are considered a functional adaptation (Hatschek 1888–1891). Other authors, on the other hand, emphasize the hydraulic advantages of serially arranged, paired, fluid-filled cavities for animals that burrow (Clark 1964) or exhibit undulatory locomotion (Gutmann 1972). In particular, the "hydroskeleton theory" of Wolfgang Friedrich Gutmann (1935–1997) attempts to explain segmentation and its variations in detail, based on biotechnologically functional plausibility considerations (Gutmann 1972). Gutmann develops hypothetical intermediate forms (Figure 1.6) that are construction-dependent and follow a principle of economy.

Structure-oriented theories of evolutionary segmentation begin with the historical question as to the original structures and the transformation processes that change them. As a rule they follow two patterns, which can be summarized by the conceptual pair: multiplication versus subdivision. The multiplication concepts assume that the parts of an originally uniform, short body are copied and then repeated in series like links in a chain. This can be vegetative budding when the multiplied bodies detach, as occurs in a series of flatworms and annelids, but also in the budding of jellyfish from polyps (Figure 1.7). According to the theory, the individuals generated in this way do not separate; instead they lose their autonomy in the course of evolution and are transformed into segments. This idea, referred to as the "corm or fission theory," was supported, varying in details, by many influential 19th-century zoologists including Ernst Haeckel (1834–1919), Carl Gegenbaur (1826–1903), and Berthold Hatschek (1854–1941) (Haeckel 1866; Gegenbaur 1870; Hatschek 1878, 1888–1891). This view was recently revived through mathematic modeling of processes of pattern formation by the theoretical biologist Hans Meinhardt (1938–2016) (Meinhardt 2015) (Figure 1.7). Similarities between the corm theory and the pattern of sequential ontogenetic segment formation is evidently not a coincidence, but instead is marked by ideas of a parallelism of ontogenetic and phylogenetic processes as formulated by Haeckel's theory of recapitulation (Haeckel 1866).

The theories based on subdivision as the mechanism for explaining segmentation view a wormlike form that already has some repeated structures along the body axis as the point of departure for the evolution of segments (Figure 1.8). These segments are ultimately formed in the course of evolution by integrating other repeating elements. The most elaborate theory along these lines is the "trophocoel theory," which was developed by Arnold Lang (1855–1914) on the basis of the

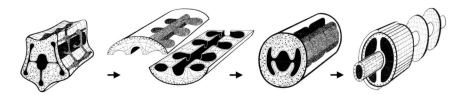

FIGURE 1.6 The "hydroskeleton theory" according to Gutmann (modified after Scholtz 2017). Sequence from left to right. From a gelatinous wormlike ancestor with a central intestine and small lateral intestinal diverticula, an organism with hydrostatic pressure opposing the external muscular system is formed by constricting the serial, fluid-filled gastric pouches. These diverticula ultimately detach from the intestine and form paired serial coelom cavities (secondary body cavities). This hydroskeleton serves to improve undulatory movement.

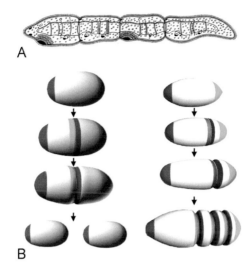

A

B

FIGURE 1.7 The "corm theory." A. A flatworm (*Microstomum lineare*) that generates complete new worms through budding, which form a chain (modified after Scholtz 2017). B. The modern version of the "corm theory" according to Meinhardt (modified after Meinhardt 2015, with permission from Elsevier). Left: Schematic depiction of budding, as illustrated in the top sketch. The colored stripes represent the expression of the genes that determine the anteroposterior axis of the animal. Right: Depiction of the hypothesis as to how segmentation might have evolved from this mechanism. This occurs through modification of the gene (suggested by the switch from violet to red) with the budded sections remaining together.

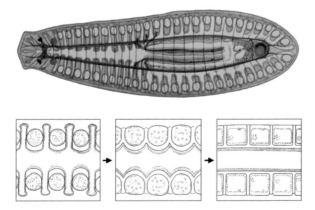

FIGURE 1.8 Arnold Lang's "trophocoel theory" (modified after Scholtz 2017). Top: Sketch of a flatworm (*Gunda segmentata*), with serial organs (nervous system [red], blind-ended diverticula [light gray], and gonads [grainy, light, oval structures]), the head is facing to the left. Bottom: A hypothetical sequence (left to right) of the evolutionary development of segmented coelom cavities from serial sexual organs. The longitudinal axis of the body is oriented horizontally. Between every two rounded gonads is in each case a short blind-ended diverticulum. These serial diverticula withdraw in the course of evolution, the gonads grow, and finally a series of large, adjacent coelom spaces forms to the left and right of the now-smooth intestines. The vascular system runs between the coelom spaces.

body organization of free-living (nonparasitic) flatworms (Lang 1881, 1903). These animals have no external rings, but some internal organs indicate repeated sections with a corresponding arrangement and rhythm. This pertains in particular to the central nervous system with numerous interconnections, the series of repeated sexual organs along the longitudinal body axis, and the numerous lateral intestinal diverticula (Figure 1.8).

Some theories combine the aspects of subdivision and duplication in that they assume an initial small number of metameres (segments), which then are supplemented and multiplied through replication. One example of this is the "enterocoel theory" (Figure 1.9) proposed by several authors in the 19th century (e.g., Sedgwick 1884; Salvini-Plawen 1998). Adolf Remane (1898–1976) took up these older ideas in the mid-20th century, developing and refining them further (Remane 1950). He assumed there are initially three serially arranged coelom spaces (the anterior one unpaired, the two posterior ones paired), which are derived from the intestinal diverticula (gastric pouches) as they occur in radially symmetrical jellyfish. As bilaterally symmetrical animals evolved, these gastric pouches transformed into coeloms that functioned as a hydroskeleton. This body structure, called archimetamerism, is first transformed into a deutometamere by reducing the two anterior metameres through minor replication of the third section. In a third step this leads to tritometamerism with the actual segments in the trunk of annelids and arthropods, and possibly also chordates.

All these scenarios are necessarily speculative. Also, they are sometimes based on different basic assumptions with respect to the phylogenetic relationships of the major animal groups and the original bilaterian body organization. There are nevertheless clear-cut differences in how well they correspond to existing organismic forms of organization and to today's view of phylogenetic relationships. Furthermore, the quality of the homology assumptions differ, since not all of the compared structures exhibit sufficient complexity to make a multiple independent evolution improbable. In his hydroskeleton theory, Gutmann (1972), with his notion of construction-dependent transformation series, distanced himself too far from actually existing organismic organizational forms. Plausibility scenarios alone obviously do not describe the course of evolution.

As tempting as the corm theory might be due to its proximity to observable ontogenetic processes, a key weakness of it lies in the fact that in no case in the process of segmentation can the anlage of replicated complete organisms be verified. The inner germ layer that forms the digestive tract also does not show a serial ontogenetic subdivision. In addition, it is obvious that in all groups in which such chain formations appear, these depict evolutionarily secondary mechanisms of reproduction. The enterocoel theory, too, suffers under a large share of untestable premises and speculations. It is simply assumed that the last common ancestor of the bilaterians had a serial coelom, which still is a debated issue. In many animal groups, unverifiable structures such as protometameres or other coelom spaces must have been lost over the course of evolution, although there is no evidence whatsoever that they were ever present in the first place. The homology between such unspecific structures as the gastric pouches of cnidarians and coelomic

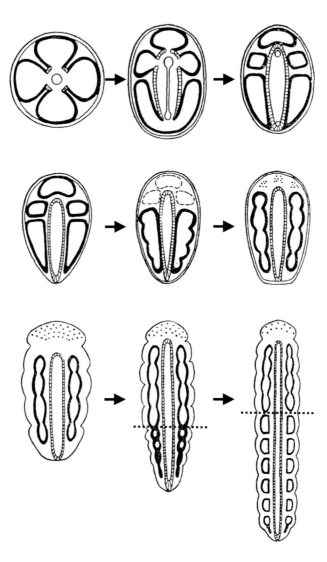

FIGURE 1.9 The "enterocoel theory" according to Remane (modified after Scholtz 2017). Top row: From left to right this top view shows the evolution of bilateral symmetry from a radially symmetrical jellyfish-like organism. The intestines show four diverticula, the mouth is shown as a small circle (left). As one axis is extended simultaneous to the elongation of the mouth, precursors to the coelom spaces (center) form along the longitudinal axis. Finally, the coelom spaces detach and form the three protometameres (the anterior one unpaired, the two posterior ones paired). The original mouth opening divides into mouth and anus, with a continuous intestine. Middle row: This depicts the stepwise loss of the protometameres (dashed lines, point clouds) and the formation of deutometameres (secondary subdivisions of the posterior coelom cavities). Bottom row: This shows the formation of the tritometameres by means of a budding zone posterior to the deutometameres (dotted line). The segmentation also becomes externally visible.

cavities of some bilaterally symmetrical animals is not well founded, and the distinction between deuto- and tritometameres lacks any foundation (Dohle 1979). The trophocoel theory, as well, is today no longer tenable in the form presented by Lang. There was certainly no linear transformation from a type of flatworm with numerous derived traits to an annelid or other segmented animal. Another weakness is the implicit assumption that the evolution of the coelom was associated with segmentation. All animals with nonserially arranged coeloms would then have experienced a reduction or in some other way would have evolved corresponding structures independently. In a certain way, the subdivision theories on the evolution of segments shift the problem to the level of the development of the original serial structure. In contrast to the multiplication theories, however, this initially has to do with individual serial structures or organs that occur recurrently and not integrated segments that have already been put together from a set of substructures.

Nevertheless, the theory of the origin of segmentation through the stepwise integration of serial structures is most compatible with the different organizational forms on the basis of the major bilaterally symmetrical animal groups. This concept can be easily reconciled with today's perspective on the evolution of animals, and thus it has also been recently advocated (Budd 2001; Scholtz 2003). Supporting the view of the origin of segments from an initially small number of repeated structures is the fact that a bilaterally symmetrical, wormlike body is not at all conceivable without serial elements. This is already true at the cell level, but also the muscular system and, even more so, the nervous system requires an inherent serial structure with precise positioning in order to interact specifically along a body axis (see Deutsch and Le Guyader 1998). It seems understandable that based on this, other—and, depending on the animal group, different—organ systems would follow this series (see also Chipman 2019). Ultimately, these organs must also be controlled by the nervous system. Also supporting these theories, there are in fact structures that evolved later and which are additionally integrated into the serial rhythm of already existing repeated structures. A good example of this are the vertebrae of vertebrates, which first appear within the segmented chordates, and whose serial arrangement is based on serial coeloms, which evolved far earlier. The same is true regarding the external rings and serial nerve ganglia of arthropods. These structures evidently first appeared in the line to the arthropods, as they are absent in the closely related *Onychophora* (velvet worms), although like arthropods these have serially arranged extremities and other serial structures (Martin et al. 2017).

Such a view of the evolution of segmentation reveals a number of implications. Similar to the process of ontogenesis, in the evolution of segmentation two processes can be distinguished. One is longitudinal growth and the other is the sequential subdivision of the body in serial structures. Longitudinal growth is certainly the evolutionarily older of the two phenomena, since also an unsegmented worm must form its longitudinal body shape from a more spherical egg through directed growth. Furthermore, the aforementioned basic premise of initially homonomous segmentation is disputable, as it obviously conflicts with the at least partial heteronomy that was always observed. If different serial structures are integrated stepwise, then initial differences in the serial arrangement can also lead to an originally heteronomous segmentation.

1.5   HOW OFTEN DID SEGMENTS EVOLVE?

Not only are the hypotheses on the evolutionary origin of segments subjects of heated debate, but also are the questions as to where and how often segments evolved. Depending on the perspective, single, twofold, or threefold independent evolutionary formation has been postulated (Balavoine and Adoutte 2003; Seaver 2003; Vellutini and Hejnol 2016). In most cases it is assumed that segmentation in chordates, annelids, and arthropods is homologous within each of the respective groups. The different hypotheses are based on varying answers to the question whether segments of these three major groups are homologous to each other and on their differing tolerance with respect to a repeated convergent loss of segmentation. Moreover, they reflect often unconscious or unclear concepts regarding segments.

For several years the hypothesis has been favored that segmentation evolved once and that the stem species of Bilateria (Urbilateria) was already segmented (see Balavoine and Adoutte 2003). Yet, recent analyses of bilaterian phylogeny led to the view that the last common ancestor of annelids, arthropods, and chordates was not Urbilateria but the stem species of the large bilaterian sub-taxon Nephrozoa (Cannon et al. 2016) (Figure 1.10). The putative sister group of Nephrozoa, the Xenacoelomorpha do not show any signs of segmentation. Hence, the discussion has been shifted to the question of whether segmentation evolved in the stem lineage of the Nephrozoa (Treffkorn et al. 2018). Due to similarities in the expression of segmentation genes, some authors view the segments of annelids, arthropods, and chordates as homologous (Carroll et al. 2001). According to this view, the stem species of the Nephrozoa was already segmented. In this case, we must assume a multiple evolutionary reduction of segmentation in the other animals, that is, the close relatives of the annelids, the arthropods, and the chordates (Figure 1.10). The same also applies if we assume a twofold development of segmentation, once in the common line of the arthropods and annelids and, independent of that, in the line leading to the chordates. It would thus be most parsimonious to assume a threefold independent origin of segmentation (Figure 1.10). This view corresponds to the conclusions drawn from the minor structural correspondence between the serial units characterized as segments in the three groups.

Also, one must ask if the term "segment" is at all appropriate for the observed phenomena. Does the previously described concept of segmentation take the given conditions into account and does it even allow us to ask the proper questions about the origins of segmentation?

The contradictions between the theories of the evolutionary formation of segments already provide an indication that the concept of segmentation might itself be problematic. This assumption is even reinforced by the following problems.

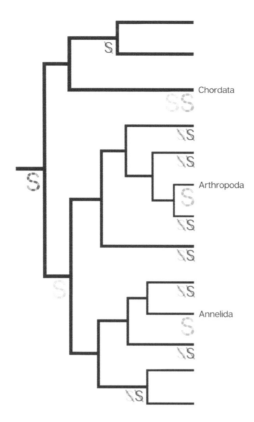

FIGURE 1.10 Greatly simplified diagram of one of several hypotheses on the phylogenetic relationships of Nephrozoa. Only the lines of the segmented animal groups are named. An S stands for the emergence of segmentation, a crossed-out S for its loss. Once the segmentation has already emerged at the base of the nephrozoans (red S), this implies an eightfold loss of segmentation. If assuming a dual evolution of segmentation in chordates and the last common ancestor of annelids and arthropods (blue S), the segmentation still undergoes a sevenfold loss. Segmentation evolving independently in the three lines of annelids, arthropods, and chordates (green S) would be the most parsimonious solution, since it assumes only three evolutionary steps.

1.6 PROBLEMS WITH STRUCTURAL DEFINITIONS OF A SEGMENT

If segments are defined according to a set of substructures that appear together, this immediately leads to the problem that only in the rarest cases are all these criteria are truly satisfied. Despite of the importance of serial coelomic cavities for the definition of segments (see earlier), adult arthropods are considered as segmented animals, although they lack these.* Yet, the problem with a structural definition of

---

* Traditionally, this absence was considered as being of minor importance, since serial coelom anlagen were described during early development. Yet, the homology of the ontogenetic anlage of coeloms is increasingly viewed critically (Koch et al. 2014). With that, the classical essential quality of segments is also lost.

segments pertains not only when different species are compared, but also within organisms, in particular when body sections would not be viewed as segments although they belong to the same series. If, for example, both nerve ganglia and extremities are absent, as is the case in the abdomen of many crustaceans, are the body sections still considered segments just based on the external rings? Other cases raise the same question: wasps lack external rings in the thorax and have no legs in the abdomen, and in many annelids, the genital organs are limited to a few specialized body sections.

1.7 PROBLEMS WITH ONTOGENETIC DEFINITIONS OF A SEGMENT

If formation through a posterior budding zone distinguishes segments from other serial structures, then the body sections that are not formed in this way are not segments. Classically, this also applies for the anteriormost (prostomium in annelids, acron in arthropods) and posteriormost (pygidium in annelids, telson in arthropods) terminal body sections of segmented animals, as they are already present prior to the differentiation of the budding zone, which then emerges between them (see Westheide and Rieger 2007). As previously discussed, however, segments are not always formed through a posterior budding zone. Frequently, several of the anterior segments are already differentiated before this budding zone becomes active. In animals such as *Drosophila*, all segments emerge almost simultaneously and independent of a budding zone. In a close reading of the classical definition, the anterior serial body sections would not be segments and *Drosophila* would not be regarded as segmented at all.

Similar problems arise if the expression of segment polarity genes is used as the basis for determining segmentation criteria. For instance, the *engrailed* gene is expressed in the forming segments of arthropods, annelids, and chordates (e.g., Scholtz and Dohle 1996; Holland et al. 1997; Prud'homme et al. 2003). Thus, this expression has been used to homologize segments and to discriminate them from other serially repeated structures. However, in arthropods, annelids, and chordates, *engrailed* and other segmentation genes are expressed in different regions of the segment anlage with respect to germ layers and segment boundaries, and sometimes they are not segmentally expressed at all (e.g., Holland et al. 2000). And differences like this can occur even within one of the segmented groups, as has been exemplified by comparative gene expression studies in Annelida (Seaver and Kaneshige 2006). Genes involved in segmentation are also expressed in the terminal regions of annelids and arthropods, and thus the anterior and posterior terminal regions were classified as segments (e.g., Starunov et al. 2015), although they have been interpreted as non-segmental body units based on structural and developmental grounds (for discussion, see Scholtz and Edgecombe 2006). Moreover, most of these segmentation genes are also expressed in the nervous system (e.g., Scholtz and Dohle 1996; Seaver and Kaneshige 2006). Thus, not every structure that is formed in which these genes are active will definitely become a segment and not all segments express corresponding genes.

There are also regions with transitory expression that do not lead to the formation of segments.

A recently published study now shows, however, that also unsegmented animals use these genes to distinguish very different morphological boundaries such as coelom spaces, brain sections, external body rings, or dorsal shell plates (Vellutini and Hejnol 2016; see Chapter 8). The serial arrangement of gene expression might not determine segment formation as such, but rather the establishment of morphogenetic boundaries as a starting point for the differentiation of additional serially arranged structures.

1.8 STRUCTURAL AND ONTOGENETIC SEGMENT DEFINITIONS LEAD TO PARADOXES

Numerous attempts have been undertaken to define segments generally or for individual groups of animals by means of a combination of developmental characteristics and/or a set of repeated substructures that occur together (Scholtz 2002). Even if such definitions are taken rather loosely, this approach leads immediately to several paradoxes. Contradictions often result between morphological, structural criteria and the criteria of developmental biology. This is the case if a body section is considered a segment if it is formed from a budding zone yet morphologically does not exhibit the necessary substructures such as extremities, coelom, or nerve ganglia, and vice versa.

1.9 SEGMENTS DO NOT FORM SPATIAL AND DIFFERENTIAL UNITS

The left and right halves of a segment evidently display a certain time-space independence of its subdivisions. This is already apparent in the different speeds of cell division on the right and left sides in crustacean embryos (Scholtz 1990). The difference is consistent throughout the entire length of the embryo. In lancelets, the larvae develop a conspicuous asymmetry in the two body halves in that the segments of the left and right halves are shifted relative to one another about the length of half a segment and the mouth develops on the left side (Hatschek 1882). The ventral and the dorsal sides do not necessarily exhibit the same series of their segmental substructures. In Notostraca among the branchiopods, the posterior segments develop not one but up to six pairs of legs per segment or body ring (Linder 1952). In the diplopods, among the myriapods (Figure 1.1) each segment or body ring is associated with two pairs of legs and nerve ganglia. Embryological studies have shown that even at the molecular level, the subdivision of the ventral side proceeds differently than on the dorsal side (Janssen et al. 2006). Furthermore, it has been shown that among insects and some crustaceans, the initial subdivision of the body through segment polarity genes does not deal with segments, but with serial structures shifted half a segment and determined by cell clones. These are called parasegments and are interpreted as the essential serially developmental units (Lawrence 1992).

1.10   SIMPLE ANOMALIES DISTURB THE PATTERN OF SEGMENTATION

Anomalies in the seriality of segments are also of interest in this context. This concerns the occurrence of half-segments and in particular the spiral or helical arrangement of segmental structures (helicomery or helicomerism) (Figure 1.11) (e.g., Morgan 1895; Leśniewska et al. 2009). Half-segments and helicomery occur through regeneration after injuries or disturbances in the embryonic development during the segmentation process. Half-segments lead to a difference in the number of serial structures on the left and right sides, showing once again the differential independence of the series formation on the two sides of the body. Helicomery, however, substantially changes the seriality. Merely a break in a segment or its embryonic anlage in combination with a slight spatial shift the length of a segment can lead to helicomery. In the normal case, most structures referred to as parts of segments, for example the legs, are arranged in bilaterally symmetrical series along the longitudinal body axis. In the case of helicomery, the segment rings that normally appear one behind the other form a continuous spiral around the body (at least in part of it) (Figure 1.11). Segmental structures on this spiral, such as the appendages, no longer appear as pairs on the respective rings, but as a continuous series. The integrity of the segments is thus eliminated. Strictly speaking, the body in the helicomerous regions is not segmented (see also Fusco et al. 2008). However, the normal seriality along the

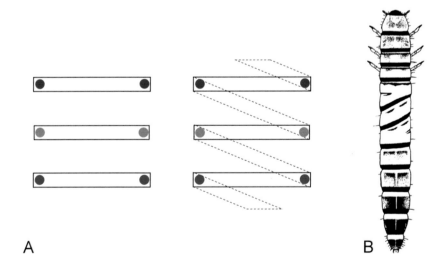

A   B

FIGURE 1.11   Anomaly: "Helicomery." A. Schematic representation. Left: The normal arrangement of segmental structures as a series along the longitudinal body axis. The colored dots represent the paired, serially arranged segmental structures, such as extremities. Right: The segmentation is eliminated due to helicomery. Segments form a continuous spiral band. The serial segmental and the paired arrangements, as well as the bilateral symmetry, are thus eliminated and replaced by a series on a spiral or screw (spiral or helical symmetry). B. The larva of a mealworm beetle with helicomery on some segments of the abdomen (modified after Balazuc 1948, ©Publications Scientifiques du Muséum national d'Histoire naturelle).

body axis is also fundamentally changed by a spiral series arrangement of the parts (Figure 1.11).

1.11 CRITICISM OF THE TERM "SEGMENT"

In summary, segments evidently do not form the assumed morphological, structural, or ontogenetically integrated units. The structures that are referred to as segments do not display much correspondence other than their seriality. This leads to the failure of any attempt at a definition. The serial coeloms also cannot be seen as essential characteristics of segments, but as one serial element among others. In terms of classification, therefore, this eliminates the distinction between segment and non-segment. Taking a look at the development does not necessarily help us any further. On the one hand, segments are formed differently, and on the other hand, it might not be the segments themselves that are laid out, but only the serial morphological boundaries. The different body halves and the ventral and dorsal sides show a certain independence in their serial differentiation. Anomalies that lead to asymmetries or helicomery break up the segmentation pattern and even the serial arrangement. Independent of one another, segmental substructures can be transformed in evolution or even completely lost. Segments probably did not evolve as a whole, but through individual steps as the eventual combination of different serially arranged substructures. From the genealogical perspective, segments and certain other serial structures make up a historical continuum. The distinction between segments and other serial body structures is evidently more a question of quantity than quality. This also blurs the boundary to unsegmented serial structures. Does this deconstruction lead the concept of "segment" to vanish? Do segments not exist, or to put it another way, are there no segmented organisms?

Some authors have in fact suggested speaking only of segmented structures or organs and no longer of segmented organisms (Budd 2001; Minelli and Fusco 2004). This choice of terminology does not resolve the problem, however, since the word "segment" was introduced in biology and in colloquial speech to mean a subdivision of a whole. This does not even apply to most of the serially arranged structures. The serially repeated nephridia of an annelid do not show any subdivisions, that is, segments of an entire kidney. Instead, they are kidneys *sui generis*, which together aim to have a greater impact, but each one can also operate on its own. Similarly, the legs of a myriapod do not constitute segments of any sort of "super leg"; instead they are individual repeated units.

At this point the concept of "series" makes sense. Structures and organs along the longitudinal axis of the body are serial but not segmented. The seriality can in fact vary between the body halves and between the dorsal and ventral sides. The concept of serial structures also does justice to a more evolutionary view, since, on the one hand, it poses the question as to the origin of segments differently, namely as an opportunity to gradually add serial structures. On the other hand, the concept of a series creates greater conceptual clarity for the controversial question as to whether the last common ancestor of the bilaterians was segmented. This question might just be worded improperly; instead it should be: Did the last common ancestor have internal or external serial structures and, if so, which ones? Were these serial structures then the point of departure for the different combinations of serial structures in the lines that led to the annelids, arthropods, and chordates? With this, the distinction between seriality and segmentation is no longer purely semantic. All in all the concept of serial structures offers greater flexibility in dealing with the observed differences and common ground between individual segments of an organism as well as between the segments of different animal groups.

Precisely by eliminating the term "segment" it might be possible to gain greater insight into what comprises segmentation as seriality and how it emerged through evolution.

ACKNOWLEDGMENTS

This text is the translated and modified version of an essay originally written in German (Scholtz 2017). I am very grateful to Allison Brown for the excellent translation.

I would like to thank Beate Witzel-Scholtz, Ariel Chipman, and Bruno Vellutini for critical reviews of the manuscript.

REFERENCES

Altenburger, A. 2016. The neuromuscular system of *Pycnophyes kielensis* (Kinorhyncha: Allomalorhagida) investigated by confocal laser scanning microscopy. *EvoDevo* 7: 25.

Auman, T., B. M. I. Vreede, A. Weiss, S. D. Hester, T. A. Williams, L. M. Nagy, and A. D. Chipman. 2017. Dynamics of growth zone patterning in the milkweed bug *Oncopeltus fasciatus*. *Development* 44: 1896–1905.

Balavoine, G., and A. Adoutte. 2003. The segmented Urbilateria: A testable scenario. *Integr. Comp. Biol.* 43: 137–147.

Balazuc, J. 1948. La tératologie des coléoptères et expériences de transplantation sur *Tenebrio molitor* L. *Mém. Mus. Nat. Hist nat. Paris (N.S.)* 25: 1–293.

Budd, G. 2001. Why are arthropods segmented? *Evol. Dev.* 3: 332–342.

Cannon, J. T., B. C. Vellutini, J. Smith, F. Ronquist, U. Jondelius, and A. Hejnol. 2016. Xenacoelomorpha is the sister group to Nephrozoa. *Nature* 530: 89–93.

Carroll, S. B., J. K. Grenier, and S. D. Weatherbee. 2001. *From DNA to Diversity*. Malden, MA: Blackwell Science.

Chipman, A. D. 2008. Thoughts and speculations on the ancestral arthropod segmentation pathway. In *Evolving Pathways – Key Themes in Evolutionary Developmental Biology*, Ed. A. Minelli, and G. Fusco, 343–358. Cambridge: Cambridge University Press.

Chipman, A. D. 2019. Becoming segmented. In *Perspectives on Evolutionary and Developmental Biology - Essays for Alessandro Minelli*, Ed. G. Fusco, 235–244. Padova: Padova University Press.

Clark, R. B. 1964. *Dynamics in Metazoan Evolution*. Oxford: Clarendon Press.

Deutsch, J. S. 2004. Segments and parasegments in arthropods: A functional perspective. *BioEssays* 26: 1117–1125.

Deutsch, J. S., and H. Le Guyader. 1998. The neural zootype. An hypothesis. *Compt. Rend. Acad. Sci. Ser. III* 321: 713–719.

Dohle, W. 1979. Vergleichende Entwicklungsgeschichte des Mesoderms bei Articulaten. *Fortschr. Zool. Syst. Evolutionsforsch.* 1: 120–140.

Dohle, W. 1999. The ancestral cleavage pattern of the clitellates and its phylogenetic deviations. *Hydrobiologia* 402: 267–283.

Dohle, W., M. Gerberding, A. Hejnol, and G. Scholtz. 2004. Cell lineage, segment differentiation, and gene expression in crustaceans. In *Evolutionary Developmental Biology of Crustacea*, Ed. G. Scholtz, 95–133. Lisse: A.A. Balkema.

Fusco, G., L. Bonato, and M. Leśniewska. 2008. Helicomery in centipede trunk: Different processes for the same pattern. *J. Morph.* 269: 1477.

Gegenbaur, C. 1870. *Grundzüge der Vergleichenden Anatomie*, 2nd ed. Leipzig: Engelmann.

Gilbert, S. F. 2003. *Developmental Biology*, 7th ed. Sunderland, MA: Sinauer Associates.

Gutmann, W. 1972. Die Hydroskelett-Theorie. *Aufs. Red. Senckenberg. Naturf. Ges.* 21: 1–91.

Haeckel, E. 1866. *Generelle Morphologie der Organismen*. Berlin: Georg Reimer Verlag.

Hannibal, R. L., and N. H. Patel. 2013. What is a segment? *EvoDevo* 4: 35.

Hatschek, B. 1878. Studien über Entwicklungsgeschichte der Anneliden. Ein Beitrag zur Morphologie der Bilaterien. *Arb. Zool. Inst. Univ. Wien* 1: 57–129.

Hatschek, B. 1882. Studien über Entwicklung des *Amphioxus*. *Arb. Zool. Inst. Univ. Wien* 4: 1–88.

Hatschek, B. 1888–1891. *Lehrbuch der Zoologie*. Jena: Gustav Fischer Verlag.

Hertwig, O., and R. Hertwig. 1881. *Die Coelomtheorie. Versuch einer Erklärung des mittleren Keimblatts*. Jena: Gustav Fischer Verlag.

Holland, L. Z., N. D. Holland, and M. Schubert. 2000. Developmental expression of *AmphiWnt1*, an amphioxus gene in the *Wnt1/wingless* subfamily. *Dev. Genes Evol.* 210: 522–524.

Holland, L. Z., M. Kene, N. A. Williams, and N. D. Holland. 1997. Sequence and embryonic expression of the amphioxus *engrailed* gene (AmphiEn): The metameric pattern of transcription resembles that of its segment-polarity homolog in *Drosophila*. *Development* 124: 1723–1732.

Janssen, R., N.-M. Prpic, and W. G. M. Damen. 2006. Dorso-ventral differences in gene expression in *Glomeris marginata* (Villers, 1789) (Myriapoda: Diplopoda). *Norw. J. Entom.* 53: 129–137.

Koch, M., B. Quast, and T. Bartolomaeus. 2014. Coeloms and nephridia in annelids and arthropods. In *Deep Metazoan Phylogeny: The Backbone of the Tree of Life*, Ed. J. W. Wägele, and T. Bartolomaeus, 173–284. Berlin: De Gruyter.

Korschelt, E., and K. Heider. 1890. *Lehrbuch der vergleichenden Entwicklungsgeschichte der Wirbellosen Tiere*. Specieller Teil, Erstes Heft. Jena: Gustav Fischer Verlag.

Lang, A. 1881. Der Bau von *Gunda segmentata* und die Verwandtschaftsverhältnisse der Plathelminthen mit Coelenteraten und Hirudineen. *Mitt. Zool. Stat. Neapel* 3: 187–251.

Lang, A. 1903. Beiträge zu einer Trophocöltheorie. *Jen. Zeitschr. Naturw. (N.F.)* 38: 1–376.

Lawrence, P. A. 1992. *The Making of a Fly*. Oxford: Blackwell.

Leśniewska, M., L. Bonato, A. Minelli, and G. Fusco. 2009. Trunk anomalies in the centipede *Stigmatogaster subterranea* provide insight into late-embryonic segmentation. *Arthropod Struct. Dev.* 38: 417–426.

Linder, F. 1952. Contributions to the morphology and taxonomy of the Branchiopoda Notostraca, with special reference to the North American species. *Proc. United States Nat. Mus.* 102: 1–69.

Martin, C., V. Gross, H.-J. Pflüger, P. A. Stevenson, and G. Mayer. 2017. Assessing segmental versus non-segmental features in the ventral nervous system of onychophorans (velvet worms). *BMC Evol. Biol.* 17: 3.

Meinhardt, H. 2015. Models for patterning primary embryonic body axes: The role of space and time. *Sem. Cell Dev. Biol.* 42: 103–117.

Minelli, A., and G. Fusco. 2004. Evo-devo perspectives on segmentation: Model organisms, and beyond. *Trends Ecol. Evol.* 19: 423–429.

Möbius, K. 1908. *Ästhetik der Tierwelt*. Jena: Gustav Fischer Verlag.

Morgan, T. H. 1895. A study of metamerism. *Quart. J. Microsc. Sci.* 37: 395–476.

Prud'homme, B., R. de Rosa, D. Arendt, J. F. Julien, R. Pajaziti, A. Dorresteijn, A. Adoutte, J. Wittbrodt, and G. Balavoine. 2003. Arthropod-like expression patterns of engrailed and wingless in the annelid *Platynereis dumerilii* suggest a role in segment formation. *Curr. Biol.* 13: 1876–1881.

Remane, A. 1950. Die Entstehung der Metamerie der Wirbellosen. In *Verhandlungen der Deutschen Zoologen vom 2. bis 6. August in Mainz*, Ed. W. Herre, 16–23. Leipzig: Geest & Portig.

Riedl, R. 1975. *Die Ordnung des Lebendigen*. Hamburg: Paul Parey Verlag.

Rieger, R. M., and G. Purschke. 2005. The coelom and the origin of the annelid body plan. *Hydrobiologia* 535/536: 127–137.

Salvini-Plawen, L. 1998. Morphologie: Haeckels Gastraea-Theorie und ihre Folgen. *Stapfia* 56: 147–168.

Schmidt-Rhaesa, A. 2007. *The Evolution of Organ Systems*. Oxford: Oxford University Press.

Scholtz, G. 1990. The formation, differentiation and segmentation of the post-naupliar germ band of the amphipod *Gammarus pulex* L. (Crustacea, Malacostraca, Peracarida). *Proc. R. Soc. London B* 239: 163–211.

Scholtz, G. 2002. The Articulata hypothesis – or what is a segment? *Org. Divers. Evol.* 2: 197–215.

Scholtz, G. 2003. Is the taxon Articulata obsolete? Arguments in favour of a close relationship between annelids and arthropods. In *The New Panorama of Animal Evolution*, Ed. A. Legakis, S. Senthourakis, R. Polymeni, and M. Thessalou-Legaki, 489–501. Sofia: Pensoft.

Scholtz, G. 2010. Deconstructing morphology. *Acta Zool.* 91: 44–63.

Scholtz, G. 2017. Segmentierung. Ein zoologisches Konzept von Serialität. In *Serie und Serialität*, Ed. G. Scholtz, 139–166. Berlin: Dietrich Reimer Verlag.

Scholtz, G., and W. Dohle. 1996. Cell lineage and cell fate in crustacean embryos - A comparative approach. *Int. J. Dev. Biol.* 40: 211–220.

Scholtz, G., and G. D. Edgecombe. 2006. The evolution of arthropod heads: Reconciling morphological, developmental and palaeontological evidence. *Dev. Genes Evol.* 216: 395–415.

Scholtz, G., and C. Wolff. 2013. Arthropod embryology: Cleavage and germ band development. In *Arthropod Biology and Evolution*, Ed. A. Minelli, G. Boxshall, and G. Fusco, 63–90. Heidelberg: Springer Verlag.

Schramke, S. 2017. Ästhetik des Serienbruchs. Versuch einer Deutung der Störung in der ornamentalen Fassadenreihung der Erweiterung des Stadtarchivs München. In *Serie und Serialität*, Ed. G. Scholtz, 59–76. Berlin: Dietrich Reimer Verlag.

Seaver, E. C. 2003. Segmentation: Mono-or polyphyletic? *Int. J. Dev. Biol.* 47: 583–596.

Seaver, E. C., and L. M. Kaneshige. 2006. Expression of 'segmentation' genes during larval and juvenile development in the polychaetes *Capitella* sp. I and *H. elegans*. *Dev. Biol.* 289: 79–194.

Sedgwick, A. 1884. On the origin of metameric segmentation and some other morphological questions. *Quart. J. Microsc. Sci.* 24: 43–82.

Starunov, V. V., N. Dray, E. V. Belikova, P. Kerner, M. Vervoort, and G. Balavoine. 2015. A metameric origin for the annelid pygidium? *BMC Evol. Biol.* 15: 25.

Tautz, D. 2004. Segmentation. *Dev. Cell* 7: 301–312.

Toepfer, G. 2017. Serialität als natürliches Phänomen. Beschreibungsmodell der Biologie und Evolutionsprodukt. In *Serie und Serialität*, Ed. G. Scholtz, 11–30. Berlin: Dietrich Reimer Verlag.

Treffkorn, S., L. Kahnke, L. Hering, and G. Mayer. 2018. Expression of NK cluster genes in the onychophoran *Euperipatoides rowelli*: Implications for the evolution of NK family genes in nephrozoans. *EvoDevo* 9: 17.

Vellutini, B. C., and A. Hejnol. 2016. Expression of segment polarity genes in brachiopods supports a non-segmental ancestral role of *engrailed* for bilaterians. *Sci. Rep.* 6: 32387.

Westheide, W. 1997. The direction of evolution within the Polychaeta. *J. Nat. Hist.* 31: 1–15.

Westheide, W., and R. Rieger (ed.). 2007. *Spezielle Zoologie, Teil 1: Einzeller und Wirbellose Tiere*, 2nd ed. München: Spektrum.

Williams, T. A., B. Blachuta, T. A. Hegna, and L. M. Nagy. 2012. Decoupling elongation and segmentation: Notch involvement in anostracan crustacean segmentation. *Evol. Dev.* 14: 372–382.

2 Diversity in Segmentation Mechanisms

Ariel D. Chipman

CONTENTS

2.1 INTRODUCTION

Segmented body plans are extremely common when measured in terms of the number of species that display them. However, in terms of their phylogenetic distribution, fully segmented body plans are only found in three phyla: arthropods, chordates, and annelids (Minelli and Fusco, 2004; Chipman, 2010). Of course, various degrees of repetitive units and metameric organization are found in numerous other taxa (see Chapter 9 for a detailed discussion). Nonetheless, regardless of the specific definition used for "truly" segmented organisms or for a segment (see Chapter 1), almost all authors would agree that the aforementioned three phyla display the most complete and consistent examples of segmented body plans.

This chapter serves as a general introduction to the themes discussed in detail in the subsequent chapters. In it, I will discuss the many different ways segments are generated and segmented body plans are arranged throughout the developmental process, and point out where in the book these points are elaborated upon. But, before doing so, it is worth asking what—if anything—segmented body plans have in common. If they have little or nothing in common, it is not surprising that they display a great diversity in developmental mechanisms. However, if there are common organizational principles to segments across taxa, and these are achieved through a diversity of mechanisms, we have an interesting phenomenon that is worth exploring, providing justification for the current book.

So what do segmented body plans have in common? Segmented body plans are made up of repeated body units, and these units contain components from a number of different organ systems, with coelomic sacs often considered to be the most important repeated units. This is the accepted definition of "true" segments (Scholtz, 2002; Minelli and Fusco, 2004; Hannibal and Patel, 2013). Beyond this, one of the hallmarks of segmented body plans, and the one that is most pertinent to the current discussion, is that during their development, segments pass through a transient stage wherein they are already morphologically distinct from segments anterior or posterior to them, but have not yet differentiated to give rise to their constituent organ systems (Scholtz, 2002; Chipman, 2019). Thus, individual segments are specified in ontogeny (embryonic or post-embryonic) as undifferentiated but distinct units, which undergo subsequent differentiation to the repeated elements of the different organ systems that make up the mature segment. The diverse mechanisms described in this chapter all give rise to such an undifferentiated transient segment.

2.2   THE SEQUENCE OF SEGMENT FORMATION

The plesiomorphic mode of segment addition is sequential addition from a subterminal domain in the posterior of the organism (Scholtz, 2002). This is true for all three of the fully segmented phyla, but there are significant differences among them and within them.

2.2.1   EMBRYONIC OR POST-EMBRYONIC SEGMENTATION

The most noteworthy variability is in whether some or all segment addition occurs during embryonic development or post-embryonically. Within annelids, the primitive and most common pattern is embryonic development to a trochophore larva, which is essentially unsegmented but includes two to three ciliary bands. A posterior region of this larva, the area between the metatroch and the telotroch (the middle and posterior ciliary bands, respectively) becomes the segment addition zone, and segments are patterned sequentially from the posterior in a post-embryonic stage (Bleidorn *et al.*, 2015). In many cases (e.g., in the model species *Platynereis dumerilii*; Fischer *et al.*, 2010), several anterior segments form during the early larval stage. This gives rise to a nectochaete larva, which grows through posterior segment addition (see Chapter 4 for more details on the diversity of annelid segmentation modes). Clitellates (leeches, earthworms, and their relatives), unlike most other annelids, are direct developers and generate all segments during embryogenesis through specialized stem cells known as teloblasts (see Chapter 7).

Segmentation in chordates is almost exclusively embryonic (cephalochordates being an exception). Segments are generated from an embryonic domain, known as the pre-somitic mesoderm (PSM), through a cellular oscillator that leads to the organization of distinct mesodermal units, known as somites (Palmeirim *et al.*, 1997). All trunk and tail segments are formed using the same mechanism, during embryogenesis, with no evidence of unusual anterior segments.

Arthropods present a much more complex picture. There is some debate as to whether direct development is the plesiomorphic state in arthropods or not, and there have been arguments raised both for embryonic generation of all segments as primitive (Chipman, 2015) and for a biphasic life cycle being primitive (Wolfe, 2017). Either way, both direct development and indirect development are found in most clades within arthropods. Insects and arachnids are almost exclusively direct developers, with all segments formed in embryogenesis (but one pair of legs differentiating only post-hatching in mites, and three segments added post-embryonically in proturans). Crustaceans are mixed, with many taxa developing into a nauplius larva, which adds segments sequentially post-embryonically, and some taxa (e.g., isopods, amphipods) generating all segments during embryonic development. In myriapods again both modes are found. Millipedes add segments post-embryonically, whereas in centipedes, post-embryonic segmentation is plesiomorphic and generation of all segments embryonically is derived—including in the geophilomorphs, which have the highest segment numbers among centipedes (Minelli, 2001).

Among fossil arthropods, there are highly detailed descriptions of post-embryonic segment addition in trilobites (Hughes *et al.*, 2006; Hopkins, 2017). Recently, there have been reports of post-embryonic segment addition in the stem group arthropod *Fuxianhuia* (Fu *et al.*, 2018).

2.2.2 SIMULTANEOUS OR SEQUENTIAL SEGMENTATION

A second variability in mode of segment formation is the extent to which segments are patterned sequentially or simultaneously. In chordates, again the picture is the simplest. All chordates generate all segments sequentially. In annelids, the formation of the nectochaete segments is almost simultaneous, but all other segments are formed sequentially, whether they are embryonic or post-embryonic (Bleidorn *et al.*, 2015).

Arthropods present great diversity in this aspect as well. While most segments in most species appear sequentially, even in direct developing species, this is not true of all segments. The anteriormost segments (the pre-gnathal segments, or naupliar segments) often appear separately from other segments, through a distinct mechanism about which we know relatively little (Scholtz and Edgecombe, 2006; Posnien *et al.*, 2010; Janssen *et al.*, 2011). Segments of the post-gnathal head and sometimes of the trunk/thorax are often patterned simultaneously (Stahi and Chipman, 2016). The most extreme cases of simultaneous segmentation are found in many holometabolous insects, including the best studied model species, *Drosophila melanogaster*, where all segments are patterned simultaneously, and there is no posterior addition of segments at all (Hartenstein and Chipman, 2015; Stahi and Chipman, 2016). Amphipod crustaceans are unusual among crustaceans in having an almost simultaneous assembly of the segments, without a phase of posterior addition (Scholtz and Wolff, 2002; Hannibal *et al.*, 2012).

2.3 CYCLICAL PROCESSES IN SEGMENTATION

The process of generating segments sequentially is by nature a cyclical process, with a sequence of events occurring repeatedly for every segment generated. The nature of this cyclical process has been studied in detail in some cases, whereas for others we have only a cursory understanding of what is involved.

The first system where cyclical segmentation was described, and the one for which we still have the largest body of data, is vertebrate somitogenesis (Pourquié, 2003). In this process, a traveling wave of gene expression moves over a field of undifferentiated mesodermal cells. At some point, the traveling wave is fixed in space, giving a repeating pattern of cell identities, which is translated into the reiterated pattern of mesodermal somites. This process is very similar to the process predicted by one of the earliest theoretical models for how repeated units can be patterned during development, the clock and wavefront model (Cooke and Zeeman, 1976). We now know that the traveling wave is maintained through entrainment of adjacent cells using Notch–Delta signaling, and a cell autonomous Hairy-based clock, and it is fixed in space through interactions with a Wnt-signaling gradient (Aulehla and Herrmann, 2004; Dubrulle and Pourquié, 2004).

Evidence of traveling waves involved in generating a segmental pattern also exists in several groups of arthropods, although the molecular mechanism behind them is probably very different, suggesting they have evolved convergently. In the centipede *Strigamia*, a traveling wave of expression of *delta* and several other segmental genes sweeps across the large posterior undifferentiated disk (Chipman and Akam, 2008). The nature of the original cycler and the mode of propagation of the wave remain unknown. Similar traveling waves have also been shown in spider development (Schoppmeier and Damen, 2005a). In the spider example there is also an involvement of the Notch pathway (Stollewerk *et al.*, 2003; Schoppmeier and Damen, 2005b), but again, the exact details are unclear. A very different type of cycler and traveling wave has been reported from the flour beetle *Tribolium*. Here, the cycler is based on a negative feedback loop between three genes, which are orthologs of *Drosophila* pair-rule genes. The negative feedback generates a reiterated pattern, with each repeat generating a two-segment unit (Choe *et al.*, 2006). In contrast, in the hemimetabolous insect *Oncopeltus*, there is no evidence of any type of traveling wave, and the nature of the repeating process, which generates single-segment units, is unknown (Auman *et al.*, 2017; Auman and Chipman, 2018).

A cyclical signal moving across cells is not the only way to generate a repeated pattern in development. An alternative way to make a repeated pattern is to link patterning with the cell division cycle. This is indeed what happens in malacostracan crustaceans (Chapter 6) and in clitellate annelids (Chapter 7). In both these cases, a single cycle of cell division is linked to the generation of the precursors of a single segmental unit. A series of dedicated stem cells, known as teloblasts, divide asymmetrically, giving rise to one daughter cell that will serve as a segmental precursor cell, and one daughter cell that maintains stem properties. Essentially, the segmentation process is regulated by the same factors that regulate the cell cycle. It remains unclear at what level the ancestral segmentation gene regulatory network comes in to the process.

2.4 PROLIFERATION VERSUS CELL REARRANGEMENT AS DRIVERS OF SEGMENT FORMATION

As new segments form from the posterior, there is an addition of tissue to the segmented portion of the embryo (e.g., the segmented germband in arthropods or the somites in vertebrates). Where is this tissue coming from? Is it new tissue that results from localized cell proliferation or is it existing tissue that has been shifted into the nascent segment through cell rearrangement? Answering these questions requires developing experimental approaches that can trace individual cells over time and provide a detailed description of cell proliferation and cell movement (see Chapter 8 for some examples of these approaches).

Cell proliferation has a central role in segment formation in the cases where the segmentation process is linked to the cell cycle, as described earlier for malacostracan crustaceans (with the exception of amphipods) and for clitellate annelids. In both these cases, we have fairly detailed data on the pattern and sequence of cell division from early development and up to the formation of distinct segmental precursors. There is no preexisting tissue that contributes to the teloblast-derived nascent segments, and all growth arises from cell proliferation within the teloblasts and their descendants.

Conversely, in a typical arthropod growth zone, there is a pool of undifferentiated cells, which are gradually recruited into the segmented germband. Throughout the process of germband extension, the growth zone becomes smaller, as the growth zone cells are recruited into nascent segments. This has been shown most dramatically in the growth zone of the geophilomorph centipede *Strigamia maritima* (Chipman *et al.*, 2004), where the growth zone starts out as a very large disk, covering nearly half of the egg, and shrinks as segments are added to the germband. However, in reality the process is somewhat more complex. Even when the tissue added to the extending germband is recruited from existing cells, there is some cell proliferation in the growth zone. Auman *et al.* (2017) showed that in the milkweed bug *Oncopeltus fasciatus,* there is a constant low level cell proliferation in the posterior of the growth zone, presumably serving to replenish the pool of undifferentiated cells. There is relatively little known about cell proliferation in segmentation in other arthropods that do not have teloblasts (see Chapter 3; Williams and Nagy, 2017).

As for vertebrates and non-clitellate annelids, to my knowledge there has been no detailed work following the relative contributions of cell division and cell rearrangement to the process of segment addition.

2.5   TYPES OF CELLS IN SEGMENTATION PROCESSES

Cells involved in the segmentation process are by nature pluripotent cells. Nascent segments go through a stage characterized by an undifferentiated morphological unit, which is of course composed of undifferentiated cells. These cells will go on to form precursors of multiple different organs and structures. As with other aspects of segmentation, our understanding of the specific identity of these cells is fairly limited, and there is almost no work dividing these cells into different cell types.

The exceptions to this gap in our understanding are the stem cells in malacostracan crustaceans and in clitellate annelids, already mentioned earlier. In these cases, there are detailed fate maps, and the individual cells are known and identified (Dohle and Scholtz, 1988; Lans *et al.*, 1993; Gerberding and Scholtz, 1999; Kuo and Shankland, 2004). In malacostracans they include a number of ectodermal stem cells (ectoteloblasts) and a smaller number of mesodermal stem cells (mesoteloblasts). The ectoteloblasts are all roughly equivalent in behavior, and in many cases there is a fixed number of them. They are symmetrically arranged on two sides of the midline, with one cell being median and unpaired. There is no median mesoteloblast. The first two rounds of division of the daughter cells generate four rows of cells for each division of the ectoteloblasts and the mesoteloblasts. These cells remain morphologically identical, and only start to differentiate in subsequent divisions.

In clitellates, there are ten teloblasts, two of these are mesoteloblasts and eight are ectoteloblasts. Each of these cells generates a distinct lineage (with symmetrical cells being identical), with different cell fates. Thus, compared with the situation in malacostracans, the teloblasts are not as pluripotent.

Vertebrate somites are already dedicated mesodermal structures, which differentiate into a series of distinct cell fates, in a regulated regionalized manner, shortly after the somites are formed. The source of the somite cells is a series of cells known as neuromesodermal progenitors (NMPs; see Chapter 5 for a detailed discussion). These cells originate in the tailbud, and they are unusual in not being committed to a specific germ layer, even after gastrulation. As their name suggests, some of them become committed to ectodermal fate and become part of the neural tube, whereas others become committed to mesodermal fate and form the somites. The NMP cells behave like stem cells, and although they cannot be said to be totipotent, they have a range of developmental fates.

Stem cells are involved in post-traumatic segmentation in organisms that can regenerate axial structures (most notably annelids; see Chapter 10). These stem cells can come either from reserve stem cells that remain in the organisms' tissues postembryonically or from dedifferentiation of somatic cells. They proliferate and differentiate to regenerate all of the required adult structures (Bely *et al.*, 2014; Zattara and Bely, 2016).

Cell lineage studies of segmental cells have mostly been performed on those species that have stereotypical stem cell division patterns (e.g., malacostracan crustaceans and leeches; Lans *et al.*, 1993; Gerberding *et al.*, 2002; Wolff and Scholtz, 2002; Alwes *et al.*, 2011). There are almost no studies following the fates of cells within the undifferentiated proto-segment or the pre-segmental domains. Although it is clear that the cells in the arthropod growth zone end up in the differentiated segments (Nakamoto *et al.*, 2015; Auman *et al.*, 2017), there is no mapping of which cells end up where. Thus, there is currently no way to identify differing cell fates within the arthropod growth zone or nascent segment. The same is true for most other taxa.

2.6 COMMONALITIES IN SEGMENTATION PROCESSES: ARE THERE ANY?

I will end with a similar question to the one I started with: What—if anything—do diverse segmentation processes have in common? A detailed comparison reveals that in fact, there is very little in common, and almost no generalizations one can make about segmentation as a developmental process. This is in contrast with segments as morphological units, which while probably convergent, do have enough in common to justify being discussed together. Nonetheless, there are numerous restricted developmental commonalities, found in specific taxa. The phylogenetic distribution of these commonalities in different aspects of the process allows a reconstruction of the evolutionary history of segmentation, both as a developmental and as a morphological phenomenon. This makes it possible to identify specific nodes in animal phylogeny where there were significant changes in development that led to segments in the first place (Chipman, 2019) and nodes where there were changes in the development of segments once they had evolved (Balavoine, 2014; Stahi and Chipman, 2016). Coupled with information from the fossil record, and with data about morphology and life history, we can begin to reconstruct the success story of segmented animals and to understand how and why they evolved.

REFERENCES

Alwes, F., B. Hinchen, and C. G. Extavour. 2011. Patterns of cell lineage, movement, and migration from germ layer specification to gastrulation in the amphipod crustacean *Parhyale hawaiensis*. *Dev. Biol.* 359: 110–123.

Aulehla, A., and B. G. Herrmann. 2004. Segmentation in vertebrates: Clock and gradient finally joined. *Genes Dev.* 18: 2060–2067.

Auman, T., and A. D. Chipman. 2018. Growth zone segmentation in the milkweed bug *Oncopeltus fasciatus* sheds light on the evolution of insect segmentation. *BMC Evol. Biol.* 18: 178.

Auman, T., B. M. I. Vreede, A. Weiss, S. D. Hester, T. A. Williams, L. M. Nagy, and A. D. Chipman. 2017. Dynamics of growth zone patterning in the milkweed bug *Oncopeltus fasciatus*. *Development* 144: 1896–1905.

Balavoine, G. 2014. Segment formation in Annelids: Patterns, processes and evolution. *Int. J. Dev. Biol.* 58: 469–483.

Bely, A. E., E. E. Zattara, and J. M. Sikes. 2014. Regeneration in spiralians: Evolutionary patterns and developmental processes. *Int. J. Dev. Biol.* 58: 623–634.

Bleidorn, C., C. Helm, A. Weigert, and M. T. Aguado. 2015. Annelida. In *Evolutionary Developmental Biology of Invertebrates*, Vol. 2, Ed. A. Wanninger, 193–230. Wien: Springer-Verlag.

Chipman, A. D. 2010. Parallel evolution of segmentation by co-option of ancestral gene regulatory networks. *Bioessays* 32: 60–70.

Chipman, A. D. 2015. An embryological perspective on the early arthropod fossil record. *BMC Evol. Biol.* 15: 285.

Chipman, A. D. 2019. Becoming segmented. In *Perspectives on Evolutionary Developmental Biology*, Ed. G. Fusco, 235–244. Padova: Padova University Press.

Chipman, A. D., and M. Akam. 2008. The segmentation cascade in the centipede *Strigamia maritima*: Involvement of the Notch pathway and pair-rule gene homologues. *Dev. Biol.* 319: 160–169.

Chipman, A. D., W. Arthur, and M. Akam. 2004. Early development and segment formation in the centipede *Strigamia maritima* (Geophilomorpha). *Evol. Dev.* 6: 78–89.

Choe, C. P., S. C. Miller, and S. J. Brown. 2006. A pair-rule gene circuit defines segments sequentially in the short-germ insect *Tribolium castaneum*. *Proc. Natl. Acad. Sci. USA* 103: 6560–6564.

Cooke, J., and E. C. Zeeman. 1976. Clock and wavefront model for control of number of repeated structures during animal morphogenesis. *J. Theoret. Biol.* 58: 455–476.

Dohle, W., and G. Scholtz. 1988. Clonal analysis of the crustacean segment - The discordance between genealogical and segmental borders. *Development* 104: 147–160.

Dubrulle, J., and O. Pourquié. 2004. Coupling segmentation to axis formation. *Development* 131: 5783–5793.

Fischer, A. H. L., T. Henrich, and D. Arendt. 2010. The normal development of *Platynereis dumerilii* (Nereididae, Annelida). *Front. Zool.* 7: 31.

Fu, D., J. Ortega-Hernandez, A. C. Daley, X. Zhang, and D. Shu. 2018. Anamorphic development and extended parental care in a 520 million-year-old stem-group euarthropod from China. *BMC Evol. Biol.* 18: 147.

Gerberding, M., W. E. Browne, and N. H. Patel. 2002. Cell lineage analysis of the amphipod crustacean *Parhyale hawaiensis* reveals an early restriction of cell fates. *Development* 129: 5789–5801.

Gerberding, M., and G. Scholtz. 1999. Cell lineage of the midline cells in the amphipod crustacean *Orchestia cavimana* (Crustacea, Malacostraca) during formation and separation of the germ band. *Dev. Genes Evol.* 209: 91–102.

Hannibal, R. L., and N. H. Patel. 2013. What is a segment? *Evodevo* 4: 35.

Hannibal, R. L., A. L. Price, and N. H. Patel. 2012. The functional relationship between ectodermal and mesodermal segmentation the crustacean, *Parhyale hawaiensis*. *Dev. Biol.* 361: 427–438.

Hartenstein, V., and A. D. Chipman. 2015. Hexapoda: A *Drosophila*'s view of insect development. In *Evolutionary Developmental Biology of Invertebrates*, Vol. 5, Ed. A. Wanninger, 1–91. Vienna: Springer.

Hopkins, M. 2017. Development, trait evolution, and the evolution of development in trilobites. *Int. Comp. Biol.* 57: 488–498.

Hughes, N. C., A. Minelli, and G. Fusco. 2006. The ontogeny of trilobite segmentation: A comparative approach. *Paleobiology* 32: 602–627.

Janssen, R., G. E. Budd, and W. G. M. Damen. 2011. Gene expression suggests conserved mechanisms patterning the heads of insects and myriapods. *Dev. Biol.* 357: 64–72.

Kuo, D. H., and M. Shankland. 2004. A distinct patterning mechanism of O and P cell fates in the development of the rostral segments of the leech *Helobdella robusta*: Implications for the evolutionary dissociation of developmental pathway and morphological outcome. *Development* 131: 105–115.

Lans, D., C. J. Wedeen, and D. A. Weisblat. 1993. Cell lineage analysis of the expression of an *engrailed* homolog in leech embryos. *Development* 117: 857–871.

Minelli, A. 2001. A three-phase model of arthropod segmentation. *Dev. Genes Evol.* 211: 509–521.

Minelli, A., and G. Fusco. 2004. Evo-devo perspectives on segmentation: Model organisms, and beyond. *Trends Ecol. Evol.* 19: 423–429.

Nakamoto, A., S. D. Hester, S. J. Constantinou, W. G. Blaine, A. B. Tewksbury, M. T. Matei, ..., T. A. Williams. 2015. Changing cell behaviours during beetle embryogenesis correlates with slowing of segmentation. *Nat. Commun.* 6: 6635.

Palmeirim, I., D. Henrique, D. Ish-Horowicz, and O. Pourquié. 1997. Avian *hairy* gene expression identifies a molecular clock linked to vertebrate segmentation and somitogenesis. *Cell* 91: 639–648.

Posnien, N., J. B. Schinko, S. Kittelmann, and G. Bucher. 2010. Genetics, development and composition of the insect head - A beetle's view. *Arthropod Struct. Dev.* 39: 399–410.

Pourquié, O. 2003. The segmentation clock: Converting embryonic time into spatial pattern. *Science* 301: 328–330.

Scholtz, G. 2002. The Articulata hypothesis - or what is a segment? *Org. Divers. Evol.* 2: 197–215.

Scholtz, G., and G. D. Edgecombe. 2006. The evolution of arthropod heads: Reconciling morphological, developmental and palaeontological evidence. *Dev. Genes Evol.* 216: 395–415.

Scholtz, G., and C. Wolff. 2002. Cleavage, gastrulation, and germ disc formation of the amphipod *Orchestia cavimana* (Crustacea, Malacostraca, Peracarida). *Contrib. Zool.* 71: 9–28.

Schoppmeier, M., and W. G. M. Damen. 2005a. Expression of Pax group III genes suggests a single-segmental periodicity for opisthosomal segment patterning in the spider *Cupiennius salei. Evol. Dev.* 7: 160–169.

Schoppmeier, M., and W. G. M. Damen. 2005b. Suppressor of hairless and Presenilin phenotypes imply involvement of canonical Notch-signalling in segmentation of the spider *Cupiennius salei. Dev. Biol.* 280: 211–224.

Stahi, R., and A. D. Chipman. 2016. Blastoderm segmentation in *Oncopeltus fasciatus* and the evolution of arthropod segmentation mechanisms. *Proc. R. Soc. Lond. B* 283: 20161745.

Stollewerk, A., M. Schoppmeier, and W. G. M. Damen. 2003. Involvement of *Notch* and *Delta* genes in spider segmentation. *Nature* 423: 863–865.

Williams, T. A., and L. M. Nagy. 2017. Linking gene regulation to cell behaviors in the posterior growth zone of sequentially segmenting arthropods. *Arthropod Struct. Dev.* 46: 380–394.

Wolfe, J. M. 2017. Metamorphosis is ancestral for crown euarthropods, and evolved in the Cambrian or earlier. *Integr. Comp. Biol.* 57: 499–509.

Wolff, C., and G. Scholtz. 2002. Cell lineage, axis formation, and the origin of germ layers in the amphipod crustacean *Orchestia cavimana. Dev. Biol.* 250: 44–58.

Zattara, E. E. and A. E. Bely. 2016. Phylogenetic distribution of regeneration and asexual reproduction in Annelida: Regeneration is ancestral and fission evolves in regenerative clades. *Invertebr. Biol.* 135: 400–414.

Section II

Cellular Mechanisms
of Segmentation

3 Cell Division, Movement, and Synchronization in Arthropod Segmentation

Lisa M. Nagy and Terri A. Williams

CONTENTS

3.1 INTRODUCTION

A segmented body plan has been a key feature of arthropod diversification and evolution. Most arthropods build their segmented bodies during embryogenesis or early larval development, and do so by the sequential addition of segments from the posterior (reviewed in Sander, 1975; Davis and Patel, 2002; Peel et al., 2005). The contrast between this mode of segmentation and the extensively studied model arthropod, *Drosophila*, has sparked numerous studies in diverse taxa examining the regulation of sequential segmentation (reviewed in Davis and Patel, 2002; Seaver, 2014; Minelli, 2004; Peel, 2004; Liu and Kaufman, 2005; Peel et al., 2005; Damen, 2007; Blair, 2008; Couso, 2009; Chipman, 2010; also see Chapter 2). Studying segment patterning in arthropod model species outside of flies has brought focus to the cellular behaviors that elongate the body and accommodate the progressive addition of new segments (Freeman, 1995; Benton et al., 2013; Nakamoto et al., 2015; Cepeda et al., 2017; Benton, 2018; Hemmi et al., 2018). New insights into cell behaviors in the posterior have emerged from careful temporal analyses, fate mapping, live imaging, and functional genetic studies. It is clear that one variable feature of the evolution of arthropod segmentation has involved changing the degree to which embryos rely on cell division *versus* cell rearrangement to elongate.

　　Here we review our current understanding of the relative roles of cell division and cell rearrangement, discuss other possible contributors to elongation, and briefly discuss how these processes relate to cell fate determination. We summarize current findings about growth and elongation in sequentially segmenting arthropods. Key elements are a minor, but essential, role for posterior cell division that is both temporally and spatially controlled by the posterior organizer, as well as a role for convergent extension driven by diverse mechanisms.

3.1.1 SEGMENTATION AND ELONGATION: THE EVOLVING ROLES OF CELL DIVISION AND CELL REARRANGEMENT

Among arthropod embryos, the size of the initial embryonic anlage from which segments form is a highly variable feature. Consequently, the degree of growth needed to add new segments and elongate a fully segmented embryo or larva is also highly variable. This variability in initial size of the anlage is characterized in insects by the designation "long germband" *versus* "short germband" development (Krause, 1939; reviewed in Sander, 1975; Davis and Patel, 2002). While these categories were originally used for the comparative description of insects, embryos and larvae in other arthropod taxa also vary in the degree to which clusters of anterior segments are simultaneously specified as well as the number of total body segments formed prior to hatching (reviewed in Scholtz and Wolff, 2013). Despite this variability, all embryos or larvae typically show notable elongation during their early development. Elongation primarily depends on either increasing the number of cells or rearranging the relative positions of cells. Indeed, the relative degree of cell division *versus* cell rearrangement is a key variable in evolution. Oriented cell division or coordinated changes in cell shape can also influence elongation but may play more minor roles (e.g., see da Silva and Vincent, 2007, also Freeman, 1995), or may simply be less well known.

3.2   THE ROLE OF CELL DIVISION

Many species, whether they develop from a sequentially segmenting embryo or larva, appear to add their new segments from a region at the posterior of the embryo/larva. Consequently, both the addition of new segments and the driving force for elongation were originally hypothesized to result from cell division in this posterior domain (Anderson, 1973; Sander, 1975). In its simplest form, a posterior "growth zone" would supply both naïve tissue for segment patterning at its anterior border and all the additional tissue required for elongation (Figure 3.1). As our understanding of the cellular mechanisms that drive elongation have grown, we now understand that segment addition can be accomplished by posteriorly localized stem cells (e.g., malacostracans; reviewed in Dohle and Scholtz, 1997; Scholtz and Wolff, 2013; and Chapter 6) or migration of cells (Benton, 2018), or by having a population of cells that divide until they exit the posterior, presumably held in a multipotent state by a posterior signal (Shinmyo et al., 2005; McGregor et al., 2008, 2009; Chesebro et al., 2013; Constantinou et al., 2016; Williams and Nagy, 2017). In addition, elongation in many species is driven by cell division and rearrangements outside of the posteriormost region of the embryo or larva (see following discussion). This diversity of mechanism led to the suggestion to replace the classical term "growth zone" with the term "segment addition zone" (Janssen et al., 2010). As both terms imply posterior growth and both terms are essentially agnostic to mechanism, we have argued for retaining "growth zone" as a generally understood term for the area from which the germband grows—albeit using a diversity of cellular mechanisms (Auman et al., 2017).

3.2.1   CASE STUDIES SUPPORT A NEW MODEL FOR THE ROLE OF CELL DIVISION IN THE POSTERIOR

We describe three case studies examining cell division in the posterior: a non-malacostracan crustacean and two different insect species. These cases (as well as other published literature reviewed) illustrate the role of cell division in our

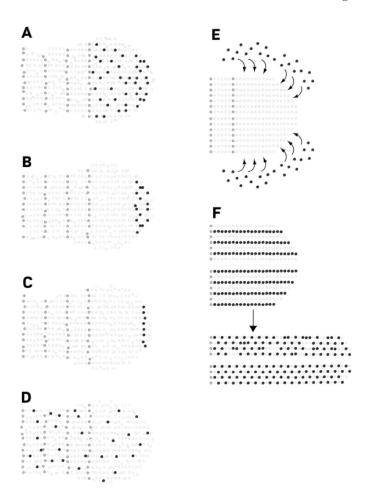

FIGURE 3.1 Various cell behaviors that could account for elongation during segmentation in sequentially segmenting arthropods. Classical models of a growth zone (A and B) have assumed high rates of cell division in the posterior. Cell division could be limited to stem cells (C), although that has been found in only one crustacean clade (Malacostraca). When actually measured, mitoses appear much more widespread within the embryo (D), usually without a posterior bias. Other known methods of elongation include cell migration (E) and intercalation (F). (Mitotic cells, red; Engrailed/Invected expressing cells, green; see text for references).

proposed model and suggest some common principles of elongation. First, cell division promoting elongation is distributed throughout the embryo or larvae, and is not a consequence of a continually or even highly proliferative posterior growth zone. Second, in most species, the growth zone shrinks over the course of segmentation, thus does in fact contribute to elongation of the embryo. However, its depletion is not sufficient to account for all the tissue of the newly added segments. Proliferation in the posterior growth zone is required for elongation but each cell in the growth zone need only divide a few times to supplement the contributions of growth from other regions of the embryo/larva. Third, the rate of cell division can vary over the course of elongation and cell cycle duration can be regulated, at least at times, by controlling the duration of cell cycle. Fourth, the region tradition- ally called the "posterior growth zone" is subdivided into a posterior domain of multipotent cells that divide infrequently, and an anterior domain of cells initiating segmental specification that rarely divide. All of these features have not yet been systematically analyzed in each case study, but in total provide a framework for future interspecies comparisons.

3.2.1.1 *Thamnocephalus*

Our first case study examining the degree to which cell division is required for posterior growth is the branchiopod crustacean *Thamnocephalus platyurus.* *Thamnocephalus* hatch as free-swimming larvae with three pairs of head append- ages and an undifferentiated trunk. Sequential segment addition and progressive dif- ferentiation gradually produce the adult morphology of eleven limb-bearing thoracic segments and eight abdominal segments, the first two of which are fused to form the genital region (Figure 3.2; Linder, 1941; Anderson, 1967; Freyer, 1983; Møller et al., 2003). Precise staging of developmental cohorts allowed for quantification of the changing dimensions of the growth zone as well as cell cycling behaviors (Constantinou et al., 2020).

Growth Zone Mitoses Are Few with More Extensive Cell Cycling in the Segmented Region

In *Thamnocephalus*, the initial growth zone has about 40% of the total tissue needed to add post-hatching trunk segments (14). During segment addition, the growth zone shrinks. Without mitosis, the initial growth zone would be quickly depleted, used up after adding four segments, despite the fact that the size of new segment anlage also decreases over development. However, mitosis does occur in the growth zone and, since the total cell number in the growth zone is quite large, we estimate that the population of cells in the initial hatchling growth zone need only divide approximately 1.5 times to produce the tissue necessary to make all segments (Constantinou et al., 2020). Consistent with this estimate, direct counts of mitosis (using Hoechst staining or an anti-phosphorylated histone 3 [pH3], which marks cells with mitotic chromosome condensation), show surprisingly little mito- sis in the growth zone: less than 5% of the cells are in mitosis at any given stage (Constantinou et al., 2020). This shows that not only are there very few mitotic cells in the posterior but, given the relatively large numbers of cells in the growth zone, not much division may be required.

What then accounts for the remaining growth? When larvae are exposed to a nucleotide analog (5-ethynyl-2'-deoxyuridine, EdU, which is incorporated into cells during S-phase) for 30 minutes, few EdU positive cells are found in the growth zone; instead many more are detected in the regions anterior to the growth zone, or *after* segment specification. Thus, it is growth in regions of the larvae anterior to the growth zone—the already specified segments—and not the growth zone itself that accounts for most of the elongation (Constantinou et al., 2020).

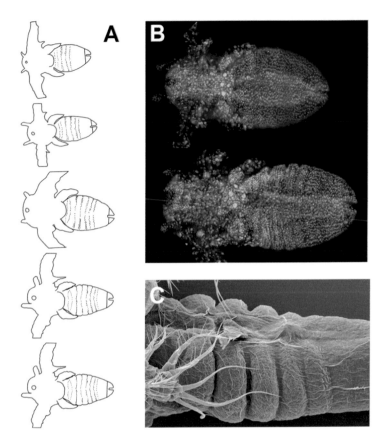

FIGURE 3.2 Sequential addition of segments from the posterior region in larval fairy shrimp, *Thamnocephalus platyurus*. A. Outlined drawings of successive one-hour time points indicating the addition of segments (as measured by Engrailed/Invected stripes) beginning at hatching to 4 hours post-hatching (graphics by P. Storrer). B. DAPI-stained larvae at hatching (top) and 4 hours post-hatching (bottom), and C. SEM of older larva showing in more detail segments beginning to undergo morphogenesis.

Growth Is Regulated through Both Synchronization of
S-Phase Domains and Slowing the Rate of S-Phase

In *Drosophila*, discrete clusters of cells specified for common cell fates undergo mitosis together, resulting in repeatable, progressive patterns of mitosis (Foe, 1989). We, and others, had anticipated finding similar mitotic domains of some sort in sequentially segmenting arthropods, but as of yet, have failed to find domains as intricate as the 25 distinct mitotic domains described in *Drosophila* (Nagy et al., 1994; Rosenberg et al., 2014) or failed to find them at all (Brown et al., 1994; Liu and Kaufman, 2009). Interestingly, EdU incorporation in *Thamnocephalus* larvae unexpectedly revealed discrete S-phase domains, demonstrating a spatial coordination in cell cycling not captured by examining mitosis (Figure 3.3). The cells in the most posterior growth zone are variably in S-phase. Anterior to this, two stable

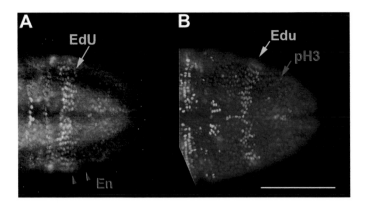

FIGURE 3.3 Spatially discrete domains of EdU incorporation in the posterior of *Thamnocephalus* larvae indicate coordinated cell cycling. A. An early larval stage showing band of EdU cells in the newest specified segment (flanked by En/In stripes, red arrows). This band is followed by a region in the anterior GZ of with very few EdU incorporating cells and then a region in the posterior GZ with scattered EdU incorporating cells. B. Similar larvae showing that the EdU band (green) excludes cells in M-phase (pH3 staining, pink). (After Constantinou et al., 2020.)

S-phase domains are associated with segmentation: a band of cells not undergoing S-phase in the anterior growth zone and a synchronized band of cells undergoing S-phase in the most recently specified segment. The stable domains indicate a synchronized transition into S-phase in the newly specified segments. In addition, during a 30 minute exposure to EdU, cells in the growth zone undergoing DNA replication incorporate less EdU than replicating cells in the segmented region of the larvae. This suggests that the rate of cell division in the posterior is regulated, at least in part, by controlling the rate of DNA synthesis, in this case by a slower the cell cycle.

Wnt/Caudal Gene Expression Are Linked to the Temporal Control of S-Phase

Strikingly, the S-phase domains correlate with expression domains of specific Wnt paralogs (Figure 3.4): *WntA* mRNA expression is restricted to the anterior growth zone and *Wnt4* mRNA is expressed only in the posterior growth zone (Constantinou et al., 2016). The boundary between *WntA* and *Wnt4* corresponds to the posterior boundary of no EdU incorporation (Constantinou et al., 2020). The broad posterior expression domain of *Wnt4* unfortunately does not provide clues to understand which 5% of the growth zone cells are given the cue to enter either S- or M-phase. Additionally, the anterior GZ boundary where cells transition to synchronized S-phase corresponds to the anterior boundary of *caudal* expression. Wnt signaling functions *via caudal* to integrate patterning, metabolism, and cell division in the posterior growth zone in other arthropods (Oberhofer et al., 2014; reviewed in Williams and Nagy, 2017). These data suggest distinct roles for Wnt paralogs in regulating the cell cycle within the growth zone.

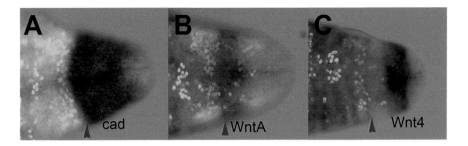

FIGURE 3.4 Expression of segment patterning genes in the growth zone of *Thamnocephalus* coincides with domains of cell cycling. A. *caudal* expression extends throughout the GZ to the border of the newest segment and abuts the band of EdU incorporating cells. B. WntA extends in a gradient through the anterior GZ, while C. Wnt4 extends in a gradient through-out the posterior GZ. In all the photos, the red arrowhead marks the boundary between the anterior GZ and the newest segment. (From Constantinou et al., 2020.)

Thus, the emerging model of the growth zone from *Thamnocephalus*, is scattered, infrequent cell division in the posterior growth zone; rare cell division in the anterior growth zone; and high numbers of cell division in newly formed segments (Figure 3.7A). Actual mitoses in the growth zone are limited. Not only are cells of the growth zone not generating a large amount of additional tissue, they are also cycling more slowly than cells in the already specified segments. Furthermore, the cell cycle in the growth zone is highly regulated both temporally and spatially. By the time cells reach the anterior growth zone, S-phase is completed (cells rarely incorporate EdU in the anterior growth zone); immediately afterward, they synchronize to cycle together in the newly specified segment (all cells are in S-phase; Constantinou et al., 2020). These cell cycle domains correlate with expression of specific Wnt paralogs. This coincident mapping of cell cycling and gene expression domains is consistent with the idea of a posterior signaling region where cells are maintained in a multipotent state until they transit anteriorly and, under new regulatory signals, move toward segment specification.

3.2.1.2 *Oncopeltus*

Our second case study focuses on the patterns of cell division underlying elongation in the hemipteran insect *Oncopeltus fasciatus*. *Oncopeltus* is classified as an intermediate germband embryo (Liu and Kaufman, 2004) and during embryogenesis develops the last nine abdominal segments sequentially from a small posterior region (Ben-David and Chipman, 2010; Birkan et al., 2011; Liu and Kaufman, 2004, 2005; Stahi and Chipman, 2016).

Posterior Cell Division, Coupled with Cell Division in the Already Segmented Region, Is Required for Normal Abdominal Segmentation

Auman et al. (2017) quantified both changes in dimensions and the distribution of cell division throughout the posterior region during the development of the abdominal segments. They compared the cell division in already specified abdominal segments (expressing *invected* stripes), and the growth zone, defined as the region posterior to

the last *invected* stripe. The growth zone shrinks over the course of abdominal segmentation, consistent with the cells in this region contributing to segment addition.
However, as in *Thamnocephalus*, its depletion is not sufficient to account for all the
tissue of the newly added segments. Thus, the growth zone must continue to proliferate, or cells must arrive from elsewhere. There is also a significant degree of growth
in regions of the embryo other than the growth zone. Consistent with this, Auman
et al. (2017) found that the length of a new segment increases as segments continue
to be added behind them.

The Anterior Growth Zone Is Characterized by a Lack of Cell Division

To more precisely define when and where cell division is taking place, Auman et al.
(2017) marked cells with pH3. This led to the discovery that differences in cell division effectively subdivide the growth zone itself into anterior and posterior domains:
the anterior growth zone has very little cell division whereas the posterior has, on
average, over twice as much (Figure 3.5). Auman et al. (2017) also found that no
more than 5% of the cells in any region were undergoing division during the time
points analyzed. The region with already specified segments showed the highest
percentage of proliferating cells. EdU incorporation to determine any patterns of
cells in S-phase was not done.

The Growth Zone Is Regionalized with Respect to Cell Cycling, and Those
Regions Correspond to Changes in Segmentation Gene Expression Patterns

Auman et al. (2017) also found a correlation with the region of reduced/no cell division and a cell's transition from undifferentiated to differentiated. In *Oncopeltus*,
the region of low cell division in the anterior of the growth zone is coincident with
striped *even-skipped* (*eve*) expression *versus* the region of higher cell division in
the posterior that is coincident with *caudal* expression and a broad *eve* expression domain (Auman et al., 2017). Again, the correlation of cell division and gene

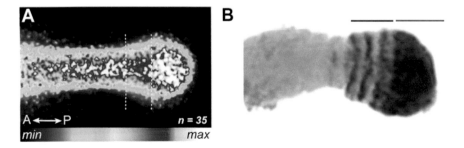

FIGURE 3.5 Rates of mitosis are different in the anterior and posterior GZ in *Oncopeltus*
and correspond to domains of *eve* expression. A. Multiple pH3-stained germbands, 46–54
hours after egg laying, merged into a heat map. The zone of low cell proliferation is indicated
by dotted lines. B. In the anterior GZ (black line), *eve* is expressed in a dynamic striped pattern. In the posterior GZ, *eve* is expressed in a continuous domain (red line). (From Auman
et al., 2017.)

expression domains is consistent with the hypothesis of a posterior signaling region where cells are maintained in a multipotent state until they transit anteriorly, toward segment specification. However, as in *Thamnocephalus*, the broad posterior expression domains of these particular segmentation genes do not provide clues to understand which 5% of the growth zone cells are given the signal to divide.

Evidence for Blastoderm Specification

This model is in contrast to an older study that showed that the cells in the posterior region might not be multipotent at early stages. Lawrence (1973) used x-ray irradiation to fate map the *Oncopeltus* embryo. When cells in the early embryo are irradiated, the irradiated cells and their progeny express a different pigment in the larval epidermis than nonirradiated cells and their progeny. Using this method, Lawrence conducted clonal analyses of abdominal segments. When individual cells were irradiated at the cleavage stage, clones contributed to multiple abdominal segments of the fifth-stage larva. However, when individual cells were irradiated at the blastoderm stage, each clone was restricted to a dorsal/ventral or left/right quadrant within a single abdominal segment of the fifth-stage larva. It should be noted though that the positions of the initial clones were not reported, making it difficult to draw precise inferences about the specificity of single-cell fates and minor contributions of a clone (Lawrence, 1973). The results are nonetheless intriguing and suggest segmental prepatterning precedes sequential segmentation of the posterior. Alternatively, it might simply indicate that the cells irradiated early undergo more cell division. If we abandon the idea that posterior cells are highly proliferative and recognize that a posterior cell in the late blastoderm only undergoes one to three division cycles and does not move around much, clones from single blastoderm cells may, in the majority of cases, ultimately lie within a single segment. However, the fact that all the clones respected segmental boundaries favors the interpretation that cells acquire some degree of segmental fate in the late blastoderm. Additional fate mapping studies would increase our understanding of when cells commit to their segmental fates.

Thus, as in *Thamnocephalus*, a model that emerges is that, at any point in time, the growth zone is shrinking but only a small fraction of cells in the posterior are dividing and additional tissue is required for elongation. The highest rates of division occur within the already segmented regions. The growth zone is regionalized with respect to cell cycling, and regionalized cell cycling corresponds with segmentation gene expression domains.

3.2.1.3 *Tribolium*

Our final case study involves the embryos of the flour beetle *Tribolium castaneum*. *Tribolium* embryos add labial, thoracic, and abdominal segments sequentially in early embryogenesis and thus fit within the category of short germband embryos. As with the other case studies, as segments are added the length and area of the posterior growth zone decreases. Thus, segment addition depletes the field of cells constituting the growth zone. The original growth zone needs to be ~18% larger to equal the total area of all segments added during extension (Nakamoto et al., 2015; note that these measurements did not account for the ingression of the mesoderm in the posterior region, which could potentially increase the percentage growth required by as much as 50%).

Does the Additional Growth Originate from Localized Cell Division in the Posterior?

Surprisingly, the answer to this question is not completely resolved. No evidence for higher rates of division in the posterior were found by counting DAPI-stained nuclei in fixed preparations (Sarrazin et al., 2012; Constantinou and Williams, unpublished observations). Similarly, when cultured early or mid-germband embryos are exposed to EdU for 60 minutes, cells in S-phase are found dispersed throughout the germband (Cepeda et al., 2017). When EdU staining is quantified in these embryos, only early germband embryos show significantly more cells in S-phase in the growth zone relative to the already segmented trunk. The growth zone or SAZ is defined by these researchers as coincident with the posterior *caudal* expression domain. That the cells in the growth zone are not dividing more frequently than trunk cells is supported by the fact that small clones marked along the anteroposterior axis of the blastoderm do not show a significant difference in the number of cell divisions undergone by the end of embryo elongation (Nakamoto et al., 2015). Oberhofer et al. (2014) come to a different conclusion and report that 38% of embryos injected with EdU (10–14 hours after egg laying; 32°C) and cultured for 3 hours show enhanced EdU uptake in the growth zone. The differences between these results are difficult to explain. Unfortunately, neither DAPI staining or EdU uptake, nor measuring division cycles in clones are ideal for quantifying proliferation. DAPI or Hoechst staining capture cells in M-phase, which spans a short segment of the cell cycle. The short time window of mitosis can be overcome with EdU labeling, where labeling records any cell that underwent some or all of S-phase during the duration of the exposure. However, the length of the cell cycle is unknown in *Tribolium* (and in most species), so interpreting how cells in S-phase translate into proliferation rates is difficult, and EdU incorporation cannot be taken as a straight metric of cell division *per se*. Another caveat is that in all these *Tribolium* studies, many, sometimes most, of the EdU or DAPI positive cells are either in the presumptive mesoderm or hindgut. Increased division in these two populations may correlate with their midline invagination or posterior internalization and may not directly contribute to segmental elongation. Without an analysis that distinguishes cells of different presumptive fates, it is difficult to confirm a posterior prevalence of cell division in the segmental anlage in these experiments. Despite these caveats, it is clear that some cell division occurs both in the growth zone and throughout the developing germband at all stages analyzed.

Tribolium *Embryos Undergo Temporal Bursts of Cell Division*

A small pulse of cell division during early segmentation was found by counting DAPI-stained mitotic nuclei in cohorts of staged embryos (Sarrazin et al., 2012; Constantinou and Williams, unpublished observations). Interestingly, when the mitotic measures of staged animals are mapped to developmental stages (Figure 3.6), the relatively higher levels of growth zone mitoses are found when the thoracic segments are being added. Whether this is to supplement thoracic segmental anlage with cells in preparation for limb outgrowth or to pad the growth zone itself for the

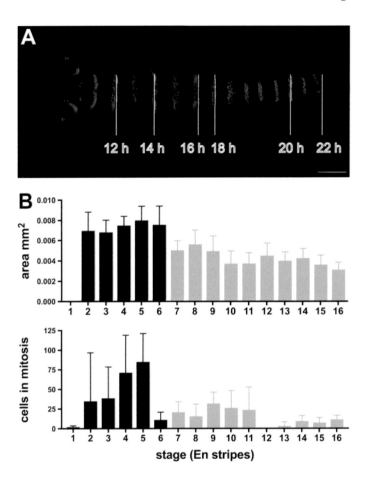

FIGURE 3.6 Higher numbers of mitosis during embryogenesis correspond to the addition of thoracic segments in *Tribolium*. A. Fully extended *Tribolium* germband stained with the 4D9 antibody that recognizes both Engrailed and Invected proteins, with position and numbers of segments added in the previous 2 hours (after egg laying). B. The highest numbers of cells in mitosis occur just prior to thoracic segments are being added (En stripes 4–6). (Panel A from Nakamoto et al., 2015.)

upcoming rapid production of abdominal segments is unknown in the absence of lineage tracing. These counts also do not distinguish between mesodermal *versus* ectodermal fates (as the aforementioned data).

Functional Approaches Confirm a Requirement for Cell Division in Elongation

Other approaches to determining the requirement for cell division in *Tribolium* elongation are consistent with a requirement for a mid-segmentation burst in cell division. To functionally test a requirement for cell division, Cepeda et al. (2017) injected embryos with a cocktail of hydroxyurea and aphidicolin, two inhibitors of DNA replication (and thereby cell division) that work by distinct mechanisms. Hydroxyurea inhibits ribonucleotide reductase, inhibiting DNA replication once nucleotide stores are depleted (Timson, 1975), while aphidicolin inhibits DNA polymerase alpha (Krokan et al., 1981). Embryos treated at an early stage when the first Engrailed/Invected (En/In) stripe is visible, only add an additional four to five En/In stripes, albeit at a slower pace than control embryos. Extension to a roughly similar stage was also seen in experiments that exposed embryos to microtubule inhibitors, expected to arrest the progress of cell division (Macaya et al., 2016). Arrest at this stage correlates well with the hypothesis that the pulse of cell division during thorax formation (En/In stripes 4–6) is required for the subsequent rapid addition of abdominal segments (Nakamoto et al., 2015). It would be interesting to extend the postblock examination of these embryos to see whether En/In stripes past the thorax (T6) were eventually added.

An alternative functional approach to testing a requirement for cell division was reported in the PhD thesis of Fu (2014). Fu injected adult females with dsRNA to knockdown either *cyclin D* or *string* function. Cyclin D (G1-cyclin) initiates the transition from G1 to S-phase (reviewed in Massagué, 2004), while *string* (cdc25) regulates the transition from G2 to M-phase (reviewed in Mailer, 1991) and regulates the three post-blastoderm cell cycles underlying the *Drosophila* mitotic domains (Edgar and O'Farrell, 1990). While the resulting embryonic phenotypes from these dsRNA injections were quite variable, in one distinct class of cuticular phenotypes the abdomen was truncated after the third thoracic segment. In addition to truncated abdomens, additional cuticular phenotypes included tiny heads, small malformed appendages, defects in dorsal closure, and asymmetric segment formation. While still preliminary, these experiments to functionally disrupt the cell cycle confirm a requirement for cell division in elongation, particularly of the abdominal segments.

Is the Tribolium Growth Zone Regionalized?

A subdivision of the *Tribolium* growth zone into distinct regions is apparent from the cycling domains of the pair-rule genes (Patel et al., 1994; Choe et al., 2006; Sarrazin et al., 2012; El-Sherif et al., 2012) as well as the recent mapping of *Dichaete* and *odd paired* expression showing separate stable domains within the posterior *Caudal* domain (Clark and Peel, 2018). But whether the anterior and posterior regions of the growth zone vary in cell cycle regulation has not been reported. Maybe cell cycle domains don't exist or perhaps they haven't been discovered. Regional synchronicity in cell cycles between individuals is easier to detect if collection times are short (e.g., 10 minute collections were used in *Thamnocephalus* hatchlings), at least short enough to not span more than the duration of a single-cell cycle. At the moment, cell cycle duration at any point in development is not known for any species other than *Drosophila*, and collecting large clutches of eggs in short amounts of time isn't feasible in many species.

Computational Modeling Points to Cell Rearrangements As Key Motor behind Elongation

In another approach to determined a requirement for cell division in *Tribolium* elongation, Nakamoto et al. (2015), modeled the rapid phase of abdominal elongation of *Tribolium* embryos (18–20 h 30°C; Engrailed/Invected stripes 7–12). The initial growth zone in the model mirrored length and cell counts in an embryo just prior to abdominal segment addition. Cells in the model did not divide but underwent directional movements paralleling those well documented in an earlier stage embryo (Sarrazin et al., 2012). The simulated elongation mimicked elongation in the embryo. Although this simulation did not account for loss of the midline mesodermal cells, it is consistent with a model in which cell rearrangements rather than a highly proliferative posterior growth drive axial elongation in the abdomen after a pulse of cell division during thoracic segmentation.

Thus, like *Thamnocephalus* and *Oncopeltus*, cell division occurs throughout the germband during *Tribolium* embryo elongation. The size of the growth zone is not sufficient to account for the additional tissue added during segmentation, and cell division is required in both the growth zone and the trunk to elongate the embryo. However, there doesn't appear to be a difference between the number of divisions a cell in either the trunk or growth zone undergoes during elongation. It is not known whether proliferation or DNA replication are restricted in the anterior growth zone, nor whether the cell cycle is regulated through control of S-phase. Novel to this species is the hypothesis that a temporal burst of cell division may fuel abdominal elongation.

3.2.2 WHAT HAVE THESE CASE STUDIES REVEALED?

3.2.2.1 The Growth Zone Typically Requires Mitosis But Only at Low Rates

One thing we can conclude from all three case studies, as well as previous studies in other arthropod species, is that cell division in the posterior is low. Regardless of the original size of the germband or larva, cell division occurs both in the growth zone and throughout the embryo/larva (Figure 3.7; Sarrazin et al., 2012; Auman et al., 2017; Cepeda et al., 2017; Constantinou et al., 2020). In the cockroach *Periplaneta americana*, Pueyo et al. (2008) point out the growth zone has a number of cells in mitosis as do the already formed segments; posterior mitosis is present but is not markedly greater than mitosis throughout the embryo (Chesebro et al., 2013). This pattern has also been observed in the velvet worm *Euperipatoides rowelli* (Mayer et al., 2010) and spiders (Stollewerk et al., 2003). These results are consistent with species that use teloblastic growth from the posterior. Although one row of cells arises from posterior ectoteloblasts to form the initial segmental anlage, two subsequent rounds of cell division of this initial row (intercalary divisions) occur prior to the differential medial/lateral cleavages that occur within each anlage (reviewed in Scholtz and Wolff, 2013; also see Chapter 6). Thus, for each new segment, the cell divisions outside the teloblastic growth zone account for much of the initial elongation. Our case studies are the first to quantify the requirements for cell division in non-teloblastic species, and they report that fewer than 5% of the cells are dividing in the growth zone (Auman et al., 2017; Constantinou et al., 2020). Similarly, where

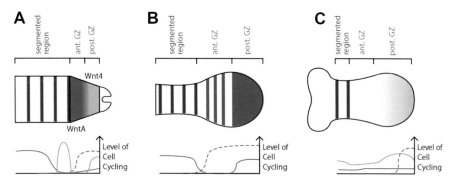

FIGURE 3.7 Overview of cell cycling in the growth zone of three arthropods. At the top, each diagramed embryo shows the posterior region, including specified segments (red stripes) and the growth zone, partitioned into an anterior and posterior region. Expression domains of differ- ent regulators at this stage of early elongation are shown in each embryo. Below each diagram are graphical representations of cell cycle behaviors. Predicted (dotted blue lines) *versus* actual (solid blue lines) levels of mitosis, and cells in S-phase (solid green lines) were known. A. In *Thamnocephalus*, mitosis levels are low (solid blue line) relative to traditional growth zone expectations (dotted blue line). S-phase falls in discrete domains (solid green lines) that map onto Wnt4 (light green) and WntA (blue green) expression domains in the growth zone. The synchronization of cells in S-phase in the last added segment (solid green line) maps to cyclin A expression (not shown). B. In *Oncopeltus*, mitosis levels are low (solid blue line) relative to clas- sical expectations (dotted blue line). Nonetheless they are regionalized within the growth zone, with more mitoses in the posterior domain in a region that corresponds to the broad domain of *eve* expression (purple). C. In *Tribolium*, mitoses are low and do not appear differentially distributed in the growth zone (solid blue line). In young germband embryos, cells in S-phase are more abundant in the posterior, in the region corresponding to the expression of *cad* (shaded blue), than in the mid-trunk region. In no species is the classical expectation of relatively high levels of mitosis borne out. Patterns of cell cycling revealed by S-phase do not directly map to mitoses (green versus solid blue lines). S-phase is likely not a simple index of cell division but instead carefully regulated as cells transit from the posterior growth zone to the anterior growth zone to the newly specified segment. References for data can be found in text.

it has been calculated, the estimated requirement for additional tissue in the posterior is less than expected (Nakamoto et al., 2015; Auman et al., 2017; Constantinou et al., 2020). There is, nevertheless, a demonstrated need for additional cells to account for all the segmental anlage, and cell division in a small percentage of cells in the posterior is required to meet this need.

3.2.2.2 The Anterior Growth Zone Functions As a Transition Zone in Segmental Specification

Both *Oncopeltus* and *Thamnocephalus* show different cell cycle behaviors in the anterior *versus* posterior growth zone, in particular, the anterior appears devoid of cycling (Figure 3.7A, B). Interestingly, in a comparison of cell death to normal mito- sis during knockdown of pair-rule genes in an intermediate germband dermestid beetle, Xiang et al. (2017) used pH3 staining to reveal a region of mitosis in the pos- terior germband separated from the rest of the germband by a region lacking mitosis.

The dermestid, like other species examined, also had cell division throughout the germband. In *Oncopeltus* and *Thamnocephalus,* the differential cell cycling behaviors that subdivide the growth zone map to expression domains of the segmentation genes. In *Thamnocephalus,* WntA is expressed in the anterior growth zone where no DNA replication is observed and could function to inhibit S-phase in the anterior growth zone. Similarly, both WntA and *caudal* have boundaries coincident with the activation of a synchronized S-phase. However, the gene expression domains in the posterior growth zone of both *Oncopeltus* and *Thamnocephalus* cover the entire posterior part of the growth zone and do not map directly to the populations of cells in M- or S-phase in the growth zone. This is distinct from the well-characterized segmentation gene expression and mitotic domains in *Drosophila* (Foe, 1989; Edgar and O'Farrell, 1989; Bianchi-Frias et al., 2004).

3.2.2.3 Cell Division Is Highly Regulated and Regionalized in Both the Growth Zone and Trunk

Although cell proliferation is not abundant in the growth zone of many species, it is required and it can be highly regulated either spatially or temporally (Figure 3.7; Auman et al., 2017; Sarrazin et al., 2012; Constantinou et al., 2020). We and others have observed a temporal pulse of cell division in the *Tribolium* embryo. In vertebrates, recent studies have shown that a precisely regulated pattern of proliferation, arrest, and proliferation is necessary during elongation for proper differentiation of somites (Bouldin et al., 2014). Similarly, the posterior cells in *Thamnocephalus* undergo a highly regulated cell cycle as cells transit from the posterior to the anterior growth zone and finally become specified as a new segment. We predict that dynamic temporal regulation of cell cycling will be found in many arthropods.

Rosenberg et al. (2014) were the first to document segmentally iterated mitotic domains that appear progressively in the developing trunk segments of hymenopteran *Nasonia*: growth appears to be fueled by segmentally iterated mitotic domains that appear in an anteroposterior progression (Rosenberg et al., 2014). These domains correlate with the iterated bands of *Nasonia eve* expression. The synchronized band of cells undergoing S-phase in the most recently specified *Thamnocephalus* trunk segments also correlate with segmentation gene expression (Constantinou et al., 2020). This provides support for the hypothesis of Rosenberg et al. (2014) that use of coordinated cell cycle domains is a strategy that seems to have evolved multiple times. This apparent coordination of mitotic or replication domains and segmentation gene expression is likely an ancestral feature of arthropod development.

3.3 THE ROLE OF CELL REARRANGEMENT

The original designations of egg types of insects (Krause, 1939) focused on the length of the germ anlage relative to the anteroposterior axis of the egg and the number of presumptive segments that form from the growth zone (or from an alternative perspective, the number of segments specified in the germ anlage before the onset of gastrulation). The presumption then was that segments not yet specified originate from a short posterior growth zone, which undergoes substantial growth as additional segments are defined. As discussed earlier, this presumption fails in a number of species, and proliferative growth underlying elongation in both short and intermediate germband insects occurs throughout the length of the germband. This failure to observe the logically predicted growth in the posterior, coupled with the observation that many early germ anlage undergo substantial changes in their shape—for example, the short and wide *Bombyx* embryo undergoes a complete change in aspect ratio, or the bulbous posterior of a *Tribolium* embryo extends and narrows—led researchers to focus their attention more on the role of orchestrated cell rearrangements as an essential mechanism underlying segmentation and elongation.

The well-studied *Drosophila* embryo has been the arthropod model for elongation *via* cell rearrangement, or convergent extension. Despite the fact that the primordia for all *Drosophila* segments are visible in the late blastoderm, the germband increases in length more than twofold (Hartenstein and Campos-Ortega, 1985). This extension post-segmentation has been the main lens through which to consider other species that elongate while adding segments. Recently, however, advances in live cell imaging have radically improved our ability to describe movements of cells during germband elongation in other arthropod species. We begin our discussion of cell rearrangements with a brief, general summary of cell rearrangement and provide an overview of convergent extension in *Drosophila*. We then describe what is known about cell rearrangements in sequentially segmenting arthropods, focusing on the best-studied case, *Tribolium*. We predict that, as more species are characterized, the cellular mechanisms underlying elongation in arthropod embryos and larvae will be diverse.

3.3.1   CELL REARRANGEMENTS FROM DIVERSE TAXA HAVE SOME COMMON FEATURES

In animals as different as frogs and flies, and in processes as distinct as gastrulation and germband extension, intercalary cell rearrangements show many commonalities (reviewed in Walck-Shannon and Hardin, 2014; Keller et al., 2000; Keller, 2006; Shindo, 2017; Huebner and Wallingford, 2018). The described cases of convergent extension all use two main cellular mechanisms to rearrange cells: bipolar basal cell protrusions and junctional remodeling, although they vary in the degree to which these two mechanisms are employed (reviewed in Zallen and Goldstein, 2017; Huebner and Wallingford, 2018). Basal cell protrusions make stable contacts with neighboring cells and use these contacts to exert traction. Protrusions are oriented along the mediolateral axis and the tissue elongates as the cells intercalate (Keller and Tibbetts, 1989; Wilson and Keller, 1991). Junctional remodeling requires the repositioning of the apical junctions that adhere cells to one another (Bertet et al., 2004; Blankenship et al., 2006). Repeated cycles of junctions contracting along the mediolateral axis and extending along the anteroposterior axis function to elongate the *Drosophila* embryo (Bertet et al., 2004; Blankenship et al., 2006). Most of the described cases of convergent extension also share common regulatory features, relying on planar cell polarity (PCP) signaling, a conserved protein network that establishes cell polarity in many different types of tissues (reviewed in Butler and Wallingford, 2017). Loss of PCP function disrupts junctional remodeling (Lienkamp et al., 2012; Nishimura et al., 2012; Shindo and Wallingford, 2014; Williams et al., 2014) and may be indirectly involved in basal protrusive activity during convergent extensions (see Huebner and Wallingford, 2018). Although PCP function in tissue polarity was first discovered in *Drosophila,* the PCP pathway is not required for *Drosophila* germband extension (Zallen and Wieschaus, 2004). Here we briefly describe the mixture of junctional remodeling and bipolar basal cell protrusions used during *Drosophila* germband extension, then discuss what is known from the best-studied sequentially segmenting insect, *Tribolium.*

3.3.2 CONVERGENT EXTENSION DRIVES ELONGATION IN DROSOPHILA

3.3.2.1 Elongation in *Drosophila* Occurs Primarily by Junctional Remodeling

In *Drosophila*, cells in the blastoderm acquire both anteroposterior (AP) and dorsoventral (DV) patterning prior to germband elongation. The elongating field of cells occupies almost the entire egg, along both the AP and DV axes, and the cells are relatively uniform in size and shape. During elongation, rows of cells along the DV axis simultaneously intercalate, collapsing the cell field along the DV axis and lengthening it along the AP axis (Figure 3.8A; Irvine and Wieschaus, 1994). This intercalation is the primary mode of elongation, although oriented cell division occurs in the posterior of the embryo during late elongation (da Silva and Vincent, 2007). Recently, basal–lateral protrusions have been discovered to be required as an additional and, to some degree, independent mechanism for germband elongation (Sun et al., 2017). Prior to junctional remodeling at the apical surface of the cells, basolateral protrusions arise in DV cells. These actin-rich protrusions require Rac (a Rho GTPase) for DV cells to actively migrate toward and sometimes past one another as they intercalate. Blocking apical junctional remodeling does not halt basolateral protrusions and when protrusions are blocked alone, germband extension is compromised, suggesting that the two mechanisms operate in parallel but somewhat independently to extend the germband (Sun et al., 2017).

3.3.2.2 Intra- and Intercellular Effectors of Cell Movements Are Polarized in *Drosophila*

In *Drosophila*, the contractile cell intercalation driving elongation apically depends on the asymmetrical distribution of proteins associated with adherens junctions: myosin II and F-actin concentrate between anterior and posterior faces of adjacent cells, while DE-cadherin, armadillo/β-catenin, and Bazooka/PAR-3 concentrate in the DV interfaces. This polarization within cells is required for differential junctional remodeling between neighboring cells: membrane in the DV axis is reduced, while membrane contacts along the AP axis increase in a fashion that depends directly on myosin II (Bertet et al., 2004; Zallen and Wieschaus, 2004). Differential junctional remodeling results in active cell–cell intercalation along the DV axis, which can be coordinated between quartets of cells or larger multicellular rosettes (Bertet et al., 2004; Blankenship et al., 2006).

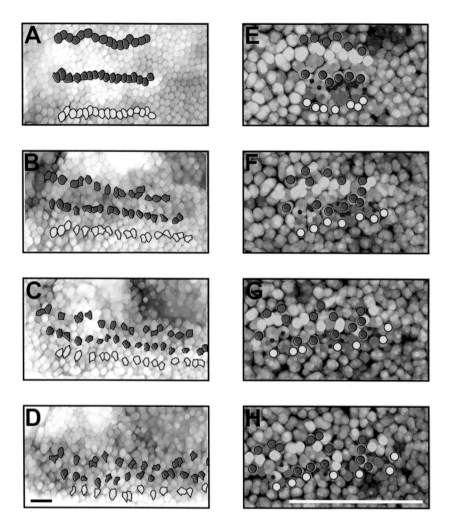

FIGURE 3.8 Comparison of cell interaction in *Tribolium* versus *Drosophila* shows less ordered cell movements. A–D. Traced cells in the extending *Drosophila* germband move almost in register together along the DV axis and spread in a correspondingly equidistant way along the AP axis. E–H. Traced cells in the extending *Tribolium* germband, while clearly intercalating, show unequal movement in both the DV and AP axes. (Panels are modified from Irvine and Wieschaus, 1994, and Benton, 2018.)

3.3.2.3 Pair-Rule Genes Drive Periodic Expression of the Toll Receptors Required for Convergent Extension

Germband elongation relies on inputs from the segmentation genes that pattern the AP axis, specifically, the pair-rule genes. In *Drosophila*, pair-rule genes form a series of stripes alternating along the AP body axis; these genes provide positional information to intercalating cells. Pair-rule gene mutants (*even-skipped* [*eve*], *runt*, *odd paired* [*odd*]) lose polarized distribution of intracellular effector molecules that drive cell intercalation (Zallen and Wieschaus, 2004; Blankenship et al., 2006; Simões et al., 2010) and fail to elongate normally (Irvine and Wieschaus, 1994; Blankenship et al., 2006). Conversely, mutants that ectopically express the pair-rule genes *eve* and *runt* cause local reorientation of the polarized effector molecules (Irvine and Wieschaus, 1994; Zallen and Wieschaus, 2004). Paré et al. (2014) identified three Toll family receptors as downstream targets of the pair-rule genes during embryo elongation by comparing the transcriptomes of mutant embryos lacking the pair-rule genes *eve* and *runt* to wild-type *Drosophila* embryos. These leucine-rich receptors are differentially expressed in stripes of cells along the AP axis. Based on *Toll* mutant phenotypes, they propose a model in which heterotypic binding between adjacent stripes drives the differential accumulation of intracellular effector molecules, e.g., MyoII, to those cell interfaces (Paré et al., 2014). There is evidence that basolateral protrusions also depend on proper AP axial patterning since basolateral rosette formation was suppressed in *eve* and *Toll* RNAi embryos (Sun et al., 2017).

3.3.3   CELL REARRANGEMENTS IN SEQUENTIALLY SEGMENTING ARTHROPODS: THE *TRIBOLIUM* MODEL

How comparable is this model from *Drosophila* to what is known in other arthropods? *Tribolium* is one of the few sequentially segmenting arthropods that has been analyzed with any molecular detail in regard to cell rearrangement. The first notable movements of cells in the *Tribolium* blastoderm are of a population of cells that condense in the posterior of the egg to form the germband and become distinct from more anterior blastoderm cells destined to form extraembryonic tissues (see blastoderm fate map in Benton, 2018). Thus, the embryonic cells have already migrated significantly from a relatively uniform blastoderm field prior to elongation of the germband *per se*. Subsequently, elongation occurs progressively, in the posterior region as new segments are specified. In addition to this apparent progressive elongation, live imaging of embryos shows periodic whole germband contractions, suggestive of myosin contractility. Overall the process of elongation in *Tribolium* is phenomenologically distinct from *Drosophila* in that it occurs in a changing cell field, the boundaries of which are changing as new segments form, and has dynamic regulatory inputs, as we describe in the following.

3.3.3.1   Live Imaging Shows Clear Convergent Extension in the *Tribolium* Germband

The dynamic behavior of cells during *Tribolium* embryo elongation has recently been analyzed using live imaging and fluorescent reporters. By tracing a subset of cell movements over one hour in the early germband (during the transition from pair-rule stripe 3 to pair-rule stripe 4), Sarrazin et al. (2012) provided evidence that cells in the early germband undergo medially directed movements that drive convergent extension: cells in the anterior of the germband move medially and anteriorly, cells in the posterior of the germband move medially and posteriorly as the germband narrows. Subsequently, Benton et al. (2013) tracked seven rows of cells during a slightly earlier stage, in what appear to be the anterior region of the lateral ectoderm (location in the embryo was not indicated but it appears to be the region that will express pair-rule stripes 1 and 2). They showed cells rearrange mediolaterally as the tissue lengthens during the hour of filming, providing clear evidence for convergence and extension. More recently, Benton (2018) also tracked cells in the posterior growth zone during abdominal segment formation. His rows of marked cells clearly mix and elongate during the 3 hours (25–28°C) of filming.

3.3.3.2 *Tribolium* Convergent Extension Does Not Appear to Arise from a *Drosophila*-Like Neighbor-Sliding Mechanism

Although these *Tribolium* studies show clear convergent extension behavior, what is occurring mechanistically and how comparable it is to *Drosophila* is not clear. One way to consider this is to compare marked cells in the *Drosophila* germband to those in the elongating *Tribolium* germband (Figure 3.8). Cell tracking in *Drosophila* clearly shows the orderly movement of cells (Irvine and Weischaus, 1994) by the coordinated junctional remodeling and basolateral interdigitation described earlier. Rows of marked cells extend in the AP direction as regularly spaced unmarked cells move past them. Rows remain roughly parallel as they move toward one another along the DV axis (Irvine and Weischaus, 1994) and the endpoint shows each of the marked cells separated by a single unmarked cell—the classic model of cell intercalation. By contrast, while rows of cells along the AP axis intermingle in *Tribolium*, they do not appear to do so based on the kind of orderly underlying mechanism documented in *Drosophila*. Some cells change neighbors during elongation, others do not. Overall, there appears to be differential movement both along the AP and DV axes (Figure 3.8B). In addition, in neither live imaging study were the rosettes of cells characteristic of junctional remodeling or basal protrusions reported, nor have polarized effectors of cell movement been documented in *Tribolium* (see later). Thus, while both species undergo convergent extension, the tracked movements of the cells show marked differences.

3.3.3.3 How Do Posterior Cells in *Tribolium* Converge and Extend?

Following clonal fates over long time periods reveals additional data about cell rearrangements in *Tribolium*. By following small clones of cells marked with a caged dye at the blastoderm stage, we found that clones marked in the posterior blastoderm elongate disproportionately compared with clones from more anterior regions of the blastoderm (Nakamoto et al., 2015). Posterior clones undergo extensive rearrangement and form narrow, elongated clones that extend through four or five segments. By contrast, anterior clones extend through only one or two segments. This switch in clone distribution occurs at the boundary between addition of thoracic and abdominal segments, but is not due to cell division since the doubling of anterior and posterior clones was not significantly different. Interestingly, the clones rarely mixed with nonlabeled cells (although they sometimes did), even in the posterior when they undergo extreme elongation to form very narrow clones (Figure 3.9A, B). In contrast, cells in the mesodermal layer undergo clear intercalary movements (Figure 3.9C, D). Despite extensive cell rearrangement, posterior clones do not show evidence of arising from repeated rounds of the kind of neighbor-sliding intercalation described in *Drosophila* (as shown in Figure 3.8A). At present, we do not know how the descendants of the clones maintain connection. The same kinds of contiguous, elongated

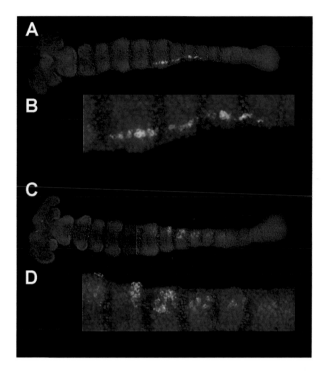

FIGURE 3.9 Blastoderm clones scored after completion of elongation. All panels show clones of roughly 20 cells marked with a photoactivated dye in the late blastoderm stage between 50–60% egg length. The fate of the clones was scored in fully elongated embryos stained with the 4D9 antibody that detects the Engrailed/Invected proteins. Because DV position cannot be determined on the blastoderm, some clones mark ectodermal precursors (A; B, higher magnification of clone in A), while others mark mesodermal precursors (C; D, higher magnification of clone in C). The ectodermal clones form remarkably elongated clusters of cells that are only occasionally interrupted by nonlabeled cells. The mesodermal clones behave quite differently and distribute across several segments, restricted to non-Engrailed/Invected expressing cells.

clones were found by Benton (2018, *e.g.* his Supplemental Figures 3 and 4). In some cases, a region of nonactivated cells breaks the integrity of the clone, however, never in the regular interdigitating fashion seen in the *Drosophila* germband. In sum, all studies indicate that cells undergo some kind of convergent extension to elongate, but the mechanism remains elusive.

3.3.3.4   Role of Pair-Rule Genes in Elongation in *Tribolium*

Despite distinct cell movements, *Tribolium* shares with *Drosophila* some regulatory control of elongation: in both species, elongation depends on pair-rule gene function. During *Tribolium* elongation, pair-rule expression is dynamic in the posterior, with cells undergoing oscillatory expression of pair-rule genes while they undergo the movements that constitute convergence and extension (Figure 3.3B; Choe et al., 2006; Sarrazin et al., 2012; El-Sherif et al., 2012). Cellular oscillations of pair-rule gene expression are regulated by a posterior gradient of Caudal protein (El-Sherif et al., 2014) and produce a wave of expression that travels from the posterior anteriorly during elongation. When it reaches the anterior, the wave stabilizes, then splits into two stripes. The result is the progressive formation of stable, segmentally iterated stripes, first, of pair-rule genes, subsequently of segment polarity genes (e.g., *engrailed*; Sarrazin et al., 2012; El-Sherif et al., 2012). Loss of *caudal* function results in severely truncated embryos that fail to undergo the wild-type cell movements that restructure the blastoderm into the initial germband (Copf et al., 2004; Schoppmeier et al., 2009; Benton et al., 2013). Parental and embryonic RNAi knockdowns of the *Tribolium* pair-rule genes *eve*, *odd*, and *runt* also result in truncated embryos: arrest occurs just after the initial condensation of the blastoderm cells into the germband (Choe et al., 2006; Nakamoto et al., 2015). Moreover, we have demonstrated that clones marked in the blastoderm do not elongate in *eve* knockdowns in *Tribolium*, consistent with a role for *eve* in regulation of elongation (Figure 3.4; Nakamoto et al., 2015). Thus, both flies and beetles show a similar failure to elongate in pair-rule gene mutants. However, the timing of the onset of cell movement relative to the expression of pair-rule genes differs dramatically between the two species. While the *Drosophila* pair-rule genes undergo dynamic expression patterns, those dynamics have stabilized in all but the most posterior segments prior to the onset of germband elongation (Kuhn et al., 2000; Clark, 2017; Clark and Peel, 2018). In *Tribolium*, dynamic pair-rule expression patterns accompany elongation. Clark and Peel (2018) argue that the upstream regulation of segment patterning is essentially the same in both species but varies mostly in temporal readout in long germband versus short germband insects. Whether the same can be hypothesized about the regulation of elongation *per se* remains unknown.

3.3.3.5   Role of Toll Receptors in *Tribolium*

Knockdown of two Toll family receptors in *Tribolium* disrupts cell rearrangements in the early embryo (Benton et al., 2016), although cell movements in later embryos were not examined. These experiments led to the hypothesis that, like *Drosophila*, the pair-rule genes are regulatory intermediates in a cascade that regulates cell rearrangements at least in part through the activity of Toll family receptors (Benton et al., 2016). There are, however, significant differences in how and when the Toll stripes are expressed relative to cell movements in both species. In *Drosophila*, each segmental anlage is four cells wide and each cell row has a distinct combination of Toll receptor expression, with proposed heterophilic interactions between cell rows. The differential expression is established prior to any of the movements of germband elongation (Pare et al., 2014). In *Tribolium*, a stripe of double segment periodicity (between five and six cells wide) of Toll7 is flanked by a stripe of double segment periodicity of Toll10, thus very few rows of cells would experience heterophilic interactions. In addition, these stripes are added sequentially from the posterior, while the embryo is elongating. Whether cells express different Toll receptors prior to the onset of their cell movements or whether Toll receptor interactions drive differential accumulation of actomyosin is not known. In sum, whether the same cellular mechanisms underlie the loss of function phenotypes in both species remains unclear.

3.3.3.6   Polarized Effectors of Cell Movement Have
Not Been Documented in *Tribolium*

In *Drosophila*, subcellular polarized distribution of motor or adhesion proteins is required for normal apical junctional remodeling (reviewed in Zallen, 2007; Zallen and Blankenship, 2008): myosin II, F-actin, DE-cadherin, armadillo/β-catenin, and Bazooka/PAR-3. Similarly, actin and lipid phosphatidylinositol 3,4,5-trisphosphate (which promotes actin polymerization) are found in basolateral protrusions (Sun et al., 2017). In *Tribolium*, polarized candidates have yet to be reported. Furthermore, RNAi and CRISPR knockdowns of the *Tribolium* homolog of E-cadherin are reported to have no effect on either elongation or segmentation, although they disrupt dorsal closure (Gilles et al., 2015). Interestingly, the mutant larva illustrated publication are significantly shorter than the wild-type.

In *Drosophila* germband extension, PCP pathway genes are not required (Zallen and Wieschaus, 2004). However, in *Tribolium,* preliminary knockdowns of PCP (*flamingo, fat,* and *dachsous*) and JNK (*basket* and *hemipterous*) pathway regulators produce larval cuticle phenotypes that include truncation and failure of dorsal closure (Fu, 2014). Since many cuticles were empty, Fu (2014) also analyzed embryonic phenotypes and found evidence that PCP pathway genes were required for proper embryo condensation from the blastoderm. While preliminary, these data suggest a possible role for PCP genes in convergent extension in *Tribolium* at both early and late stages, in contrast to the lack of PCP regulation in *Drosophila* germband extension.

3.3.3.7   Possible Hypotheses for Mechanisms of
Convergent Extension in *Tribolium*

Novel mechanisms may underlie the cellular rearrangement during posterior segmentation in *Tribolium*. One hypothesis is that tissue-level forces moving cells medially, coupled with differential adhesive behaviors could create both cell mixing and elongated clones. The ectodermal tissue that elongates in *Tribolium* is a quite narrow tissue, bounded medially by the invaginating mesoderm and laterally by the edge of the germband. The medial ingression of mesoderm and the geometry of the tissue borders could be a "tissue-level polarizer" that directs polarized elongation of the embryo. Boundary effects have been proposed to play a role in the elongating ascidian notochord (Weliky et al., 1991) as well as more generally in migrating tissues (Vedula et al., 2013). Backes et al. (2009) argued that the physical edge of an adjacent tissue can lead to an elongated arrangement of cells, even in the absence of polarized cell behaviors. If boundaries that run along the lateral and medial edges are significant mechanical players in elongation, we might predict DV genes play a significant role. The Bmp gradient of the zebrafish gastrula guides migrating lateral cells by regulating cell–cell adhesion (von der Hardt et al., 2007). However, we have found that modeling cell movements in the *Tribolium* germband using forces that drive cells medially and posteriorly (mimicking those tracked by Sarrazin et al., 2012) in the absence of differential adhesion can create clones that mix and interdigitate without an active neighbor-sliding mechanism (Hester, Williams, and Nagy, unpublished).

Although not yet fully explored or characterized, it is interesting to note that the knockdown of Toll7 and Toll10 in *Tribolium* germbands (Benton et al., 2016) not only halts intercalation but also results in distinct changes in cell shape: cells constrict globally causing a shrinking of the tissue. Cell shape change has not been a typical focus of live imaging as yet but may be significant. For example, Pechmann (2016) describes the formation of the germ disc in *Parasteatoda tepidariorum* arising from a combination of changes in cell shape that drive tissue movements. The loose pyramidal cells of the early blastoderm condense on one side of the egg forming a cuboidal epithelium while the extraembryonic cells spread and flatten. Condensation of the germ disc is disrupted by an actin polymerization inhibitor (cytochalasin D) but not a microtubule polymerization inhibitor (colchicine).

3.4 SYNCHRONIZING CELL DIVISION, CELL REARRANGEMENTS, AND CELL FATE

Because the available studies of elongation have revealed highly regulated temporal and spatial sequences of cell cycling and/or cell movements, an obvious question to ask is how the progressive specification of cell fates during segmentation is linked to the cell behaviors described earlier. Where and when are segmental progenitor cells restricted to their fates in different arthropod species? Are cells in the growth zone initially multipotent? Are cell fates and cell movements mechanistically linked?

Most species share the same general behaviors: some cells in the ventral epithelium that comprises the posterior growth zone move out of the epithelium and become mesodermal, while the remaining cells move through the anterior growth zone and transition to their segmental fates. We can propose a generic model for cell fate specification, based on inferences from multiple species. That the growth zone functions to maintain cells in a multipotent state has been previously proposed as a consequence of posterior Wnt signaling (McGregor et al., 2008, 2009; Chesebro et al., 2013; Oberhofer et al., 2014). In the emerging models of arthropod growth, cells experience a temporal sequence of transcription factors and/or signals, set up by a posterior Wnt signal, that provide a roadmap for both segmental fate, growth, and cell movement. Cells in the posterior region of high Caudal are presumed to be held in an undifferentiated state. These cells already have information as to their DV position, established in the blastoderm (as shown in Stappert et al., 2016 for *Tribolium*). Dynamic waves of determinants of cell fate, e.g., pair-rule genes (El-Sherif et al., 2012; Sarrazin et al., 2012) or the Notch ligand Delta (Chesebro et al., 2013), pass through this region, but do not stabilize, or presumably influence cell fate, until they are outside the region of high Caudal expression. When released from the posterior Caudal signal, cells begin their journey down a transcriptional regulatory path that allows progressive determination of fate: position within a segment, neural vs. ectodermal, tagmatic position, etc., ultimately resulting in the complete differentiation of the insect body plan. These cell fate decisions are realized as cells rearrange their positions within the embryo. Based on the temporal dynamics of gene expression, Cad (and Dichaete [Clark and Peel, 2018] or Cad and Delta [Chesebro et al., 2013]) expressing cells would be at least multipotent (their ability to differentiate into serosa, amnion, mesoderm may have already ended).

Interestingly, some experimental data suggests posterior blastoderm cells in insects may not be multipotent, e.g., the *Oncopeltus* experiments described earlier

(Lawrence, 1973) or the *Bombyx* blastoderm irradiation experiments (Myohara, 1994). While there may be alternative explanations for these nettling results, in truth, no experimental data has verified that cells in the arthropod posterior growth zone are multipotent. In vertebrates, progenitor cells that lie upstream of somite formation use Cdx (the vertebrate *caudal* ortholog) expression to maintain Wnt and Fgf signaling in the growth zone. In mice, the Cdx/Wnt signaling forms a feedback circuit which, if broken, leads to premature differentiation through a failure to clear retinoic acid from the posterior (Young et al., 2009)—strong evidence that the feedback loop maintains pluripotency. Support for a role for the Cdx/Wnt feedback loop in maintaining pluripotency in the presomitic mesoderm is also confirmed in experiments that replicate somitogenesis from induced pluripotent stem cells in culture. Similar feedback loops have been demonstrated in cockroach embryos, where there is evidence for a clocklike mechanism based on Notch signaling instead of a pair-rule circuit: the Notch ligand Delta acts both up- and downstream of Wnt signaling during segmentation (Chesebro et al., 2013). But unlike the vertebrate embryo, loss of function of the components of the feedback loop results in a failure to elongate, but do not appear to result in premature differentiation of tissue: premature appearance of segment polarity stripes of *wingless* in the posterior were not detected.

Ideally, we would experimentally test a cell's determination through transplantation to different locations and score their subsequent development for segment specific markers. Unfortunately, such direct measures of cell determination are impossible in most arthropod embryos. However, in the current era of single-cell RNA-seq, comparing single-cell transcriptome profiles in anterior and posterior growth zones over time may clarify patterning signals that both limit and promote the fate of posterior cells, and propose mechanistic links between cell fate, cell division, and cell movement.

3.5 SUMMARY

In the preceding discussion, we focused our attention on the role of cell division and cell rearrangement in embryo elongation. Both types of cell behaviors figure prominently in arthropod embryo elongation, and the relative degree to which an embryo relies on one or the other is a key variable in evolution. Recent quantitative studies on cell division allow us to confirm that most arthropods do not have dedicated posterior stem cells that fuel elongation, nor has a canonical region of high posterior proliferation been discovered. Whether the initial embryonic primordia is small or large, embryo elongation is fueled by cell division throughout the embryo in all cases analyzed. Despite the fact that actual mitoses appear much less frequently than predicted in the growth zone, cell division is nonetheless present and plays a role in supplying new tissue to the embryo. One notable, and currently less explored, characteristic of cell cycling is that it is highly regulated in both space and time. The growth zone has anterior and posterior domains of cell cycling and even the speed of the cell cycle varies consistently, at least in *Thamnocephalus*. One exciting avenue of future research will be to document cell cycling, both M- and S-phase, more carefully in more species; uncover the regulators of cell cycling; and examine how that regulation is coordinated with segmental patterning in sequentially segmenting arthropods.

Beyond cell division, simple observations of development across arthropods make it clear that cell movements that drive convergent extension play a significant role in elongation but are currently less well characterized across arthropods (Benton et al., 2013, Benton 2018; Sarrazin et al., 2012; Nakamoto et al., 2015). In *Tribolium*, convergent extension plays an important role in elongation, and interestingly, cell movements are temporally variable at least in extent and, possibly, in their mechanistic basis. Elements of a *Drosophila* model—in which pair-rule genes drive the differential expression of multiple leucine-rich receptors, which promote differential adhesion through heterophilic interactions, that in turn affect intracellular myosin localization—has some traction as a conserved mechanism to link segmental patterning to cell movement (Benton et al., 2016). Future studies to link interactions among these receptors to the mechanical forces that move cells in more species will be important. We predict that detailed studies of embryo elongation in other species are likely to reveal additional combinations of cell division and cell movements.

ACKNOWLEDGMENTS

The authors would like to thank Susan Hester, Savvas Constantinou, and Ayaki Nakamoto for many interesting discussions about the work reviewed here; Paula Storrer and Netta Kasher for figure graphics, and Ariel Chipman for his dedication to getting this volume completed. The authors research was supported by the National Science Foundation [NSF-IOS 1024220 and 1322350 to T. Williams; NSF-IOS 1024446 and 1322298 to L. Nagy].

REFERENCES

Anderson, D. T. 1967. Larval development and segment formation in the branchipod crustaceans *Limnadia stanleyana* King (Conchostraca) and *Artemia salina* (L.) (Anostraca). *Aust. J. Zool.* 15: 47–91.

Anderson, D. T. 1973. *Embryology and Phylogeny in Annelids and Arthropods*.

Auman, T., B. M. I. Vreede, A. Weiss, S. D. Hester, T. A. Williams, L. M. Nagy, and A. D. Chipman. 2017. Dynamics of growth zone patterning in the milkweed bug *Oncopeltus fasciatus*. *Development* 144: 1896–1905.

Backes, T. M., R. Latterman, S. A. Small, S. Mattis, G. Pauley, E. Reilly, and S. R. Lubkin. 2009. Convergent extension by intercalation without mediolaterally fixed cell motion. *J. Theor. Biol.* 256: 180–186.

Ben-David, J., and A. D. Chipman. 2010. Mutual regulatory interactions of the trunk gap genes during blastoderm patterning in the hemipteran *Oncopeltus fasciatus. Dev. Biol.* 346: 140–149.

Benton, M. A. 2018. A revised understanding of *Tribolium* morphogenesis further reconciles short and long germ development. *PLoS Biol.* 16: e2005093.

Benton, M. A., M. Akam, and A. Pavlopoulos. 2013. Cell and tissue dynamics during *Tribolium* embryogenesis revealed by versatile fluorescence labeling approaches. *Development* 140: 3210–3220.

Benton, M. A., M. Pechmann, N. Frey, D. Stappert, K. H. Conrads, Y. T. Chen, …, S. Roth. 2016. Toll genes have an ancestral role in axis elongation. *Curr. Biol.* 26: 1609–1615.

Bertet, C., L. Sulak, and T. Lecuit. 2004. Myosin-dependent junction remodeling controls planar cell intercalation and axis elongation. *Nature* 429: 667–671.

Bianchi-Frias, D. A. Orian, J. J. Delrow, Vazquez J, A. E. Rosales-Nieves, S. M. Parkhurst, and M. Parkhurst. 2004. Hairy transcriptional repression targets and cofactor recruitment in *Drosophila*. *PLoS Biol.* 2: E178.

Birkan, M., N. D. Schaeper, and A. D. Chipman. 2011. Early patterning and blastodermal fate map of the head in the milkweed bug Oncopeltus fasciatus. *Evol. Dev.* 13: 436–447.

Blair, S. S. 2008. Segmentation in animals. *Curr. Biol.* 18: R991–R995.

Blankenship, J. T., S. T. Backovic, J. S. P. Sanny, O. Weitz, and J. A. Zallen. 2006. Multicellular rosette formation links planar cell polarity to tissue morphogenesis. *Dev. Cell* 11: 459–470.

Bouldin, C. M., C. D. Snelson, G. H. Farr, and D. Kimelman. 2014. Restricted expression of cdc25a in the tailbud is essential for formation of the zebrafish posterior body. *Genes Dev.* 28: 384–395.

Brown, S. J., N. H. Patel, and R. E. Denell. 1994. Embryonic expression of the single *Tribolium engrailed* homolog. *Dev. Genet.* 15: 7–18.

Butler, M. T., and J. B. Wallingford. 2017. Planar cell polarity in development and disease. *Nat. Rev. Mol. Cell Biol.* 18: 375–388.

Cepeda, R. E., R. V. Pardo, C. C. Macaya, and A. F. Sarrazin. 2017. Contribution of cell proliferation to axial elongation in the red flour beetle *Tribolium castaneum*. *PLoS One* 12: e0186159.

Chesebro, J. E., J. I. Pueyo, and J. P. Couso. 2013. Interplay between a Wnt-dependent organiser and the Notch segmentation clock regulates posterior development in *Periplaneta americana*. *Biol. Open* 2: 227–237.

Chipman, A. D. 2010. Parallel evolution of segmentation by co-option of ancestral gene regulatory networks. *Bioessays* 32: 60–70.

Choe, C. P., S. C. Miller, and S. J. Brown. 2006. A pair-rule gene circuit defines segments sequentially in the short-germ insect *Tribolium castaneum*. *Proc. Natl. Acad. Sci. U.S.A.* 103: 6560–6564.

Clark, E. 2017. Dynamic patterning by the *Drosophila* pair-rule network reconciles long-germ and short-germ segmentation. *PLoS Biol.* 15: e2002439.

Clark, E., and A. D. Peel. 2018. Evidence for the temporal regulation of insect segmentation by a conserved sequence of transcription factors. *Development* 145: dev155580.

Constantinou, S. J., N. Duan, L. M. Nagy, A. D. Chipman, T. A. Williams. 2020. Elongation during segmentation show as axial variability, low mitotic rates and synchronized cell cycle domains in the crustacean, *Thamnocephalus platyurus*. *EvoDevo* 11: 1.

Constantinou, S. J., R. M. Pace, A. J. Stangl, L. M. Nagy, and T. A. Williams. 2016. Wnt repertoire and developmental expression patterns in the crustacean *Thamnocephalus platyurus*. *Evol. Dev.* 18: 324–341.

Copf, T., R. Schröder, and M. Averof. 2004. Ancestral role of caudal genes in axis elongation and segmentation. *Proc. Natl. Acad. Sci. U.S.A.* 101: 17711–17715.

Couso, J. P. 2009. Segmentation, metamerism and the Cambrian explosion. *Int. J. Dev. Biol.* 53: 1305–1316.

da Silva, S. M., and J. Vincent. 2007. Oriented cell divisions in the extending germband of *Drosophila*. *Development* 134(17): 3049–3054.

Damen, W. G. M. 2007. Evolutionary conservation and divergence of the segmentation process in arthropods. *Dev. Dyn.* 236: 1379–1391.

Davis, G. K., and N. H. Patel. 2002. Short, long, and beyond: Molecular and embryological approaches to insect segmentation. *Annu. Rev. Entomol.* 47: 669–699.

Dohle, W., and G. Scholtz. 1997. How far does cell lineage influence cell fate specification in crustacean embryos? *Semin. Cell Dev. Biol.* 8: 379–390.

Edgar, B. A., and P. H. O'Farrell. 1989. Genetic control of cell division patterns in the *Drosophila* embryo. *Cell* 57: 177–187.

Edgar, B. A., and P. H. O'Farrell. 1990. The three postblastoderm cell cycles of *Drosophila* embryogenesis are regulated in G2 by string. *Cell* 62: 469–480.

El-Sherif, E., M. Averof, and S. J. Brown. 2012. A segmentation clock operating in blastoderm and germband stages of *Tribolium* development. *Development* 139: 4341–4346.

El-Sherif, E., X. Zhu, J. Fu, and S. J. Brown. 2014. Caudal regulates the spatiotemporal dynamics of pair-rule waves in *Tribolium*. *PLoS Genet.* 10: e1004677.

Foe, V. E. 1989. Mitotic domains reveal early commitment of cells in *Drosophila* embryos. *Development* 107: 1–22.

Freeman, J. A. 1995. Epidermal cell cycle and region-specific growth during segment development in *Artemia*. *J. Exp. Zool.* 271: 285–295.

Freyer, G. 1983. Functional ontogenetic changes in *Branchinecta ferox* (Milne-Edwards) (Crustacea: Anostraca). *Philos. Trans. R. Soc. London B Biol.* 303: 229–343.

Fu, J. 2014. Contribution of the canonical Wnt pathway in Tribolium anterior-posterior axis patterning. PhD Thesis. Kansas State University.

Gilles, A. F., J. B. Schinko, and M. Averof. 2015. Efficient CRISPR-mediated gene targeting and transgene replacement in the beetle *Tribolium castaneum*. *Development* 142: 2832–2839.

von der Hardt, S., J. Bakkers, A. Inbal, L. Carvalho, L. Solnica-Krezel, C. P. Heisenberg, and M. Hammerschmidt. 2007. The Bmp gradient of the Zebrafish gastrula guides migrating lateral cells by regulating cell-cell adhesion. *Cur. Biol.* 17(6): 475–487.

Hartenstein, V., and J. A. Campos-Ortega. 1985. Fate-mapping in wild-type Drosophila *melanogaster*. *Wilhelm Roux Arch.* 194: 181–195.

Hemmi, N., Y. Akiyama-Oda, K. Fujimoto, and H. Oda. 2018. A quantitative study of the diversity of stripe-forming processes in an arthropod cell-based field undergoing axis formation and growth. *Dev. Biol.* 437: 84–104.

Huebner, R. J., and J. B. Wallingford. 2018. Coming to consensus: A unifying model emerges for convergent extension. *Dev. Cell* 46: 389–396.

Irvine, K. D., and E. Wieschaus. 1994. Cell intercalation during *Drosophila* germband extension and its regulation by pair-rule segmentation genes. *Development* 120: 827–841.

Janssen, R., M. Gouar, M. Pechmann, F. Poulin, R. Bolognesi, E. E. Schwager, …, A. P. McGregor. 2010. Conservation, loss, and redeployment of Wnt ligands in protostomes: Implications for understanding the evolution of segment formation. *BMC Evol. Biol.* 10: 374.

Keller, R. 2006. Mechanisms of elongation in embryogenesis. *Development* 133: 2291–2302.

Keller, R., L. Davidson, A. Edlund, T. Elul, M. Ezin, D. Shook, and P. Skoglund. 2000. Mechanisms of convergence and extension by cell intercalation. *Philos. Trans. R. Soc. London B Biol.* 355: 897–922.

Keller, R., and P. Tibbetts. 1989. Mediolateral cell intercalation in the dorsal, axial mesoderm of *Xenopus laevis*. *Dev. Biol.* 131: 539–549.

Krause, G. 1939. Die Eitypen der Insekten. *Biol. Zentr.* 59:495–536.

Krokan, H., E. Wist, and R. H. Krokan. 1981. Aphidicolin inhibits DNA synthesis by DNA polymerase alpha and isolated nuclei by a similar mechanism. *Nucleic Acids Res.* 9: 4709–4719.

Kuhn, D. T., J. M. Chaverri, D. A. Persaud, and A. Madjidi. 2000. Pair-rule genes cooperate to activate en stripe 15 and refine its margins during germ band elongation in the *D. melanogaster* embryo. *Mech. Dev.* 95: 297–300.

Lawrence, P. A. 1973. A clonal analysis of segment development in *Oncopeltus* (Hemiptera). *J. Embryol. Exp. Morphol.* 30: 681–699.

Lienkamp, S. S., K. Liu, C. M. Karner, T. J. Carroll, O. Ronneberger, J. B. Wallingford, and G. Walz. 2012. Vertebrate kidney tubules elongate using a planar cell polarity–dependent, rosette-based mechanism of convergent extension. *Nat. Genet.* 44(12): 1382–1387.

Linder, F. 1941. Contributions to the morphology and the taxonomy of the Branchiopoda Anostraca. *Zool. Bidrag. Uppsala* 20: 101–302.

Liu, P., and T. C. Kaufman. 2009. Morphology and husbandry of the large milkweed bug, *Oncopeltus fasciatus*. *Cold Spring Harb. Protoc.* 2009: pdb.emo127–pdb.emo127.

Liu, P. Z., and T. C. Kaufman. 2004. Krüppel is a gap gene in the intermediate germband insect *Oncopeltus fasciatus* and is required for development of both blastoderm and germband-derived segments. *Development* 131: 4567–4579.

Liu, P. Z., and T. C. Kaufman. 2005. Short and long germ segmentation: Unanswered questions in the evolution of a developmental mode. *Evol. Dev.* 7: 629–646.

Macaya, C. C., P. E. Saavedra, R. E. Cepeda, V. A. Nuñez, and A. F. Sarrazin. 2016. A *Tribolium castaneum* whole-embryo culture protocol for studying the molecular mechanisms and morphogenetic movements involved in insect development. *Dev. Genes Evol.* 226: 53–61.

Mailer, J. L. 1991. Mitotic control. *Curr. Opin. Cell Biol.* 3: 269–275.

Massagué, J. 2004. G1 cell-cycle control and cancer. *Nature* 432: 298–306.

Mayer, G., C. Kato, B. Quast, R. H. Chisholm, K. A. Landman, and L. M. Quinn. 2010. Growth patterns in Onychophora (velvet worms): Lack of a localised posterior proliferation zone. *BMC Evol. Biol.* 10: 1–12.

McGregor, A. P., M. Hilbrant, M. Pechmann, E. E. Schwager, N. M. Prpic, and W. G. M. Damen. 2008. *Cupiennius salei* and *Achaearanea tepidariorum*: Spider models for investigating evolution and development. *Bioessays*: 30(5): 487–498.

McGregor, A. P., M. Pechmann, E. E. Schwager, and W. G. M. Damen. 2009. An ancestral regulatory network for posterior development in arthropods. *Commun. Integr. Biol.* 2: 174–176.

Minelli, A. 2004. Evo-devo perspectives on segmentation: Model organisms, and beyond. *Trends Ecol. Evol.* 19: 423–429.

Møller, O. S., J. Olesen, and J. T. Høeg. 2003. SEM studies on the early larval development of *Triops cancriformis* (Bosc)(Crustacea: Branchiopoda, Notostraca). *Acta Zool.* 84: 267–284.

Myohara, M. 1994. Fate mapping of the silkworm, *Bombyx mori*, using localized UV irradiation of the egg at fertilization. *Development* 120: 2869–2877.

Nagy, L., L. Riddiford, and K. Kiguchi. 1994. Morphogenesis in the early embryo of the lepidopteran *Bombyx mori*. *Dev. Biol.* 165: 137–151.

Nakamoto, A., S. D. Hester, S. J. Constantinou, W. G. Blaine, A. B. Tewksbury, M. T. Matei, …, T. A. Williams. 2015. Changing cell behaviours during beetle embryogenesis correlates with slowing of segmentation. *Nat. Commun.* 6: 6635.

Nishimura, T., H. Honda, and M. Takeichi. 2012. Planar cell polarity links axes of spatial dynamics in neural-tube closure. *Cell* 149: 1084–1097.

Oberhofer, G., D. Grossmann, J. L. Siemanowski, T. Beissbarth, and G. Bucher. 2014. Wnt/β-catenin signaling integrates patterning and metabolism of the insect growth zone. *Development* 141: 4740–4750.

Paré, A. C., A. Vichas, C. T. Fincher, Z. Mirman, D. L. Farrell, A. Mainieri, and J. A. Zallen. 2014. A positional Toll receptor code directs convergent extension in *Drosophila*. *Nature* 515: 523–527.

Patel, N. H., B. G. Condron, and K. Zinn. 1994. Pair-rule expression patterns of even-skipped are found in both short- and long-germ beetles. *Nature* 367: 429–434.

Pechmann, M. 2016. Formation of the germ-disc in spider embryos by a condensation-like mechanism. *Front. Zool.* 13: 1–13.

Peel, A. 2004. The evolution of arthropod segmentation mechanisms. *Bioessays* 26: 1108–1116.

Peel, A. D., A. D. Chipman, and M. Akam. 2005. Arthropod segmentation: Beyond the *Drosophila* paradigm. *Nat. Rev. Genet.* 6: 905–916.

Pueyo, J. I., R. Lanfear, and J. P. Couso. 2008. Ancestral Notch-mediated segmentation revealed in the cockroach *Periplaneta americana*. *Proc. Natl. Acad. Sci. U.S.A.* 105: 16614–16619.

Rosenberg, M. I., A. E. Brent, F. Payre, and C. Desplan. 2014. Dual mode of embryonic development is highlighted by expression and function of *Nasonia* pair-rule genes. *Elife* 3: e01440.

Sander, K. 1975. Specification of the basic body pattern in insect embryogenesis. *Adv. Insect Physiol.* 12: 125–238.

Sarrazin, A. F., A. D. Peel, and M. Averof. 2012. A segmentation clock with two-segment periodicity in insects. *Science* 336: 338–341.

Scholtz, G., and C. Wolff. 2013. Arthropod embryology: Cleavage and germ band development. In *Arthropod Biology and Evolution: Molecules, Development, Morphology*, 2nd ed. Alessandro Minelli, Geoffrey Boxshall, Giuseppe Fusco (eds), 97, 63–89. Berlin: Springer.

Schoppmeier, M., S. Fischer, C. Schmitt-Engel, U. Löhr, and M. Klingler. 2009. An ancient anterior patterning system promotes caudal repression and head formation in Ecdysozoa. *Curr. Biol.* 19: 1811–1815.

Seaver, E. C. 2014. Variation in spiralian development: Insights from polychaetes. *Int. J. Dev. Biol.* 58: 457–467.

Shindo, A. 2017. Models of convergent extension during morphogenesis. *WIREs Dev. Biol.* 89: e293–17.

Shindo, A., and J. B. Wallingford. 2014. PCP and septins compartmentalize cortical actomyosin to direct collective cell movement. *Science* 343: 649–652.

Shinmyo, Y., T. Mito, T. Matsushita, I. Sarashina, K. Miyawaki, H. Ohuchi, and S. Noji. 2005. Caudal is required for gnathal and thoracic patterning and for posterior elongation in the intermediate-germband cricket *Gryllus bimaculatus*. *Mech. Dev.* 122: 231–239.

Simões, S. de M., J. T. Blankenship, O. Weitz, D. L. Farrell, M. Tamada, R. Fernandez-Gonzalez, and J. A. Zallen. 2010. Rho-kinase directs Bazooka/Par-3 planar polarity during *Drosophila* axis elongation. *Dev. Cell* 19: 377–388.

Stahi, R., and A. D. Chipman. 2016. Blastoderm segmentation in *Oncopeltus fasciatus* and the evolution of insect segmentation mechanisms. *Proc. Biol. Sci.* 283: 20161745.

Stappert, D., N. Frey, C. von Levetzow, and S. Roth. 2016. Genome-wide identification of *Tribolium* dorsoventral patterning genes. *Development* 143: 2443–2454.

Stollewerk, A., M. Schoppmeier, and W. G. M. Damen. 2003. Involvement of Notch and Delta genes in spider segmentation. *Nature* 423: 863–865.

Sun, Z., C. Amourda, M. Shagirov, Y. Hara, T. E. Saunders, and Y. Toyama. 2017. Basolateral protrusion and apical contraction cooperatively drive *Drosophila* germ-band extension. *Nat. Cell Biol.* 19: 375–383.

Timson, J. 1975. Hydroxyurea. *Mutat. Res.* 32: 115–132.

Vedula, S. R. K., A. Ravasio, C. T. Lim, and B. Ladoux. 2013. Collective cell migration: A mechanistic perspective. *Physiology (Bethesda)* 28: 370–379.

Walck-Shannon, E., and J. Hardin. 2014. Cell intercalation from top to bottom. *Nat. Rev. Mol. Cell Biol.* 15: 34–48.

Weliky, M., S. Minsuk, R. Keller, and G. Oster. 1991. Notochord morphogenesis in *Xenopus laevis*: Simulation of cell behavior underlying tissue convergence and extension. *Development* 113: 1231–1244.

Williams, M., W. Yen, X. Lu, and A. Sutherland. 2014. Distinct apical and basolateral mechanisms drive planar cell polarity-dependent convergent extension of the Mouse neural plate. *Dev. Cell* 29: 34–46.

Williams, T. A., and L. M. Nagy. 2017. Linking gene regulation to cell behaviors in the posterior growth zone of sequentially segmenting arthropods. *Arthropod Struct. Dev.* 46: 380–394.

Wilson, P., and R. Keller. 1991. Cell rearrangement during gastrulation of *Xenopus*: Direct observation of cultured explants. *Development* 112: 289–300.

Xiang, J., K. Reding, A. Heffer, and L. Pick. 2017. Conservation and variation in pair-rule gene expression and function in the intermediate-germ beetle *Dermestes maculatus*. *Development* 144: 4625–4636.

Young, T., J. E. Rowland, C. Ven, M. Bialecka, A. Novoa, M. Carapuco, …, J. Deschamps. 2009. Cdx and Hox genes differentially regulate posterior axial growth in mammalian embryos. *Dev. Cell* 17: 516–526.

Zallen, J. A. 2007. Planar polarity and tissue morphogenesis. *Cell* 129: 1051–1063.

Zallen, J. A., and J. T. Blankenship. 2008. Multicellular dynamics during epithelial elongation. *Semin. Cell Dev. Biol.* 19: 263–270.

Zallen, J. A., and B. Goldstein. 2017. Cellular mechanisms of morphogenesis. *Semin. Cell Dev. Biol.* 67: 101–102.

Zallen, J. A., and E. Wieschaus. 2004. Patterned gene expression directs bipolar planar polarity in *Drosophila*. *Dev. Cell* 6: 343–355.

4 Cellular and Molecular Mechanisms of Segmentation in Annelida
An Open Question

Eduardo E. Zattara and David A. Weisblat

CONTENTS

4.1   INTRODUCTION TO THE ANNELIDA

Annelida comprises a large and diverse set of worms inhabiting marine, freshwater, and terrestrial habitats. Most annelids share a stereotypical elongated vermiform body plan formed by a usually large number of repeated segmental units bound at the anterior and posterior ends by caps of terminal non-segmental tissues (Figure 4.1A, C). They have a through gut, blood vessels, and repeated excretory organs running along the body, traversing a large coelomic cavity frequently compartmentalized by segmental septa. Their body wall is muscular, with variably thick bands of longitudinal muscle running along and finer rings of circular muscle, oriented transversal to the main body axis. Most species have segmentally iterated bundles of bristles that emerge directly from the body wall or from lateral appendages called parapodia. A ventral nerve cord runs from the posterior to the anterior end, where it is connected to a dorsal brain by pairs of nerves that go around the foregut.

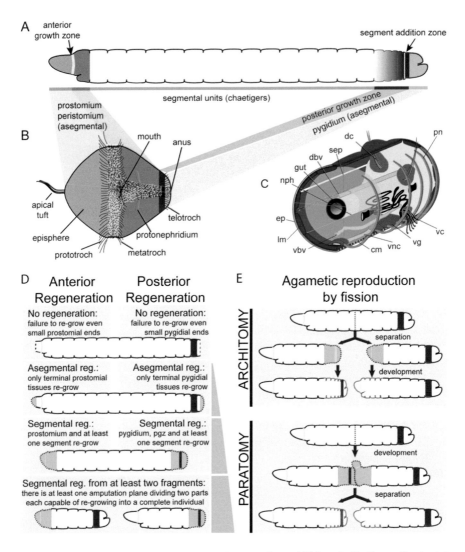

FIGURE 4.1 Annelid body plan and developmental capabilities. A–B: Generalized adult (A) and larval (B) annelid body plans. A subterminal segment addition zone in the trochophore larva develops a posterior growth zone that intercalates segmental units between the terminal anterior (prostomium and peristomium) and posterior (pygidium) regions. C: Generalized body segment, as seen in clitellates. Abbreviations: cm, circular muscle; dbv, dorsal blood vessel; dc, dorsal chaetal sac; ep, epidermis; lm, longitudinal muscle; nph, metanephridium; pn, peripheral nerve; sep, septum; vbv, ventral blood vessel; vc, ventral chaetal sac; vg, ventral ganglion; vnc, ventral nerve cord. Polychaetes often bear their chaetae in lateral appendages called parapodia. D: Levels of regenerative ability (anterior and posterior) found across annelids. Regenerated tissues represented in green; growth zones shown in gray. E: Types of agametic reproduction by fission found in annelids. New tissues shown in green; growth zones shown in gray. A, D, and E modified after Zattara and Bely (2016); B modified after Nielsen (2005); C modified after Zattara and Bely (2015).

Although this basal body plan is well conserved across the phylum, several groups show significant divergence, to the point of completely losing the segmental organization, like sipunculans, echiurans, or orthonectids.

More than 18,000 annelid species have been described. Many of them serve important ecological roles, being key prey or predators, ecosystem engineers, or even unwelcome invaders (Brusca and Brusca 1990; Zhang 2013). Annelids are ancestrally marine; most marine segmented worms are usually referred to as polychaetes. Recent phylogenetic analyses support a number of annelid lineages (including Oweniidae, Mageloniidae, Chaetopteridae, Amphinomida, and Sipuncula) branching basally to two major clades: Errantia and Sedentaria (Struck et al. 2011; Weigert et al. 2014). Within Sedentaria arose a lineage that moved out of the sea into freshwater and terrestrial habitats. This group is known as the Clitellata, reflecting one of their adaptations to these novel environments. Clitellata diversified greatly, giving rise to a series of oligochaete lineages and the one giving rise to leeches (former class Hirudinea).

Most annelids reproduce sexually; several lineages have also evolved asexual reproduction (Schroeder and Hermans 1975; Zattara and Bely 2016). Most polychaete annelids have separate sexes; their gonadal tissue tends to be diffuse and spread out along the body. Males and females gather to spawn, and fertilized eggs develop indirectly into a variety of larval forms: planktotrophic larvae feed on other organisms in the plankton; lecithotrophic larvae continue development nourished by yolk inherited from the egg. Eventually most larvae metamorphose and settle down, adopting a more benthic habit. Different lineages have evolved a diversity of parental care strategies, like brooding in pouches or tubes, or provisioning of nurse eggs. For example, the lineage leading to the Clitellata evolved a suite of strategies to adapt to freshwater and terrestrial environments, including hermaphroditism, internal fertilization, laying their eggs within a protective cocoon secreted by a specialized epidermal band, the clitellum, and direct development of the eggs to hatch small juveniles. Correlated with these traits, gonadal development in clitellates became restricted to a few specific segments of the body.

Many annelid species also exhibit agametic asexual reproduction by fission to quickly increase their numbers. There are two main modes of fission (Figure 4.1E): In architomic fission, a worm fragments itself into two or more pieces; each piece then regenerates any missing body part resulting in a completely functional worm. In paratomic fission, the development of new anterior and posterior regions within the axis of the parental animal precedes physical separation of the daughter individuals. Despite the heterochronic difference in development and separation between architomy and paratomy, both modes of fission rely heavily on developmental processes associated with regeneration (Figure 4.1D), and most likely represent a co-option of regeneration for reproductive purposes (Chapter 10; Zattara and Bely 2011, 2016).

4.2 SEGMENTATION IN ANNELIDS

The basic annelid body plan (see previous section) follows a clear pattern of segmental metamerism. Here, metamerism and segmentation are considered as two nested levels of organization. Metamerism, the more basic level, is defined as the organization of an animal body plan into repeating morphological units along the anterior–posterior (A–P) axis within any of the three germ layers (ectoderm, mesoderm, or endoderm). To rise to the level of segmentation, metamerism should meet two additional criteria: (1) morphological repeats should be present in two or more of the germ layers; and (2) repeats should have the same spatial frequency across layers. By this definition, metamerism is seen in a range of taxa including flatworms, mollusks, nematodes, onychophorans, and tardigrades (see Chapter 8), but segmentation is confined to annelids, arthropods, and chordates. The broad phylogenetic distribution of these body plan features has been interpreted in two ways. One is that metamerism or even segmentation was present at the base of the bilaterians and has been lost or modified in many different lineages. Alternatively, it could be that the ancestral bilaterian was neither segmented nor metameric, and that these features have evolved independently in various lineages.

Given the current level of knowledge in the field of evolutionary developmental biology (evo-devo), either of these scenarios may be true (Davis and Patel 1999; Patel 2003; Couso 2009; Chipman 2010). More important, in either case, the broad and phylogenetically intermingled distribution of non-metameric, metameric, and segmented body plans across the three superphyla of bilaterally symmetric animals (Deuterostomia, Ecdysozoa, Lophotrochozoa/Spiralia) calls for a much greater degree of evolutionary plasticity in developmental processes than was assumed to be the case when phylogenetic trees were constructed using morphological comparisons, as exemplified in the Articulata hypothesis (Scholtz 2002). In constructing such trees, segmentation often weighed heavily in organizing taxa, based on two assumptions: (1) that a segmented body plan would be difficult to evolve; and (2) that once evolved it would not be lost because of the selective advantages it confers. To more fully understand the evident evolutionary plasticity of developmental pathways leading to the gain or loss of segmentation and metamerism, it is necessary to compare axial growth and patterning among diverse models, including both segmented and unsegmented taxa representing different branches of the phylogenetic tree.

The phylum Annelida has traditionally been considered one of the three truly segmented phyla, the other two being the ecdysozoan Arthropoda and the deuterostome Chordata. Several annelid groups, especially those more widely familiar to the general public like earthworms or lugworms, are often presented as the epitome of metameric organization: they have a long body composed of numerous, externally and internally similar compartments, and show little to no regional specialization, save for the terminal regions. Despite including some quintessential homonomously segmented groups, many annelids lineages show regional specializations grading from subtle differences of size and organ distribution along the anteroposterior body axis, to strong tagmatization similar to that seen in most arthropod groups. Furthermore, Annelida as currently understood based on molecular phylogenies (Struck et al. 2011; Weigert et al. 2014) also includes several taxa that show little to no trace of segmentation as adults, like peanut worms (sipunculans), spoon worms (echiurids), or beard worms and giant tube worms (siboglinids). The fact that traditional, morphologically driven phylogenies had classified these taxa as independent phyla, based in large part on their unsegmented body plans, is evidence of the evolutionary plasticity of developmental mechanisms in general and segmentation in particular. Thus, out of the "big three" truly segmented phyla, Annelida shows the largest diversity of evolutionary elaborations of the basic (perhaps, even ancestral) homonomously metameric body plan.

Another fundamental difference between annelid segmentation and that of arthropods and chordates is that segment development in annelids is not restricted to embryonic and larval development in most groups. Furthermore, segment number is usually variable, with segments forming throughout the whole life of the worms (see Chapter 7 for exceptions to these generalizations). In addition, many groups can replace lost segments by regeneration, and several can redeploy regenerative abilities during asexual reproduction. As such, annelid segment development can be considered the most robust yet flexible pathway to metamery among Metazoans.

Despite the undoubted relevance of annelid segmentation to our understanding of the developmental and evolutionary mechanisms leading to metameric organization in animals, we still know surprisingly little about its cellular and molecular underpinnings, relative to the state of knowledge in arthropods (Chapter 3) and chordates (Chapter 5). Most insights into how cells that will form segmental tissues are born and fated comes from studies on embryos of leeches, a specialized group within the clitellate annelids, which are already derived relative to their marine counterparts (Chapter 7). While what we have learned so far from leeches provides enormous insight into how cells can become segmental tissues, the mechanisms themselves are unlikely to adequately represent the segmentation mechanisms across most annelids, just as segmentation mechanisms described for *Drosophila* flies have been found to be quite distinct from those later found in other arthropod groups (Peel, Chipman, and Akam 2005). Fortunately, the last two decades have seen the emergence of new and powerful annelid models, spearheaded by the nereid *Platynereis dumerilii* and the capitellid *Capitella teleta*, which are likely to be followed by several other species with the help of the next generation of functional genomic tools and the rise of evo-devo approaches.

4.3 AN OVERVIEW OF ANNELID DEVELOPMENT

All annelids born by gametic reproduction, be it sexual or parthenogenetic, undergo similar phases of early cleavage, blastulation, and gastrulation. Further development is highly variable depending on lineage and life history strategy. The diversity of annelid life histories can be grouped in two initial categories: indirect and direct developers. Indirect developers hatch as some type of trochophore larva, which comprises the rudiments of the anterior and posterior terminal regions of the adult worm (Figure 4.1B). During post-embryonic development, this larva intercalates a variable number of segments between these rudiments, and then undergoes a metamorphic process into a juvenile form, usually associated with a habitat shift. Post-metamorphic juveniles go on adding segments from a posterior subterminal region known as the segment addition zone (SAZ), which together with the developing new segments, form the posterior growth zone (PGZ). In direct developers, embryonic development eschews the larval stage, instead generating a juvenile with many segments at hatching time (Anderson 1973). Most annelids continue adding segments at their PGZ throughout their life, but a few lineages have evolved determinate growth and fixed segment numbers. In leeches, which have a fixed segment number, hatchlings are born with their full complement of segments and have no active post-embryonic PGZ (Anderson 1973).

4.3.1 EARLY EMBRYONIC DEVELOPMENT

Early embryonic development in annelids provides classic examples of spiralian development (Anderson 1973). The fertilized zygote has an intrinsic animal–vegetal axis along which the first and second cleavage planes are extended, and which corresponds roughly to the future anteroposterior axis. In contrast to mollusks (Freeman and Lundelius 1992), most annelids that have been studied exhibit unequal first and second cleavages (Dohle 1999). These unequal cleavages yield a four-cell stage with identifiable cells, which thus defines the second axis of the embryo, corresponding roughly to the future dorsoventral axis. By convention, these first four blastomeres are designated as A, B, C, and D macromeres, with D giving rise primarily to dorsal* progeny and B to ventral progeny in the early embryo; A and C contribute left- and right-lateral fates, respectively (Figure 4.2A, E). The next four cleavages are usually unequal and occur with obliquely equatorial planes of division, yielding successive quartets of smaller micromeres in animal territory that are offset in either a clockwise or anticlockwise manner from the parent macromeres (Figure 4.2B–C, F–G). Between consecutive rounds of division, the location of the spindles rotates 90 degrees in alternating clockwise and counterclockwise direction (as seen from either of the animal–vegetal poles). This series of (usually unequal) cleavages eventually results in a 64-cell embryo showing a pattern reminiscent of a spiral when seen from the anterior animal pole and is thus termed "spiral cleavage" (Figure 4.2D, H). Spiral cleavage is a landmark of spiralian development, is often highly stereotyped, and has been found to be quite conserved across many groups of protostome animals, including annelids, mollusks, nemerteans, and flatworms (Henry 2014).

Spiral cleavage is difficult to visualize and to describe, as attested both by early attempts to describe this developmental process during the 19th century, and by students past and present trying to understand it (Anderson 1973). At the close of the 19th century, E.B. Wilson (1892) and E.G. Conklin (1897) developed a nomenclature to name individual blastomeres at each step of cleavage, thus allowing for more precise descriptions and more meaningful comparisons across species. This system was developed initially for species showing unequal rounds of cleavage that left smaller blastomeres toward the anterior, animal pole of the embryo and larger blastomeres toward the posterior, vegetal pole, and thus the nomenclature suggests a stem cell–like nature of the posterior blastomeres. While investigation in further species revealed this not to be accurate, the naming system was already deeply embedded in most descriptive and comparative work and is still widely used.

As described earlier, the zygote initially undergoes two rounds of meridional cleavage resulting in an embryo whose four cells are designated A, B, C, and D. The third cleavage results in animal and vegetal quartets of cells. Each of the blastomeres

---

* It is important to note that the "dorsality" of D quadrant descendants is only true for even-numbered quartets (2d, 4d; Henry and Martindale 1998; Shankland and Seaver 2000).

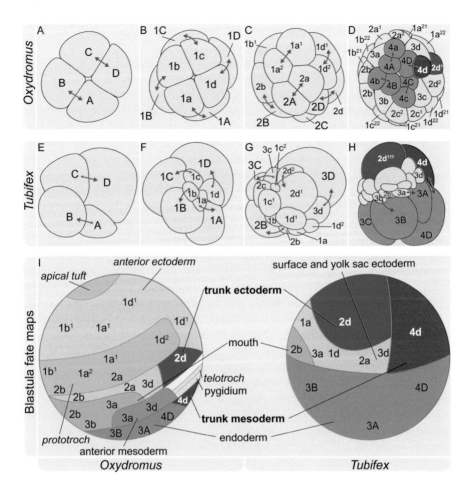

FIGURE 4.2   Early annelid development: cleavage and blastula fate maps. A–H: Comparison of early cleavage of a small oligolecithal egg of an indirect-developing polychaete and the large yolky egg of a direct developing clitellate. Green double arrows show daughter cells; degree of cleavage inequality is approximated by relative arrowhead size. Animal pole views except where otherwise indicated. A–D: Early cleavage of *Oxydromus obscurus* (Errantia: Hesionidae), a polychaete with small eggs (~60 μm) and little yolk. A: 4-cell embryo resulting from two rounds of equal cleavage. B: 8-cell embryo after an equal third cleavage. C: 16-cell embryo after a roughly equal fourth cleavage. D: 32-cell embryo, showing the trunk ectoderm precursor 2d^1 (dark blue), trunk mesoderm precursor 4d (red), and endoderm precursors (purple); vegetal pole view. E–F: Early cleavage of *Tubifex tubifex* (Clitellata: Naididae), a clitellate with large, yolky eggs (~450 μm). E: 4-cell embryo resulting from two rounds of unequal cleavage. F: 8-cell embryo after unequal third cleavage. G: 17-cell embryo, after subsequent cleavage. H: 22-cell embryo, showing trunk ectoderm precursor 2d^{111} (dark blue), trunk mesoderm precursor 4d (red), and endoderm precursors (purple); lateral pole view. I: Comparison of blastula fate maps of *Oxydromus* and *Tubifex*, as seen from left lateral view. A–D after Treadwell (1901) and Anderson (1973); E–H after Penners (1922) and Anderson (1973); I modified after Anderson (1973).

inherits the letter from their progenitor cell, but the animal blastomeres are designated by the lowercase letter corresponding to their quartet of origin, while the vegetal ones retain the capital letter designation; all get a number 1 in front. Thus, an 8-cell embryo has an animal quartet composed of micromeres 1a, 1b, 1c, and 1d, and a vegetal quartet composed by macromeres 1A, 1B, 1C, and 1D (Figure 4.2B, F). After the fourth cleavage, the descendants of the micromeres retain the prefix 1, while the animal and vegetal descendants of the vegetal quartet get the prefix 2; for example, macromere 1D generates an animal micromere and a vegetal macromere, designated 2d and 2D, respectively. As the micromeres divide, the animal-facing daughter of each micromere is distinguished by a superscript 1, and the vegetal-facing progeny gets a superscript 2. For example, the animal and vegetal daughter cells of blastomere 1d are $1d^1$ and $1d^2$, respectively. Further divisions add another superscript 1 or 2, e.g., the daughters of $1d^2$ are $1d^{21}$ and $1d^{22}$ (Figure 4.2C–D, G–H). In further rounds of division, descendants of the vegetal-most quartet of cells, the macromeres, keep incrementing their prefix, while progeny of the remaining cells increment their superscript as described earlier. This system, while somewhat cumbersome, provides a helpful framework that helps to trace the cell lineage for each cell of a 64-cell embryo, and to enable cell fate comparisons across species. On the other hand, the assumption that equivalently named cells are homologous can become problematic, for example, when evolutionary changes in cell division patterns or difficulties in catching certain divisions lead to ambiguity as to specific cell identities.

Comparisons of cleavage patterns across Annelida have shown both notable conservation and substantial deviations of the stereotypical pattern (Anderson 1973; Seaver 2014). In general, early development can be seen as a process that sequentially segregates cytoplasmic determinants—the molecular information that will eventually lead each cell and its progeny toward a specific developmental pathway. Such information might act in a cell-autonomous manner, or it might prime the cells to act in response to inductive interactions with other cells. Strategies to partition determinants differ across species, and the most obvious evidence is the equality or inequality in sister cell size at each cleavage. Cleavage equality is determined primarily by the amount and distribution of yolk, which in turn is strongly dependent both on the current life history strategy of each species, and the developmental biases imposed by the evolutionary history of its lineage.

4.3.2   CELL FATE MAPS: TROCHOPHORE LARVAE AND DIRECT DEVELOPMENT

In the context of our current knowledge of annelid phylogeny (Struck et al. 2011; Weigert et al. 2014), indirect development with an intermediate trochophore larval form is considered the ancestral mode for this phylum. A stereotypical trochophore (Figure 4.1B) is spheroid to fusiform in shape, with its long axis aligned with the embryonic anteroposterior (or animal–vegetal) axis (Rouse 2006). A characteristic ciliated band called the *prototroch* encircles the larva at an approximately equatorial position. Tissues anterior to the prototroch form the *episphere* and develop initially as larval epithelium and organs (in species with little yolk and plankton-feeding larvae) or directly as adult prostomial tissues (in species with larger, yolkier embryos and more abbreviated development). The mouth is usually located midventrally, posteriorly adjacent to the prototroch. Sometimes a second ciliated band called *metatroch* is located at a subequatorial position. Tissues posterior to the prototroch/metatroch form the hyposphere, which often bears a second, smaller and subterminal ciliated band known as the *telotroch*, and a longitudinal ciliated band running along the ventral midline from the mouth to the telotroch known as the *neurotroch*. Additional bands of ciliated cells are present in some groups; however, no lineage has a larval form possessing all known types of bands (Rouse 2006). The hyposphere becomes incorporated into the adult pygidium (Figure 4.1A–B).

The animal-most quartet of cells, descended from the primary micromere quartet (that is, cells $1a^1$, $1b^1$, $1c^1$, $1d^1$) develop to become the episphere and part of the prototroch, while their vegetal sister cells ($1a^2$, $1b^2$, $1c^2$, $1d^2$), together with the second micromere quartet (2a, 2b, 2c, 2d) contribute to the prototroch, post-trochal ectoderm, and larval muscles (Figure 4.2I, Figure 4.3). In most species, all trunk ectoderm will derive from descendants of 2d, while all trunk mesoderm is derived from cell 4d in the fourth micromere quartet (Figure 4.3). Several major groups present slight to extreme deviations from this generic pattern (Meyer et al. 2010), but an important aspect to keep in mind is that a large proportion of the initial set of blastomeres are fated to build the component tissues of the trochophore. In turn, most of the trochophore (i.e., tissues located anterior to the telotroch) will be either reabsorbed or become part of the prostomium, the anterior, non-segmental terminal region of the adult worm. Larval tissues posterior to the telotroch develop into the pygidium, the posterior, non-segmental terminal region of the worm (Nielsen 2005). Thus, most of the body of an adult, indirect-developing annelid—the segmented trunk—is composed of tissues derived from a very small fraction of the initial embryonic blastomeres (Figure 4.2I, Figure 4.3).

Trochophore larvae eventually metamorphose into juvenile worms; the degree of morphological change involved in this process varies depending on the life history strategies of each species and group. Along with ecological changes associated to shifts in habitat and/or feeding mode, metamorphosis is made evident by the developmental activation of the segmental region of the body, which effectively converts a spheroidal larva into a long, segmented worm (Figure 4.4). This growth by addition of segments is achieved through the proliferative activity of a PGZ (Figure 4.1A–B). This PGZ derives from a ring of ectodermal stem cells (termed ectoteloblasts) enclosing a pair of large mesodermal stem cells (the M cells), just in front of the telotroch (Anderson 1973). In groups where unequal cleavage leads to large 2D and 2d blastomeres, this ectoteloblast ring derives from descendants of the 2d blastomere; when 2d is relatively small, progeny of 3c and 3d also contribute to this region. In most groups, equal division of the 4d cell of the fourth quartet along a parasagittal plane generates the pair bilateral of M cells, also known as mesoteloblasts. Together, the ectoteloblasts and the mesoteloblasts will proliferate to form segmental tissues (see next section).

Among indirectly developing groups, variation in the amount of egg yolk and larval ecology correlates with both the initial size of the ecto- and mesoteloblasts, and the timing of production of segmental tissues. Planktotrophic larvae (e.g., *Owenia*, *Chaetopterus*) developing from small eggs usually show smaller teloblasts with little

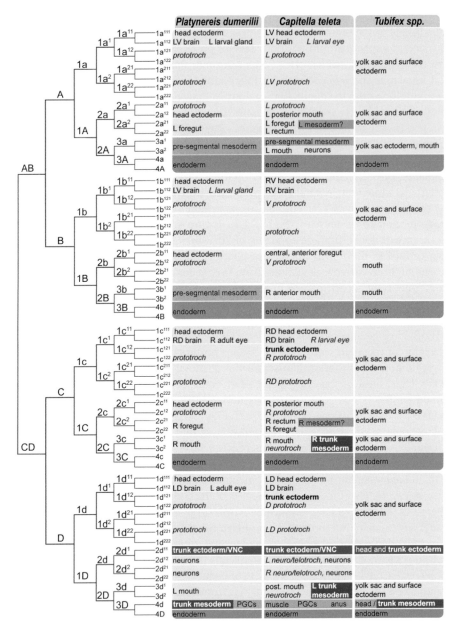

FIGURE 4.3 Comparison of embryonic cell fates between *Platynereis dumerilii* (Errantia: Nereidae), *Capitella teleta* (Sedentaria: Capitellidae), and *Tubifex tubifex* (Clitellata: Naididae). Ectodermal fates are indicated in blue (dark blue for trunk ectoderm, light blue for other ectoderm), mesodermal fates in red (red for trunk mesoderm, pink for other mesoderm), and endodermal fates in purple.

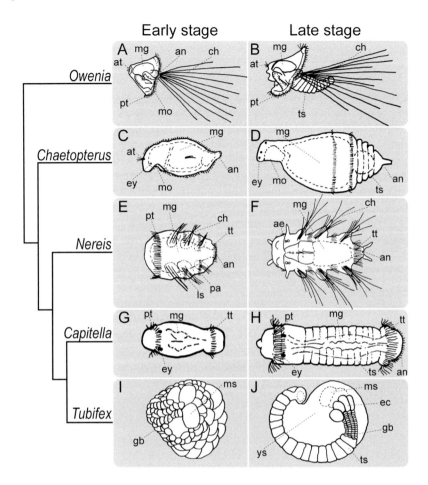

FIGURE 4.4 Developmental diversity of annelids. A–B: Protrochophore (A) and metatrochophore (B) stages of *Owenia fusiformis*; notice the developing segmented trunk in B. C–D: Larval L2 (C) and L4 (D) stages of *Chaetopterus* sp. E–F: Metatrochophore (4.5 days old) (E) and nectochaete (9 days old) (F) larvae of *Nereis pelagica*. G–H: Protrochophore (G) and metatrochophore (H) stages of *Capitella* sp. I–J: Gastrulating (I) and germband (J) stages of embryos of *Tubifex tubifex*. Abbreviations: ae, adult eye; an, anus; at, apical tuft; ch, chaetal bundle; ec, ectoteloblast; ey, larval eye; gb, germband; ls, larval segment; mg, midgut; mo, mouth; ms, mesoteloblast; pt, prototrochal band; ts, trunk segment; tt, telotrochal band; ys, yolk sac. Drawings not to scale. A–B and G–H after Lacalli (1980); C after Wilson (1883); D after Irvine *et al.* (1999); E–F after Wilson (D. P. Wilson 1932b); I after Penners (1924); J after Anderson (1973).

to no developmental activity during early larval life; activity resumes after larva begins to feed. In contrast, lecithotrophic larvae (e.g., *Platynereis*, *Capitella*) developing from larger, yolkier eggs tend to show larger teloblasts that initiate segment formation sooner. In many groups, three or more segments form before metamorphosis; in many cases, these initial segments are specialized for larval function. They develop very quickly, often almost simultaneously, and show ectodermal segmentation before any signs of mesodermal segmentation (in contrast to normal segmental development, see later). Upon metamorphosis, formation of segments is resumed, this time at a slower pace and showing a clearer anteroposterior sequential progression. These differences can be explained by heterochronic adjustments in the developmental timing of ectodermal and endodermal metamerism, relative to each other and to metamorphosis. Presence of larval segments varies not only across larger annelid taxa, but also among closer relatives that vary in their larval life history strategies (Åkesson 1967). Such lability demonstrates the flexibility of spiralian development to support adaptive changes in the ecology of a lineage.

A more dramatic demonstration of the flexibility of annelid development and the role of heterochronic adjustments is seen in the direct development of Clitellate annelids (e.g., *Tubifex*). In this group, the larval stage is completely omitted, and segment development starts early in development from very large teloblastic precursors (see later and Chapter 7). Interestingly, even though this group comprises both species with large, yolky eggs and species where yolk has been reduced and feeding switched to albumen, cleavage leads to blastulae sharing a similar basic pattern: a large dorsal 2d cell or equivalent that gives rise to presumptive ectoderm, a similarly large 4d cell or equivalent that gives rise to presumptive mesoderm, an arc of micromeres dorsal and anterior to the 2d cell that will form presumptive stomodaeum, and a set of larger vegetal cells that will become presumptive midgut (Figures 4.2I and 4.3). Such convergence is reached by different adaptations of the cleavage pattern in each lineage to deal with the reduction of yolk and the switch to albumenotrophy (i.e., embryogenesis driven by albumen reserves within the egg cocoon and yet external to the embryo itself), and greatly illustrates both the evolutionary plasticity of early development and the high conservation of specific embryonic stages.

4.4 SEGMENTATION OF THE TRUNK

Development of segmental units can be seen as the combination of three processes: elongation of the anteroposterior (AP) axis, segregation of cells into segmental units, and patterning of each segmental unit (Balavoine 2015). In most species in which segmentation has been studied, axial elongation of the body is achieved by cell proliferation at the posterior end of the body and is followed by the formation of boundaries that separate and specify fields of cells to form consecutive metamers. As detailed in other chapters, these metamers may arise sequentially (as is the case with vertebrates and short germ insects) or simultaneously (as in long germ insects) along the AP axis, and may correspond to segmental, para-segmental, or double-segmental primordia. The actual patterns of cell division leading to the formation of morphological pattern elements within the units are often variable (the variable and irregularly shaped clones of cells within *Drosophila* compartments, for example), and cell clones arising within one metamer are restricted from crossing boundaries separating them from adjacent metamers. This is analogous to creating a repeating pattern in a row of bushes by trimming them so that their branches do not intermingle, without regard for the branching patterns of the individual bushes. We refer to these processes as boundary-driven segmentation mechanisms.

Segmentation by boundary-driven mechanisms is so widespread among the commonly studied animal models that one may be forgiven in assuming that this is the only way to generate metamers and segments. However, most organisms used to study segmentation are either arthropods or vertebrates. In contrast, clitellate annelids are known to generate segments by another mechanism, in which the components of each segment arise as the clonal descendants of individual cells. We call this mechanism lineage-driven segmentation. Returning to the analogy of the row of bushes, if the bushes are carefully pruned to achieve the same branching pattern for each bush, then the row of bushes will present a repeating pattern even if the branches of adjacent bushes intermingle. In groups showing lineage-driven segmentation, axial elongation and segment specification are achieved by the same process: cell division.

Among annelids, lineage-driven segmentation has been unambiguously described for clitellate annelids, particularly in leeches (see Chapter 7). In what appears to be the ancestral mode of development for this group, large yolky embryos exhibit a modification of the D quadrant cleavage pattern so that the equivalents of micromeres 2d and 4d are large cells (designated as DNOPQ and DM, respectively). Their further divisions yield a fixed set of ten large stem cells known as teloblasts, amenable to microinjection of cell lineage markers. This approach has been used to show that the teloblast progeny also undergoes stereotyped patterns of cell division to generate segments in a lineage-dependent manner, but there is still considerable debate about whether the remaining annelid groups also show a similar mechanism of segment formation. At the heart of the issue is the question of whether segment formation in polychaete annelids segments is also lineage-driven, or boundary-driven instead. Answering this question has proven difficult because most polychaete annelids have smaller embryos than clitellates and develop indirectly to produce an intermediate larval form. Depending on the group, larval stages have from zero to several segments, which can be more or less specialized (Anderson 1973; Rouse 2006). Having a mobile, active larval stage greatly complicates tracing cell fates. Furthermore, in many groups, activation of post-larval segment development requires larvae to reach metamorphosis, a process that is often difficult to induce under artificial culture conditions. For these reasons, embryonic development of marine annelids is much better studied than development at later life stages, including post-metamorphic development of the trunk's segmental units. Despite these challenges, the last few decades have seen the emergence of two major non-clitellate annelid models: the errant nereid *Platynereis dumerilii* and the sedentary capitellid *Capitella teleta* (Zantke et al. 2014; Seaver 2016). Application of modern cell-tracing strategies are starting to clear the picture of how segmental units develop in marine annelids, although studies have given somewhat conflicting evidence.

Embryonic and larval development studies from both live specimens and histological sections were made for representatives of several families of marine annelids during the late 1800s and first half of the 1900s (Anderson 1973). Most of these studies show that the ectodermal component of trunk segments forms from proliferation of a ring of ectoteloblasts derived from the second quartet blastomere 2d, while the mesodermal component of trunk segments arises from ventrolateral bands that result from teloblastic proliferation of the paired M mesoteloblasts, descended from bilateral equal division of the fourth quartet blastomere 4d (Figures 4.2 and 4.3). However, a lack of cell-tracing tools at post-embryonic stages made it impossible to test if segment formation in these groups is lineage-driven (as seen in clitellate embryos) or else boundary-driven. While early presence of ecto- and mesoteloblasts similar to those of clitellates was evident in embryos, and developmental series showed likely clonal segment formation (Anderson 1973), most post-metamorphic annelids show no large, teloblast-like cells at their posterior growth zone (Seaver, Thamm, and Hill 2005). This conundrum, however, began to be solved with the application of a novel approach combining endogenous cell cycle reporters, *in vivo* time-lapse fluorescence imaging, and *in silico* cell tracing (Özpolat et al. 2017). This work showed that in the nereid *Platynereis dumerilii*, the M cells daughter of the 4d cell initially divide as teloblasts to generate primordial germ cells and four larval segments, but after their eighth division, they divide more or less equally to form a ring of cells located immediately in front of the pygidium and internal to the ectot-eloblast rings—becoming the mesodermal stem cells of the posterior growth zone. In other words, ectodermal and mesodermal stem cells derived from 2d and 4d, respectively, initially use lineage-driven segmentation and then transition to what currently is best described as boundary-driven segmentation (see Chapter 10).

The number of segments formed during embryonic development can range from none (in indirect-developing species with larvae without segments, e.g., Oweniidae, Chaetopteridae), a few, often specialized larval segments (e.g., Serpulidae, Nereididae, Eunicidae, Tomopteridae), to several (e.g., Capitellidae, Spionidae, Pectinariidae, non-leech clitellates), or even the complete adult complement (e.g., leeches). Below we describe in varying degree of detail what is known about segment formation in representatives of four families that cover a broad phylogenetic spectrum: (1) *Owenia fusiformis*, a member of the early branching family Oweniidae (Figure 4.4A–B); (2) *Platynereis dumerilii*, a member of the family Nereididae of errant polychaetes (Figure 4.4E–F); (3) *Capitella teleta*, a member of the family Capitellidae of sedentary polychaetes (Figure 4.4G–H); and (4) *Tubifex tubifex*, a member of the family Naididae of clitellates (Figure 4.4I–J). Segment formation in glossiphoniid leeches (*Helobdella* spp.) is treated separately in Chapter 7.

4.4.1 SEGMENT DEVELOPMENT IN OWENIA

Detailed developmental descriptions in two species of the genus *Owenia*—*O. fusi-formis* and *O. collaris*—are currently available (D.P. Wilson 1932a; Smart and Von Dassow 2009; Helm et al. 2016). *Owenia* spp. lay oligolecithal eggs of between 70 and 120 μm that cleave following the usual spiralian pattern. Embryos develop into a bell-shaped planktotrophic larva known as mitraria (Figure 4.4A). Mitraria larvae have a U-shaped larval gut whose ends are separated by a small region of ventral ectoderm. The gut is covered by a conical episphere with an apical organ on top. They harbor a pair of larval chaetal sacs that secrete two bundles of characteristic needle-like bristles. During larval life, this ventral region invaginates, forming a pocket that comes to surround the hindgut and becomes the trunk rudiment. As this rudiment grows, it protrudes from under the episphere (Figure 4.4B). Metamorphosis of larvae into juveniles is drastic and takes places over few minutes. The larval blastocoelic cavity deflates, the epithelium is orally invaginated and incorporated to the head of the juvenile, and the ventral pocket is evaginated, so that the anterior end narrows to match the width of the trunk. Ciliary bands and chaetal sacs disappear and larval bristles are shed.

Early studies of larval development in *Owenia fusiformis* mitraria (D.P. Wilson 1932a) state that trunk mesoderm derives from a single pair of M cells located adjacent to the larval anus, below the angle formed between the endoderm and the posterior ectoderm (Figure 4.5A). These cells "give rise in the manner of teloblasts" to paired mesodermal blocks with an increasing number of cells (Figure 4.5B). As each block grows in cell number, an internal cavity forms inside, giving rise to the mesodermal coelom (Figure 4.5C).* Walls of consecutive blocks fuse to form intersegmental septa. At the same time, the ring of ectoteloblasts generates a layer of ectoderm over the developing mesoderm. In contrast to the mesoderm, this ectoderm is not initially obviously divided in segmental units; segmental demarcation appears a bit later, as mesodermal segments develop (Figure 4.5C). While this suggests that modulation of ectodermal segment formation by underlying mesoderm (as described for clitellates, see Chapter 7) might be a general annelid feature, there are many examples of groups where ectodermal segmental boundaries form before mesodermal segments become evident (Anderson 1973). Since during lineage-driven segmentation mesoderm cells are born with segmental identity, it is still possible that they can modulate the overlying ectoderm before they become obviously segmented.

4.4.2 SEGMENT DEVELOPMENT IN *PLATYNEREIS DUMERILII*

In the errant nereid *Platynereis dumerilii*, the 4d cell divides bilaterally into a pair of M cells: Ml and Mr (Ackermann, Dorresteijn, and Fischer 2005; Fischer and Arendt 2013; Özpolat et al. 2017). Each undergoes two highly asymmetrical divisions, giving rise to four cells (ml^1, mr^1, ml^2, and mr^2) that become the founders of the germline (Rebscher et al. 2007; Özpolat et al. 2017). The next two also highly asymmetrical divisions yield four cells (ml^3, mr^3, ml^4, and mr^4) that undergo a limited number of symmetric divisions and then migrate toward the anterior end of the embryo, likely to form anterior non-segmental mesoderm (Özpolat et al. 2017). Each of the following four divisions yields a pair of cells that further proliferate to form a segmental unit: the clones descended from ml^5 and mr^5 form "segment 0," a cryptic segmental unit that bears no chaetae and is integrated into the head (Steinmetz et al. 2011), while the clones formed by ml^6 and mr^6, ml^7 and mr^7, and ml^8 and mr^8 respectively form larval segments 1 through 3 (Figure 4.6A–B). After that, 8Ml and 8Mr, sisters to ml^8 and mr^8, divide once more; then, while a daughter from each remains quiescent, the other undergoes two rounds of mitosis, yielding a total of five

* In a more recent study of *O. collaris* (Smart and Von Dassow 2009), the segmentally iterated structures described as mesodermal somites by Wilson (1932a) are interpreted as nephridial precursors, "derived from the ventral epidermis of the larva." While this interpretation clearly contradicts Wilson's (1932a) and Anderson's (1973) descriptions of teloblastic segmental mesoderm formation in *O. fusiformis*, no additional evidence is presented, in contrast to the detailed series of histological snapshots provided by Wilson. However, given that no cell-tracing experiments were conducted in either study, it is not possible at this time to reject either interpretation.

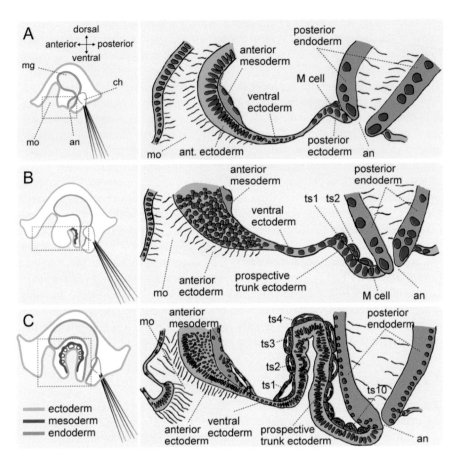

FIGURE 4.5 Development of segmental mesoderm in *Owenia* sp. Left column shows dia-grammatic sagittal sections (modified after Smart and Von Dassow, 2009); right column shows histological sagittal sections located approximately at the position indicated on the dashed boxes (modified after Wilson, 1932a). A: Mitraria larva, ~2–3 days old; a mesoblast progenitor (M cell) is located at the internal angle between posterior ectoderm and posterior endoderm. B: Mitraria larva, ~12 days old; two clusters of cells (ts1 and ts2), presumptive progeny of the posterior mesoblast, can be seen. C: Mitraria larva, ~17 days old; trunk meso-derm composed of 11 cell clusters (ts1, ts2, ts3, ts4, …, ts10) showing an anteroposterior developmental gradient. A coelomic cavity has appeared in most clusters. The prospective trunk ectoderm adjacent to the anterior-most clusters begins to show indentations with seg-mental periodicity.

descendants for each of 8Ml and 8Mr (Figure 4.6B). Those ten cells form a ring and become the mesodermal precursors of the PGZ that will generate segments during post-larval life.

Cell tracing of the 4d lineage during early development of *Platynereis* shows that segmental larval mesoderm formation clearly fulfills the expectations of a lineage-driven mechanism (Figure 4.6C). Whether post-larval segment formation is also lineage-driven remains an open question: even though mesoteloblasts shrink in size

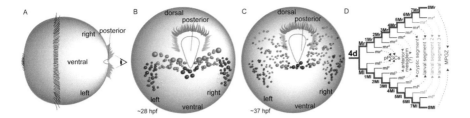

FIGURE 4.6 Origin of segmental mesoderm in *Platynereis dumerilii*. A: Simplified mid-trochophore larva, as seen in ventral view (anterior to the right). The eye symbol shows the point of view of B and C. B–C: Diagrammatic representation of cell progeny of the M mesoteloblasts in ~28 hours post-fertilization (hpf) (A) and ~37 hpf (B) mid-trochophore larvae. Each color represents descendants from the same blast cell (shown in D). D: Cell lineage tree of the progeny of the 4d cell. MPGZ, posterior growth zone mesodermal stem cells; ml^n and mr^n, blast cell produced by the nth division of the left or right mesoteloblast; nMl and nMr: left or right mesoteloblast after the nth division; pPGCs, putative primordial germ cells. A modified after Fischer *et al.* (2010)*;* B–C modified after Özpolat *et al.* (2017).

and split into several pairs as they become the founder mesodermal cells of the PGZ, whether they continue giving birth to single cells whose progeny is confined to a segment's worth of tissue or else switch to a boundary-driven mechanism (like the one described for regeneration in Chapter 10) is not yet known. Interestingly, both the rate and the degree of asymmetry of mesoteloblast divisions is gradually reduced in the M to 8M lineage (Özpolat et al. 2017): the first five rounds of mitosis happen with approximately 30 minutes intervals and then they slow down (40 minutes for 6M and 50 minutes for 7M). As 8M divides to generate the founder MPGZ cells, intervals become much longer (around 150 minutes). Cell asymmetry also changes; initial divisions are highly unequal (M being larger than m), but size difference between daughter cells decreases each round, until reaching equal size between 7M and 7m. The trend continues, so that the next division is again unequal, but this time 8M is smaller than its sister 8m.

As 4d lineage cells begin generating the mesodermal precursors of the larval segments, the 2d[112] micromere, located mid-dorsally at the embryo's hyposphere, divides bilaterally and proceeds to proliferate a pair of ectodermal sheets that converge toward the ventral midline (Ackermann, Dorresteijn, and Fischer 2005). However, whether larval segmental ectoderm formation follows a teloblastic behavior similar to that of 4d descendants, including foundation of the ectoteloblast ring at the PGZ, has not yet been determined conclusively (Gazave et al. 2013). Given our current knowledge, ectodermal segmentation could be either lineage-driven initially or boundary-driven throughout development.

4.4.3 Segment Development in *Capitella teleta*

Although the origin of segmental founders in the sedentary capitellid *Capitella teleta* is not yet described with cellular resolution, cell proliferation assays and injection of cell tracers have provided a very complete picture of proliferation dynamics and fate map in this species (Seaver, Thamm, and Hill 2005; Thamm and Seaver 2008; Meyer and Seaver 2010; Meyer et al. 2010). In contrast to many other anne-lids, cell 4d in *Capitella* has a similar size to other third and fourth quartet cells. Surprisingly, cell-tracing experiments have shown that 4d descendants generate the primordial germ cells but do not contribute to either larval or adult segmental meso-derm. Instead, left and right segmental mesoderm forms from descendants of the 3c and 3d blastomeres, respectively. Each mesodermal band is initially visible as a row of subsurface cells extending anteriorly from the telotroch. These bands thicken, becoming several cells wide and tall (except at the posterior end, which contains a single large cell), and expand initially dorsally and then ventrally following an ante-rior–posterior progression.

In *Capitella*, trunk ectoderm between the prototroch and the telotroch is formed by the progeny of $2d^{11}$. This cell divides into anterior daughter $2d^{111}$ and posterior $2d^{112}$. Interestingly, $2d^{112}$ does not contribute to the ectoderm located between the prototroch and the mouth; thus, this region's ectoderm derives from $2d^{111}$ progeny exclusively. $2d^{112}$ divides bilaterally into a right $2d^{1121}$ and left $2d^{1122}$, which further expand ventrally and dorsally to form right and left segmental ectoderm bands, including chaetal sacs and the ventral nerve cord. While both bands meet symmetri-cally at the ventral midline, they show variable asymmetry dorsally, so that a band from one side might not reach the dorsal midline or extend past the midline and into the contralateral side.

The mesodermal and ectodermal segmental plates develop into ten segments; segments from each side develop independently and eventually fuse at the mid-line. Labeled descendants of $2d^{11}$, 3c and 3d also show at a region located right in front of the telotroch. This region becomes the posterior growth zone, from which three additional segments are generated, to complete this species' 13 larval seg-ments (after about 5 days post fertilization). At this point, segment formation stops, and resumes only after metamorphosis, at a rate of about one segment every 3 days (Seaver, Thamm, and Hill 2005).

4.4.4 SEGMENT DEVELOPMENT IN *TUBIFEX TUBIFEX*

In the clitellate naidid *Tubifex tubifex*, cleavage yields a 25-cell embryo in which large $2d^{111}$, 4d, and 4D blastomeres lie along the future midline (Figure 4.7A–B). Each of these blastomeres then divides bilaterally to generate ectodermal, mesoder-mal, and endodermal precursors, respectively (Penners 1922; Penners 1924; Shimizu 1982; Goto, Kitamura, and Shimizu 1999; Nakamoto, Arai, and Shimizu 2004; Shimizu and Nakamoto 2014).

4d division yields two paired mesoteloblasts: Mr and Ml (Figure 4.7C). These mesoteloblasts divide at regular intervals (2.5 hours at 22°C) to generate small m^n primary blast cells (Figure 4.7J). Although the initial rounds of division are highly unequal, M mesoteloblasts become smaller each round, to the point that after the 35th division, M35 and m^{35} are almost the same size. Their fate after this point is unknown, but it is likely that they divide a few more times to generate mesodermal stem cells at the posterior growth zone of the juveniles. The *n*th division of the M mesoteloblasts adds an m^n blast cell posteriorly adjacent to the previous m^{n-1} blast

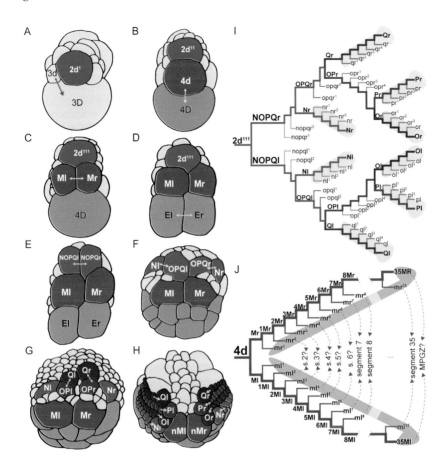

FIGURE 4.7 Development of segmental mesoderm in *Tubifex tubifex*. A–H) Origin of ecto-dermal, mesodermal and endodermal precursors. Endodermal El and Er cells and their prog-eny (lilac) derive from bilateral division of 4D; mesoteloblasts Ml and Mr and their progeny (red) derive from bilateral division of 4d. Ectoteloblast pairs N, O, P, and Q and their progeny (dark blue) derive from bilateral division of 2d^{111}. A) Formation of 3d cell after fifth cleavage; 17-cell embryo, 24 hours post-fertilization (hpf). B) Formation of 4d cell after sixth cleav-age; 22-cell embryo, 28.5 hpf. C) Formation of M mesoteloblast pair; 33 hpf. D) Formation of E cell pair; 34 hpf. E) Formation of NOPQ precursor pair; 36 hpf. F) Segregation of N teloblast pair; 48 hpf. G) Segregation of Q teloblast pair; 64 hpf. H) Elongation of germ bands (early gastrula); 96 hpf. I) Cell lineage tree of the progeny of the 2d^{111} cell. Stem cells in bold and uppercase, blast cells in lower case. J) Cell lineage tree of the progeny of the 4d cell. Abbreviations as in Figure 4.6. A–G modified after Shimizu (1982); H modified after Shimizu (1982) and Anderson (1973).

cell; in this way, Mr and Ml generate bilaterally paired columns of cells. Within each column, m^n undergoes a stereotyped series of divisions, generating a clonal cluster of cells that will generate the mesoderm of a single segment. These columns, known as mesodermal germ bands, are initially located at the dorsal side of the embryo, but as they extend, they migrate laterally and then ventrally until they converge at the ventral midline. At this point, the anterior region of the bands begins to show evident segmental organization. After ventral convergence, the germ bands begin expanding dorsally, migrating between the thin outer layer of yolk sac ectoderm and the yolky endoderm, until they meet along the dorsal midline. Blast cell proliferation, convergence, and morphological segmentation follow a clear anteroposterior gradient.

Segmental ectoderm follows a similar teloblastic pattern. $2d^{111}$ divides bilaterally to yield paired ectoteloblast precursors $2d^{1111}$ and $2d^{1112}$, which in clitellate literature are known as NOPQr and NOPQl, respectively (Figure 4.7E). Each ectoteloblast precursor divides twice very unequally producing small $nopq^1$ and $nopq^2$ cells and then divides less asymmetrically into an N ectoteloblast and an OPQ precursor (Figure 4.7F). OPQ also generates two small cells (opq^1 and opq^2) and then divides into a smaller Q ectoteloblast and a larger OP precursor (Figure 4.7G). OP divides very unequally four times (yielding op^1 through op^4) and then divides equally to yield O and P ectoteloblasts. After their birth, each of the eight large ectoteloblasts (four on each side) divides very unequally with a constant frequency (every 2.5 hours at 22°C) giving rise to bilateral columns of small cells called primary blast cells that are arranged into a bilateral pair of bandlets (Figure 4.7H). These bandlets overlie the mesodermal germ bands and show a similar behavior of ventral convergence followed by dorsal extension until meeting at the dorsal midline to enclose the yolky endoderm. Descendants from those primary blast cells proliferate and follow a stereotyped developmental program, contributing to a segment's worth of specific tissues (Figure 4.7I).

In *Tubifex*, as in other clitellates including leeches (see Chapter 7), both trunk mesoderm and ectoderm show unambiguous lineage-driven segmentation. The origin of the cell lineage present at the posterior growth zone of juveniles and adults has not yet been determined; however, the behavior of *Tubifex* M mesoteloblasts is highly reminiscent to that of *Platynereis* (see earlier), strongly suggesting that after giving rise to the last embryonic m blast cell, the now small M mesoteloblast gives rise to the founder mesodermal stem cells at the PGZ. In the same vein, it is likely that N, O, P, and Q ectoteloblasts eventually form the ring of ectodermal stem cells at the PGZ. Assuming this to be the case, another fascinating question is whether the stem cell progeny of the N, O, P, and Q progeny retain the lineage restrictions exhibited by their respective parent cells during embryonic segmentation.

4.5   MOLECULAR BASIS OF ANNELID SEGMENTATION

In contrast to the relatively good understanding of the main molecular players during the segmentation process of arthropods and vertebrates, mechanistic knowledge of the gene networks controlling segment development in annelids is very scant. The main reason for this lag, besides the fact that developmental biology of annelids is a much less populated field than that of arthropods or vertebrates, is the historical lack of strong functional molecular tools applicable to annelid models. More recently, however, the emergence of transcriptional profiling approaches, combined with powerful and broadly applicable gene editing techniques (e.g., Crispr-CAS9 gene editing) are lifting this constraint, and many research programs are currently working on providing better answers to the question of how annelids organize their tissues into segmental units.

Most approaches to uncovering the molecular genetic basis of segment formation in annelids have revolved around candidate gene approaches (see Bleidorn et al., 2015, for a summary): orthologs of genes known to be involved in the arthropod segmentation process (especially those known from *Drosophila* fruit flies) were identified in several annelid species, and their expression patterns during development were explored using *in situ* mRNA hybridization assays. Initial results were interpreted as supporting the existence of segmentation mechanisms common to both arthropods and annelids, or even to vertebrates. Then, when evidence from molecular phylogenies led to rejection of the Articulata hypothesis that Arthropoda and Annelida are closely related, these same data were interpreted as supporting the hypothesis of a truly segmented last common ancestor of all bilaterians.

However, closer examination of the data concerning segmentation gene homologs has raised doubts about this interpretation. In some cases, the spatial and temporal patterns of expression proved incompatible with the putative roles of the genes in segmentation. In other cases, it was realized that the candidate genes are broadly expressed during various aspects of development in diverse taxa, thus weakening the argument that similar patterns are indicative of shared ancestry—any gene involved in building any component of a segment will show an iterated, segmental pattern. For example, the genes *wnt1/wingless* and *engrailed* (*en*) interact in *Drosophila* to define para-segmental boundaries (Heuvel et al. 1993). Initial studies of expression of these genes in *Platynereis* showed a pattern consistent with a role in defining segment polarity in this species (Prud'homme et al. 2003): *Pdu-en* is expressed at the anterior portion of developing segments, while *Pdu-wnt1* is expressed in their posterior part. These findings were hailed as evidence of a common arthropod–annelid patterning system. However, investigations in other groups failed to support the generality of the *Platynereis* pattern: expression of *en* in *Chaetopterus, Capitella*, and *Hydroides* is not compatible with a fundamental early role in segmentation (Seaver et al. 2001; Seaver and Kaneshige 2006). In addition, experiments in the leech *Helobdella* show that ablation of *en*-expressing cells within clones of ectoteloblast progeny does not affect establishment of segment polarity (Seaver and Shankland 2001). Similar disagreements have been found for other fly segmentation genes, like *hedgehog* (Seaver and Kaneshige 2006; Kang et al. 2003; Dray et al. 2010), *even-skipped* (*eve*), *runt*, and *paired* (*prd, Pax3/7*; Song et al. 2002; de Rosa, Prud'homme, and Balavoine 2005; Seaver et al. 2012) and *hunchback* (Iwasa, Suver, and Savage 2000; Werbrock et al. 2001; Kerner et al. 2006). In most cases, a candidate gene found expressed in developing segments was found to also show broad expression at other places and times, suggesting that even if they do play a role in segmentation, then convergent co-option of a widely used toolkit gene is a more parsimonious hypothesis than developmental homology inherited from the last common ancestor between arthropods and annelids or a segmented last common bilaterian ancestor.

Perhaps a notable exception to the aforementioned disagreements is the *Notch* and *hairy* (*hes*) genes known to be involved in arthropod and vertebrate segmentation. In *Platynereis*, *Notch* expression has not been described, but 15 paralogues of *hairy* have been reported, 7 of which show expression patterns consistent with a role in segmentation (Gazave, Guillou, and Balavoine 2014). Six of them (*Pdu-Hes1/2, 4, 5, 6,* and *8*) are expressed in the ectoteloblast ring around the posterior growth zone (PGZ). *Pdu-Hes5* is also expressed in the underlying ring of mesoteloblasts derived from 4d descendants (see earlier). In all cases, *Hes* expression is associated with the established PGZ but not with earlier development of the mesoteloblast lineage. In *Capitella*, the patterns of *Notch* and *hes* expression during embryonic development do not support the hypothesis that these genes play an interacting role in segmentation; however, *CapI-hes1* shows transient expression in small bands of pre-segmental mesoderm in larvae, and later in small mesodermal domains of the PGZ of late larvae and juveniles (Thamm and Seaver 2008). In contrast, in juveniles and adults, *CapI-Delta, CapI-Notch, CapI-hes1, CapI-hes2,* and *CapI-hes3* are all expressed in the PGZ, even though their expression doesn't extend anteriorly to developing segments. Interestingly, while *Cap-hes1* expression is limited to unsegmented mesoderm, its anterior boundary is the posterior boundary of the nascent segment, which is also the posterior limit of *CapI-Delta* and *CapI-Notch* expression domains. *CapI-hes2* and *CapI-hes3* expression straddles this boundary and extends both anteriorly into the forming segment and posteriorly into the unsegmented segment addition zone. One interpretation of these data is that segmentation at the PGZ involves progenitor stem cells that maintain *CapI-hes1* expression independently of *CapI-Notch*, which instead has a role in segment boundary formation (Thamm and Seaver 2008). Involvement of *Notch* and *hes* has also been shown in leeches, where experimental simultaneous disruption of Notch signaling and *hes* expression results in somewhat weak segmentation defects of ectoteloblast progeny (Rivera and Weisblat 2009). Such treatment did not affect teloblast division and blast cell formation, the primary mechanism behind lineage-driven segmentation, nor did it interfere with clonal expansion, but the clones failed to generate regular and segmentally iterated patterns. Evidence of the generalized involvement of *Notch* and *hes* in annelid segmentation suggests this mechanism is part of the annelid ground plan and supports its presence in the last common ancestors of this phylum, arthropods and vertebrates. However, this does not imply a segmented Urbilateria: *Notch/hes* signaling might also have been involved in posterior elongation and co-opted independently during the evolution of segmentation in each phylum (Chipman 2010).

Thus, our knowledge about the molecular underpinnings of annelid segmentation is still too limited to make any broad statement about conservation or variability of genes, gene regulatory networks, or developmental processes across the segmented phyla. But there is hope that recent developments on functional molecular research tools will soon begin to shed new light on how different annelid groups organize their cells and tissues into repeated metameric units.

4.6   EVOLUTIONARY REMARKS

As defined and discussed in this chapter, the formation of true segments in a boundary-driven process comprises three steps: (1) elongation of the anteroposterior (AP) axis, (2) segregation of cells into metameric units within one or more germ layers, and (3) patterning within each segmental unit showing the same spatial frequency in two or more germ layers (Balavoine 2015). Note that the second step can be partially bypassed if patterning in one germ layer is imposed directly by inductive signaling from metamers in another germ layer. In lineage-driven segmentation, by contrast, the second step is obviated entirely, because metameric patterning arises directly from the stereotyped lineages of cells produced by the PGZ, as seen during teloblastic segmentation in clitellates and in early *Platynereis* development.

It is intriguing to consider that several annelids may transition from lineage-driven to boundary-driven segmentation mechanisms, for example, during post-embryonic segmentation after the teloblasts have exhausted their capacity for regular blast cell production, or during posterior regeneration, after complete removal of the PGZ (see Chapter 10). Under this scenario, the differences in segment development dynamics described for different annelid lineages could be explained by heterochronic shifts of this transition along developmental trajectories. This shift could also provide an evolutionary mechanism to adjust for changes in life history strategies, for example, changes in the amount of maternal yolk, switches from planktotrophy to lecithotrophy and back, evolution of direct development, and emergence of alternative maternal provisioning strategies such as albumenotrophy or adelphophagy (i.e., when egg sacs contain abortive nurse eggs that provide additional nutritional reserves to the remaining viable eggs).

Evolutionary plasticity in the timing of the switch between lineage- and boundary-driven segmentation could also provide a mechanism for evolving heteronomy, fixed segmental counts and even loss of segmentation. Heteronomy denotes the presence of larval segments whose morphology and development are proposed to be qualitatively different from that of most adult segments. In many annelids with a larval stage, embryos develop into planktonic larvae with three or more segments. In some groups these larval segments do not exhibit a strict anteroposterior sequence of development: in *Hydroides elegans* for example, segments 4 to 7 begin differentiating after segments 8 to 11 have already formed. And in myzostomid annelids, a highly modified parasitic lineage, segment number has become fixed (as in leeches); furthermore, segments do not develop following an anteroposterior sequence of development (Jägersten 1940). Still other annelid lineages have lost segmentation altogether: sipunculans, echiurids, and orthonectids show very limited evidence for segmentation during development, and almost no trace of it in adults (Hessling 2003; Kristof, Wollesen, and Wanninger 2008; Boyle and Rice 2014; Tilic, Lehrke, and Bartolomaeus 2015; Schiffer, Robertson, and Telford 2018; Zverkov et al. 2019). This enormous morphological and developmental diversity shows that the ancestral annelid segmentation mechanisms have evolved in many directions since their last common ancestor, and hint that excessive emphasis on reconstructing ancestral traits and trying to find the homologies across annelids, arthropods, and vertebrates can be misleading and can even blind us from appreciating the true evolutionary richness of diverse animal morphologies and the developmental pathways by which they arise.

REFERENCES

Ackermann, C., A. Dorresteijn, and A. Fischer. 2005. Clonal domains in postlarval *Platynereis dumerilii* (Annelida: Polychaeta). *J. Morphol.* 266(3): 258–280.

Åkesson, B. 1967. The embryology of the polychaete *Eunice kobiensis*. *Acta Zool.* 48(1–2): 142–192.

Anderson, D. T. 1973. *Embryology and Phylogeny in Annelids and Arthropods*, Vol. 50. International Series of Monographs in Pure and Applied Biology Zoology. Oxford: Pergamon Press.

Balavoine, G. 2015. Segment formation in Annelids: Patterns, processes and evolution. *Int. J. Dev. Biol.* 58(6–8): 469–483.

Bleidorn, C., C. Helm, A. Weigert, and M. T. Aguado. 2015. Annelida. In *Evolutionary Developmental Biology of Invertebrates 2: Lophotrochozoa (Spiralia)*, Ed. A. Wanninger, 193–230. Vienna: Springer Vienna.

Boyle, M. J., and M. E. Rice. 2014. Sipuncula: An emerging model of spiralian development and evolution. *Int. J. Dev. Biol.* 58(6–8): 485–499.

Brusca, R. C., and G. J. Brusca. 1990. *Invertebrates.* Sunderland, MA: Sinauer.

Chipman, A. D. 2010. Parallel evolution of segmentation by co-option of ancestral gene regulatory networks. *BioEssays* 32(1): 60–70.

Conklin, E. G. 1897. The embryology of *Crepidula*, a contribution to the cell lineage and early development of some marine gasteropods. *J. Morphol.* 13(1): 1–226.

Couso, J. P. 2009. Segmentation, metamerism and the Cambrian explosion. *Int. J. Dev. Biol.* 53(8–10): 1305–1316.

Davis, G. K., and N. H. Patel. 1999. The origin and evolution of segmentation. *Trends Genet.* 15(12): M68–M72.

Dohle, W. 1999. The ancestral cleavage pattern of the clitellates and its phylogenetic deviations. *Hydrobiologia* 402: 267–283.

Dray, N., K. Tessmar-Raible, M. Le Gouar, L. Vibert, F. Christodoulou, K. Schipany, …, G. Balavoine. 2010. Hedgehog signaling regulates segment formation in the annelid *Platynereis*. *Science* 329(5989): 339–342.

Fischer, A. H., T. Henrich, and D. Arendt. 2010. The normal development of *Platynereis dumerilii* (Nereididae, Annelida). *Front. Zool.* 7(1): 31.

Fischer, A. H. L., and D. Arendt. 2013. Mesoteloblast-like mesodermal stem cells in the polychaete annelid *Platynereis dumerilii* (Nereididae). *J. Exp. Zool. B Mol. Dev. Evol.* 320(2): 94–104.

Freeman, G., and J. W. Lundelius. 1992. Evolutionary implications of the mode of D quadrant specification in coelomates with spiral cleavage. *J. Evol. Biol.* 5(2): 205–247.

Gazave, E., J. Béhague, L. Laplane, A. Guillou, L. Préau, A. Demilly, …, M. Vervoort. 2013. Posterior elongation in the annelid *Platynereis dumerilii* involves stem cells molecularly related to primordial germ cells. *Dev. Biol.* 382(1): 246–267.

Gazave, E., A. Guillou, and G. Balavoine. 2014. History of a prolific family: The Hes/Hey-related genes of the annelid *Platynereis*. *EvoDevo* 5(1): 29.

Goto, A., K. Kitamura, and T. Shimizu. 1999. Cell lineage analysis of pattern formation in the *Tubifex* embryo. I. Segmentation in the mesoderm. *Int. J. Dev. Biol.* 43(4): 317–327.

Helm, C., O. Vöcking, I. Kourtesis, and H. Hausen. 2016. *Owenia fusiformis* – a basally branching annelid suitable for studying ancestral features of annelid neural development. *BMC Evol. Biol.* 16: 129.

Henry, J. Q. 2014. Spiralian model systems. *Int. J. Dev. Biol.* 58(6–8): 389–401.

Henry, J. Q., and M. Q. Martindale. 1998. Conservation of the spiralian developmental program: Cell lineage of the nemertean, *Cerebratulus lacteus*. *Dev. Biol.* 201(2): 253–269.

Hessling, R. 2003. Novel aspects of the nervous system of *Bonellia viridis* (Echiura) revealed by the combination of immunohistochemistry, confocal laser-scanning microscopy and three-dimensional reconstruction. *Hydrobiologia* 496(1–3): 225–239.

Heuvel, M. van den, J. Klingensmith, N. Perrimon, and R. Nusse. 1993. Cell patterning in the *Drosophila* segment: Engrailed and wingless antigen distributions in segment polarity mutant embryos. *Development* 119(Supplement): 105–114.

Irvine, S. Q., O. Chaga, and M. Q. Martindale. 1999. Larval ontogenetic stages of *Chaetopterus*: Developmental heterochrony in the evolution of chaetopterid polychaetes. *Biol. Bull.* 197(3): 319–331.

Iwasa, J. H., D. W. Suver, and R. M. Savage. 2000. The leech hunchback protein is expressed in the epithelium and CNS but not in the segmental precursor lineages. *Dev. Genes Evol.* 210(6): 277–288.

Jägersten, G. 1940. Zur kenntnis der morphologie, Entwicklung und taxonomie der Myzostomida. *Nova Acta Regiae Soc. Sci. Upsal.* 11: 1–84.

Kang, D., F. Huang, D. Li, M. Shankland, W. Gaffield, and D. A. Weisblat. 2003. A hedgehog homolog regulates gut formation in leech (*Helobdella*). *Development* 130(8): 1645–1657.

Kerner, P., F. Zelada González, M. Le Gouar, V. Ledent, D. Arendt, and M. Vervoort. 2006. The expression of a hunchback ortholog in the polychaete annelid *Platynereis dumerilii* suggests an ancestral role in mesoderm development and neurogenesis. *Dev. Genes Evol.* 216(12): 821–828.

Kristof, A., T. Wollesen, and A. Wanninger. 2008. Segmental mode of neural patterning in sipuncula. *Curr. Biol.* 18(15): 1129–1132.

Lacalli, T. 1980. *A Guide to the Marine Flora and Fauna of the Bay of Fundy: Polychaete Larvae from Passamquoddy Bay.* Canadian Technical Reports of Fisheries and Aquatic Sciences.

Meyer, N. P., M. J. Boyle, M. Q. Martindale, and E. C. Seaver. 2010. A comprehensive fate map by intracellular injection of identified blastomeres in the marine polychaete *Capitella teleta*. *EvoDevo* 1(1): 8.

Meyer, N. P., and E. C. Seaver. 2010. Cell lineage and fate map of the primary somatoblast of the polychaete annelid *Capitella teleta*. *Integr. Comp. Biol.* 50(5): 756–767.

Nakamoto, A., A. Arai, and T. Shimizu. 2004. Specification of polarity of teloblastogenesis in the oligochaete annelid *Tubifex*: Cellular basis for bilateral symmetry in the ectoderm. *Dev. Biol.* 272(1): 248–261.

Nielsen, C. 2005. Trochophora larvae and adult body regions in annelids: Some conclusions. *Hydrobiologia* 535/536(1–3): 23–24.

Özpolat, B. D., M. Handberg-Thorsager, M. Vervoort, and G. Balavoine. 2017. Cell lineage and cell cycling analyses of the 4d micromere using live imaging in the marine annelid *Platynereis dumerilii*. *eLife* 6: e30463.

Patel, N. H. 2003. The ancestry of segmentation. *Dev. Cell* 5(1): 2–4.

Peel, A. D., A. D. Chipman, and M. Akam. 2005. Arthropod segmentation: Beyond the *Drosophila* paradigm. *Nat. Rev. Genet.* 6(12): 905–916.

Penners, A. 1922. Die Furchung von *Tubifex rivulorum* Lam. *Zool. Jahrb. Abt. Für. Anat. Ontog. Tiere* 43: 323–368.

Penners, A. 1924. Experimentelle Untersuchungen zum Determinationsproblem am Keim von *Tubifex rivulorum* Lam. *Arch. Für Mikrosk. Anat. Entwicklungsmech.* 102(1): 51–100.

Prud'homme, B., R. de Rosa, D. Arendt, J. F. Julien, R. Pajaziti, A. W. Dorresteijn, …, G. Balavoine. 2003. Arthropod-like expression patterns of engrailed and wingless in the annelid *Platynereis dumerilii* suggest a role in segment formation. *Curr. Biol.* 13(21): 1876–1881.

Rebscher, N., F. Zelada-González, T. U. Banisch, F. Raible, and D. Arendt. 2007. Vasa unveils a common origin of germ cells and of somatic stem cells from the posterior growth zone in the polychaete *Platynereis dumerilii*. *Dev. Biol.* 306(2): 599–611.

Rivera, A., and D. Weisblat. 2009. And Lophotrochozoa makes three: Notch/Hes signaling in annelid segmentation. *Dev. Genes Evol.* 219(1): 37–43.

de Rosa, R., B. Prud'homme, and G. Balavoine. 2005. Caudal and even-skipped in the annelid *Platynereis dumerilii* and the ancestry of posterior growth. *Evol. Dev.* 7(6): 574–587.

Rouse, G. W. 2006. Annelid larval morphology. In *Reproductive Biology and Phylogeny of Annelida*, Vol. 4, Ed. G. W. Rouse, and F. Pleijel, 141–177. Reproductive Biology and Phylogeny 4. Enfield, NH: Science Publishers.

Schiffer, P. H., H. E. Robertson, and M. J. Telford. 2018. Orthonectids are highly degenerate annelid worms. *Curr. Biol.* 28(12): 1970.e3–1974.e3.

Scholtz, G. 2002. The Articulata hypothesis – or what is a segment? *Org. Divers. Evol.* 2(3): 197–215.

Schroeder, P. C., and C. O. Hermans. 1975. Annelida: Polychaeta. In *Reproduction of Marine Invertebrates*, Volume III, Ed. A.C. Pearse and J.S. Pearse, 1–213. New York, NY: Academic Press.

Seaver, E. C. 2014. Variation in spiralian development: Insights from polychaetes. *Int. J. Dev. Biol.* 58(6–8): 457–467.

Seaver, E. C. 2016. Annelid models I: *Capitella teleta. Curr. Opin. Genet. Dev.* 39 Developmental Mechanisms, Patterning and Evolution (August): 35–41.

Seaver, E. C., and L. M. Kaneshige. 2006. Expression of "segmentation" genes during larval and juvenile development in the polychaetes *Capitella* sp. I and *H. elegans. Dev. Biol.* 289(1): 179–194.

Seaver, E. C., D. A. Paulson, S. Q. Irvine, and M. Q. Martindale. 2001. The spatial and temporal expression of Ch-en, the *engrailed* gene in the polychaete *Chaetopterus*, does not support a role in body axis segmentation. *Dev. Biol.* 236(1): 195–209.

Seaver, E. C., and M. Shankland. 2001. Establishment of segment polarity in the ectoderm of the leech *Helobdella. Development* 128(9): 1629–1641.

Seaver, E. C., K. Thamm, and S. D. Hill. 2005. Growth patterns during segmentation in the two polychaete annelids, *Capitella* sp. I and *Hydroides elegans*: Comparisons at distinct life history stages. *Evol. Dev.* 7(4): 312–326.

Seaver, E. C., E. Yamaguchi, G. S. Richards, and N. P. Meyer. 2012. Expression of the pair-rule gene homologs *runt, Pax3/7, even-skipped-1* and *even-skipped-2* during larval and juvenile development of the polychaete annelid *Capitella teleta* does not support a role in segmentation. *EvoDevo* 3(1): 8.

Shankland, M., and E. C. Seaver. 2000. Evolution of the bilaterian body plan: What have we learned from annelids? *Proc. Natl. Acad. Sci.* 97(9): 4434–4437.

Shimizu, T. 1982. Development in the freshwater oligochaete *Tubifex*. In *Developmental Biology of Freshwater Invertebrates*, Ed. F.W. Harrison and R.R. Cowden, 286–316. New York: Alan R. Liss.

Shimizu, T., and A. Nakamoto. 2014. Developmental significance of D quadrant micromeres 2d and 4d in the oligochaete annelid *Tubifex tubifex. Int. J. Dev. Biol.* 58(6–8): 445–456.

Smart, T. I., and G. von Dassow. 2009. Unusual development of the mitraria larva in the polychaete *Owenia collaris. Biol. Bull.* 217(3): 253–268.

Song, M. H., F. Z. Huang, G. Y. Chang, and D. A. Weisblat. 2002. Expression and function of an even-skipped homolog in the leech *Helobdella robusta. Development* 129(15): 3681–3692.

Steinmetz, P. R. H., R. P. Kostyuchenko, A. Fischer, and D. Arendt. 2011. The segmental pattern of *otx, gbx*, and *Hox* genes in the annelid *Platynereis dumerilii. Evol Dev* 13(1): 72–79.

Struck, T. H., C. Paul, N. Hill, S. Hartmann, C. Hösel, M. Kube, …, C. Bleidorn. 2011. Phylogenomic analyses unravel annelid evolution. *Nature* 471(7336): 95–98.

Thamm, K., and E. C. Seaver. 2008. Notch signaling during larval and juvenile development in the polychaete annelid *Capitella* sp. I. *Dev. Biol.* 320(1): 304–318.

Tilic, E., J. Lehrke, and T. Bartolomaeus. 2015. Homology and evolution of the chaetae in Echiura (Annelida). *PLoS One* 10(3): e0120002.

Treadwell, A. L. 1901. The cytogeny of *Podarke obscura. J. Morphol.* 17: 399–486.

Weigert, A., C. Helm, M. Meyer, B. Nickel, D. Arendt, B. Hausdorf, …, T. Struck. 2014. Illuminating the base of the annelid tree using transcriptomics. *Mol. Biol. Evol.* 31(6): 1391–1401.

Werbrock, A. H., D. A. Meiklejohn, A. Sainz, J. H. Iwasa, and R. M. Savage. 2001. A polychaete *hunchback* ortholog. *Dev. Biol.* 235(2): 476–488.

Wilson, D. P. 1932a. IV. On the mitraria larva of *Owenia fusiformis* Delle Chiaje. *Philos. Trans. R. Soc. Lond. Ser. B Contain. Pap. Biol. Character* 221(474–482): 231–334.

Wilson, D. P. 1932b. The development of *Nereis pelagica* Linnaeus. *J. Mar. Biol. Assoc. U K* 18: 203–218.

Wilson, E. B. 1883. Observations on the early developmental stages of some polychaetous annelids. *Stud. Biol. Lab. Johns Hopkins Univ.* 2: 271–299.

Wilson, E. B. 1892. The cell lineage of *Nereis. J. Morphol.* 6: 361–480.

Zantke, J., S. Bannister, V. B. V. Rajan, F. Raible, and K. Tessmar-Raible. 2014. Genetic and genomic tools for the marine annelid *Platynereis dumerilii. Genetics* 197(1): 19–31.

Zattara, E. E., and A. E. Bely. 2011. Evolution of a novel developmental trajectory: Fission is distinct from regeneration in the annelid *Pristina leidyi. Evol. Dev.* 13(1): 80–95.

Zattara, E. E., and A. E. Bely. 2015. Fine taxonomic sampling of nervous systems within Naididae (Annelida: Clitellata) reveals evolutionary lability and revised homologies of annelid neural components. *Front. Zool.* 12(1): 8.

Zattara, E. E., and A. E. Bely. 2016. Phylogenetic distribution of regeneration and asexual reproduction in Annelida: Regeneration is ancestral and fission evolves in regenerative clades. *Invertebr. Biol.* 135(4): 400–414.

Zhang, Z.-Q. 2013. Animal biodiversity: An update of classification and diversity in 2013. In: Zhang, Z.-Q. (Ed.) Animal biodiversity: An outline of higher-level classification and survey of taxonomic richness (Addenda 2013). *Zootaxa* 3703(1): 5–11.

Zverkov, O. A., K. V. Mikhailov, S. V. Isaev, L. Rusin, M. D. Logacheva, A. A. Penin, …, V. V. Aleoshin. 2019. Dicyemida and Orthonectida: Two stories of body plan simplification. *Front. Genet.* 10: 443.

5 Progenitor Cells in Vertebrate Segmentation

Benjamin Martin

CONTENTS

5.1   SEGMENTATION IN VERTEBRATES

Segmentation of the vertebrate body is most noticeable from the clade's eponymous vertebrae that surround the spinal cord, as well as the ribs and skeletal muscles of the trunk. These features arise from transient embryonic structures called somites (Figure 5.1). Somites originate from the paraxial mesoderm, which forms in bilateral stripes on either side of the embryonic midline (Hubaud and Pourquie, 2014). Pairs of somites bud off from the presomitic paraxial mesoderm in a progressive fashion, with the first pair forming close to the head and the final pair at the posterior end of the body. The rate at which somites form, and the final number generated, is specific to each species (Maroto et al., 2012). In addition to ribs and vertebrae, somites also give rise to skeletal muscle, dermis, tendons, cartilage, and endothelial cells. This chapter will focus on the formation of the somites, which represent the origin of the

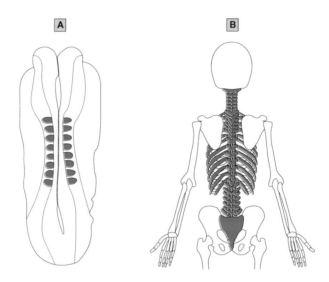

FIGURE 5.1 Segmentation in humans. (A) A dorsal view of a Carnegie stage 10 human embryo. Forming somites (red) are present on either side of the spinal cord. (B) The segmental somites later give rise to segmental structures in the adult, such as the vertebrae of the spinal column and the ribs (red).

bulk of the segmented tissue in the vertebrate body and also play a direct role in patterning adjacent segmental structures.

Other tissues in the vertebrate embryo undergo a segmental mode of development, including the nervous system, vasculature, notochord, and specific head structures. Alongside the spinal cord, sensory, motor, and autonomic nerves of the peripheral nervous system are present in a segmental organization. This pattern is organized by the somites, which express a repulsive migratory molecule exclusively in the posterior half of the somite, causing lateral migration of the nerves along the anterior half somite (Keynes, 2018). Similarly, a repulsive cue expressed throughout the somite causes intersomitic blood vessels to develop in a segmental pattern between somites (Gu et al., 2005). Recent work in zebrafish showed that the outer layer of cells in the notochord, called the sheath cells, display segmental gene expression (Lleras Forero et al., 2018). These cells later become the chordacentra of the vertebrae. The segmental pattern of the notochord sheath is influenced by the signals from the segmented somites, but also can occur in the absence of proper somite segmentation, suggesting that the axial mesoderm undergoes an autonomous segmentation program (Lleras Forero et al., 2018). Nevertheless, the developmental life history of the axial mesoderm progenitors share many similarities with the somite progenitor cells (Row et al., 2016), and thus the segmental underpinnings of both tissues may be influenced by a common molecular network while they are in the progenitor state.

There are other segmental structures in the embryo for which the segmentation mechanism is not directly influenced by the segmental somites. These include the rhombomeres of the hindbrain and the pharyngeal arches, which give rise to the majority of the musculoskeletal and neurovascular tissues of the head. Although there are some shared molecular mechanisms involved in these segmentation processes, they occur in a largely unique fashion that has been described in recent reviews (Frank and Sela-Donenfeld, 2019; Frisdal and Trainor, 2014; Wilkinson, 2018).

5.2 THE CLOCK AND WAVEFRONT MODEL OF SOMITOGENESIS

The molecular mechanism governing somite formation was first proposed as a theoretical model in 1976, called the clock and wavefront model (Cooke and Zeeman, 1976). The clock represents a molecular oscillator, where individual presomitic cells cycle between on and off states of the expression of critical genes. The wavefront marks an important activity threshold at the anterior extent of a posterior-to-anterior signaling gradient that emanates from the posterior terminus of the embryo. During posterior extension, presomitic cells are displaced anteriorly. As this occurs, the molecular oscillator within these cells slows; when it becomes sufficiently slow—and when the displaced cells intersect with the wavefront—a border forms and creates a new somite (explained in more detail in Section 5.6) (Figure 5.2). Several molecules have been discovered that fit the predicted role of either the clock or the wavefront (Aulehla and Pourquie, 2010; Hubaud and Pourquie, 2014; Maroto et al., 2012; Oates et al., 2012; Pourquie, 2018). In particular, the hairy/enhancer of split genes (referred

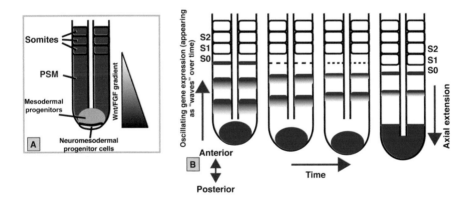

FIGURE 5.2 The clock and wavefront mechanism of somitogenesis. (A) A schematic illustrating the relative positions of the somites, the presomitic mesoderm, and the posteriorly localized neuromesodermal progenitors and mesodermal progenitors. The posterior-to-anterior gradient of Wnt and FGF signaling is also illustrated. (B) A series of four time points are shown during the formation of one somite. Individual cells have on/off cycling gene expression represented by the blue color (on) or white (off). The cycling expression is coordinated with neighbor cells, and as they cycle they give the impression of a traveling wave of gene expression through the presomitic mesoderm. Cells in the anteriormost stripe stop cycling and remain in the on state, and when these cells drop below a threshold level of wavefront signal, a new somite border is formed. In the schematic, S1 is the most recently formed somite. Somites continue to form in the posterior direction at the same time as the axis is extending in the posterior direction. The schematics are shown from a ventral view with posterior to the bottom and anterior to the top.

to as *Hes* in mouse and chick and *her* zebrafish) oscillate in the presomitic mesoderm and are a core component of the molecular clock governing somitogenesis (Mara and Holley, 2007; Oates et al., 2012; Pourquie, 2011). These genes encode bHLH transcriptional repressors that can repress their own transcription. The wavefront is produced from signals emanating from the posteriormost region of the embryo (called the tailbud), notably FGF and canonical Wnt signaling (Aulehla and Pourquie, 2010). Low levels of signal create a threshold value that initiates somite border formation at a particular phase of *her*/*Hes* oscillation. As paraxial mesoderm progenitor cells exit the tailbud, they are displaced anteriorly with respect to the tailbud and pass through the signaling gradient.

The cycling of the *her*/*Hes* genes begins in the mesodermal progenitors within the tailbud. Subsequently, as cells exit from the tailbud into the unsegmented paraxial mesoderm (called the presomitic mesoderm and abbreviated PSM), oscillations continue but become out of phase with the mesoderm in the tailbud. This cycling activity gives the impression of waves of gene expression through the presomitic mesoderm (Mara and Holley, 2007) (Figure 5.2). The oscillations of gene expression slow as cells mature in the PSM. When the wavefront of FGF and Wnt signaling passes through the PSM and intersects with a permissive oscillatory state of the clock, a new physical boundary is formed, thus creating a new somite (Aulehla and Pourquie, 2010). In the mouse, boundary formation is accomplished by the activation of the *Mesp2* gene, which establishes the anterior limit of expression of *Tbx6* (Saga, 2012; Saga et al., 1997). In cells where *Tbx6* is inactivated, border formation occurs. The situation is slightly different in zebrafish, where a quadruple knockout of all the *mesp* genes causes only mild somite defects (Yabe et al., 2016). In zebrafish, the anterior limit of *tbx6* expression also plays a critical role in border formation, but the expression domain is independent of *mesp* function (Kinoshita et al., 2018; Yabe et al., 2016).

5.3 NEUROMESODERMAL PROGENITORS AS THE CELLULAR SOURCE OF VERTEBRATE SEGMENTS

The continuous posterior addition of somites requires an influx of progenitor cells to sustain the somitogenesis process. Work in several vertebrate model systems, including mouse, chick, *Xenopus*, and zebrafish has established that cells within the posteriormost embryonic structure called the tailbud act as the progenitor pool for new presomitic mesoderm production, and subsequent somite addition (Henrique et al., 2015; Kimelman, 2016; Martin, 2016). These cells are called neuromesodermal progenitors (NMPs) and are unique in that they have not made a germ-layer decision even after gastrulation. Local signaling cues direct the fate of neuromesodermal progenitors into either neural ectoderm, where they give rise to the spinal cord, or mesoderm, the majority of which gives rise to the somites with minor contribution to the endothelial tissue of the vascular system (Henrique et al., 2015; Kimelman, 2016; Martin, 2016) (Figure 5.3).

Retrospective lineage analysis in the mouse embryo showed that single cells can give rise to both spinal cord and somites as the body axis extends, thus confirming a long-standing hypothesis that germ-layer naïve cells exist in the tailbud

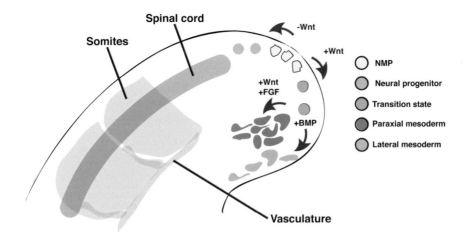

FIGURE 5.3 Neuromesodermal progenitors give rise to the somites. NMPs, here shown in the tailbud, make a germ-layer decision between ectoderm (spinal cord) and mesoderm. The mesoderm is further patterned into paraxial and lateral fates. The paraxial mesoderm gives rise to the somites, whereas the NMP-derived lateral mesoderm gives rise to vascular endothelium. The initial germ-layer decision is based on canonical Wnt signaling activity, with Wnt signaling promoting mesodermal fate. Continued Wnt and FGF signaling induces paraxial fate in the mesoderm, while BMP signaling induces the endothelial fate.

(Tzouanacou et al., 2009). This work changed our understanding of lineage relationships in the early mouse embryo, demonstrating that spinal cord cells are more closely related to somite cells than to other ectodermal derivatives such as the epidermis. Several years later, NMPs were identified in the zebrafish embryo (Martin and Kimelman, 2012). Although single zebrafish NMPs rarely give rise to daughter cells in both the spinal cord and somite lineages (Attardi et al., 2018; Martin, 2016), they are competent to differentiate into either cell type, and make that decision based on local canonical Wnt signaling (Martin and Kimelman, 2012). High Wnt signaling induces NMPs to become mesoderm, whereas low Wnt signaling causes NMPs to adopt a neural fate. NMPs are defined as cells that express both the neural-inducing transcription factor *sox2* as well as the mesoderm-inducing transcription factor *Brachyury* (discussed in more detail in Section 5.6), and a posteriorly localized *Brachyury/Sox2* population of cells has also been identified in human, chick, *Xenopus*, and axolotl (Garriock et al., 2015; Henrique et al., 2015; Martin and Kimelman, 2012; Olivera-Martinez et al., 2012; Taniguchi et al., 2017; Tsakiridis and Wilson, 2015). Thus, NMPs are evidently a conserved cell type present in all vertebrates that contribute cells to the somites throughout axis extension.

5.4 DO ALL SOMITES ALONG THE ANTERIOR–POSTERIOR AXIS COME FROM NMPS?

Vertebrate somites are present along most of the body axis, including the trunk and tail. NMPs exist in the post-gastrulation tailbud and give rise to the posterior trunk and tail of the body. NMPs have been best studied in this post-gastrulation context, but there is experimental evidence to suggest that they are present during gastrulation and give rise to anterior trunk somites. During zebrafish gastrulation, the prospective anterior spinal cord and trunk somites reside within the dorsal–lateral margin of the embryo (Kimelman and Martin, 2012). Lineage analysis in zebrafish shows that cells in the margin of the gastrula are generally mono-fated, giving rise to either spinal cord or mesoderm (Attardi et al., 2018). However, these mono-fated cells could maintain neuromesodermal plasticity, but due to their signaling environment, predominantly give rise to only one cell type. The role of Wnt signaling in post-gastrulation neuromesodermal fate decision suggests that this is the case. But the role of Wnt signaling during the neuromesoderm fate decision is stage-specific. When cells containing an inducible cell-autonomous Wnt inhibitor are transplanted into the margin of host embryos, where the future spinal cord and somitic cells reside, inhibition of Wnt signaling at the end of gastrulation causes NMPs to contribute to tail spinal cord and inhibits contribution to tail somites, whereas transplanted cells that had already made the decision to become paraxial mesoderm when Wnt signaling was inhibited joined the trunk somites normally (Martin and Kimelman, 2012). On the other hand, if Wnt signaling is inhibited in cells transplanted into the margin of wild-type host embryos at earlier stages during gastrulation, transplanted cells contribute to trunk spinal cord and fail to contribute to trunk somites (Martin and Kimelman, 2012). The continuous role of Wnt signaling in the neuromesoderm fate decision before and after gastrulation suggests that NMPs exist during gastrulation and are allocated to spinal cord or mesodermal fate depending upon their local Wnt signaling environment, although further work is required to definitively demonstrate the gastrula stage existence of NMPs.

In mouse embryos, there is single-cell clonal descendant contribution to anterior trunk mesoderm and spinal cord (Tzouanacou et al., 2009). During gastrulation, these cells originate in the epiblast, and at the completion of gastrulation, NMPs are deposited in the tailbud of the embryo (Cambray and Wilson, 2007; Rodrigo Albors et al., 2018; Wymeersch et al., 2016). These results suggest that there exists NMPs during mouse gastrulation that contribute to anterior somites, in addition to post-gastrula NMPs that give rise to the posterior somites. Thus, in both mouse and zebrafish it appears that somitic cells along the entire anterior–posterior axis originate from NMPs that exist during and after gastrulation.

5.5   ARE NMPS STEM CELLS?

NMPs clearly have a great degree of plasticity, giving rise to both neural ectoderm and mesoderm, but it is not clear whether they fit the definition of a stem cell, with the ability to indefinitely self-renew, or are rather a progenitor cell-type that have limited capacity to self-renew. Experiments in amniotes suggest that they are stem cells based on serial transplantations. Cells from the tailbud of chick or mouse embryos were removed and transplanted into younger embryos, and the host embryos were then allowed to develop to later tailbud stages, and the procedure was repeated (Cambray and Wilson, 2002; McGrew et al., 2008; Tam and Tan, 1992). In these experiments, tailbud cells self-renew and contribute to somites along the body axis over several serial transplants. These results suggest that the cells giving rise to the somites are stem cells and have the potential to self-renew their undifferentiated population while simultaneously giving rise to differentiated cells of the somites. Similar experiments have not been performed in zebrafish or *Xenopus* embryos, but evidence from zebrafish indicates that NMPs rarely divide, and instead are arrested in the G2 phase of the cell cycle at post-gastrulation stages (Bouldin et al., 2014). In the absence of significant self-renewal, NMPs in zebrafish fit better under the definition of a progenitor cell. In the zebrafish, as NMPs are induced to become mesoderm, they undergo a synchronous cell division event, which acts as a transit-amplifying step to increase the number of cells contributing to the somites (Bouldin et al., 2014).

In vitro–derived NMPs exhibit stem cell–like properties. Mouse NMPs can be efficiently induced in culture from epiblast stem cells (and can also be induced from embryonic stem cells) through the addition of FGF and the GSK-3 inhibitor Chiron, which activates canonical Wnt signaling (Edri et al., 2019a; Gouti et al., 2014; Tsakiridis and Wilson, 2015; Turner et al., 2014). Cultured NMPs can self-renew, but over several passages begin to differentiate, with a tendency to become neural tissue (Edri et al., 2019a). Mouse NMPs exist adjacent to an embryonic structure called the node, which emits signals to neighboring cells. When NMPs are induced from epiblast stem cells, node-like cells are also generated in the culture (Edri et al., 2019b). These node-like cells are depleted over time, corresponding to the loss of self-renewal among the NMPs. When cultured NMPs are passaged and then replenished with node-like cells, this extends their undifferentiated and self-renewal state (Edri et al., 2019b). Thus, when in vitro–derived NMPs are cultured in an environment that mimics their native in vivo environment, they also display stem cell–like properties. These results may also explain the sustained renewal of NMPs in the amniote serial transplant experiments (Cambray and Wilson, 2002; McGrew et al., 2008; Tam and Tan, 1992), as transplanting the cells to earlier stage embryos ensure that they are continuously exposed to neighboring node cells.

5.6 THE TRANSCRIPTION FACTOR *BRACHYURY* LINKS NMP MAINTENANCE WITH SEGMENTATION

The *Brachyury* gene plays a key role in both maintaining NMPs and promoting proper segmentation, as mutations in *Brachyury* result in truncated embryos with fewer than normal somites, and the somites that do form have irregular borders (Martin, 2016). A mutation in the mouse *Brachyury* gene (also called *T*, an abbreviation of the word "tailless") was first identified in 1927 (Dobrovolskaia-Zavadskaia, 1927). Animals that are heterozygous have the obvious phenotype of having a short tail relative to siblings, and the mutation was aptly named *Brachyury*, which is Latin for "short tail." The molecular identity of the mutated gene was discovered in 1990 (Herrmann et al., 1990), which became the founding member of the T-box transcription factor family, for which there are now 17 members identified in the mouse and human, and 26 in zebrafish (Ahn et al., 2012; Papaioannou, 2014). The *Brachyury* gene is often referred to now by the name *Tbxt*. NMPs are defined by the co-expression of *Brachyury* and the *Sox2* transcription factor, and complete loss of *Brachyury* function in the mouse or zebrafish results in severely truncated embryos, with only 8–12 of the anteriormost somites forming (Chesley, 1935; Halpern et al., 1993; Martin and Kimelman, 2008; Schultemerker et al., 1994). The truncated axis phenotype occurs because Brachyury directly regulates the expression of genes that are essential for the maintenance of the NMP population. Brachyury, along with the caudal homeobox transcription factor Cdx2, directly activates the expression of canonical Wnt signaling ligands (Amin et al., 2016; Evans et al., 2012; Martin and Kimelman, 2008; Morley et al., 2009), and Wnt signaling is involved in the specification and expansion of NMPs, as well as their maintenance. Wnt signaling in turn activates *Brachyury* and *Cdx2* expression, creating an autoregulatory loop in the NMPs (Amin et al., 2016; Martin and Kimelman, 2008; Yamaguchi et al., 1999).

Another Brachyury direct target that supports NMP maintenance is the *cyp26a1* enzyme, which is responsible for degrading the retinoic acid (RA) signaling molecule (Martin and Kimelman, 2010). RA is produced in the most recently formed somites by the enzyme Aldh1a2. After production, RA then diffuses into neighboring cells and binds to retinoic acid receptors, which act as transcription factors to regulate downstream gene expression (Duester, 2008). Retinoic acid normally functions in the mouse to induce an initial population of NMPs (although this is species-specific, as this is not observed in zebrafish), and subsequently to promote neural differentiation from the NMPs (Berenguer et al., 2018; Gouti et al., 2017). Due to this later role, it is important to keep the majority of NMPs in a low retinoic acid environment, so they can be sustained in an undifferentiated state. Mutations in the *cyp26a1* enzyme in mouse and zebrafish result in axial truncations with fewer somites than normal due to the failure to maintain the NMPs, with truncated embryos exhibiting expanded posterior neural tissue (Abu-Abed et al., 2001; Emoto et al., 2005; Martin and Kimelman, 2010). In the chick, the increased proximity of the retinoic acid source to the NMPs as the presomitic mesoderm shortens during late axial extension is hypothesized to be important for terminating axis extension through NMP depletion via neural induction (Olivera-Martinez et al., 2012). These studies reveal that an essential role of Brachyury is to create an environment of low retinoic acid through *cyp26a1* activation, and together with its activation of Wnt ligand expression, Brachyury establishes a molecular niche critical for NMP maintenance (Martin and Kimelman, 2010).

The poorly formed anterior somites present in *Brachyury* mutants indicate it also plays a role in the segmentation process. Indeed, Brachyury directly activates several key components of the segmentation process and has been implicated as a factor causing human congenital scoliosis when mutated (Ghebranious et al., 2008). Notch signaling is required for segmentation by coupling the oscillating gene expression in neighboring cells (Horikawa et al., 2006; Jiang et al., 2000; Mara et al., 2007; Okubo et al., 2012; Ozbudak and Lewis, 2008; Riedel-Kruse et al., 2007). The zebrafish *brachyury* orthologue *tbxta* (*ntla*) directly activates the expression of the Notch ligands *deltaC* and *deltaD* (Garnett et al., 2009; Jahangiri et al., 2012; Morley et al., 2009). Mutations in either of these genes result in severe somitogenesis defects (Holley et al., 2000; Julich et al., 2005; vanEeden et al., 1996). Direct regulation of Delta ligand expression by Brachyury is conserved in the mouse, where Brachyury activates *Dll1* expression. *Dll1* has t-box binding sites in its regulatory region that are essential for expression, and like the zebrafish delta mutants, mouse *Dll1* mutants have severe somitogenesis defects (Concepcion and Papaioannou, 2014; Hofmann et al., 2004; Hrabe de Angelis et al., 1997). Notch/Delta signaling coordinates the oscillations of the *Hes/her* transcription factors in neighboring cells, which are core components of the segmentation process (Horikawa et al., 2006; Jiang et al., 2000; Mara et al., 2007; Okubo et al., 2012; Ozbudak and Lewis, 2008; Riedel-Kruse et al., 2007). The *Hes/her* genes are also directly regulated by Brachyury. Zebrafish Tbxta binds to the –9kb to +3kb genomic regions of both *her1* and *her7* (Morley et al., 2009), which are together required for proper somitogenesis (Henry et al., 2002; Holley et al., 2002; Oates and Ho, 2002). There is a functional t-box binding site in the *her1* promoter that is required for *her1* expression (Brend and Holley, 2009). The *her7* gene has one of the largest clusters of t-box biding sites of any gene in the zebrafish genome, but these sites have not been tested functionally (Garnett et al., 2009). Similarly in the mouse, the *Hes7* oscillatory gene is a direct transcriptional target of Brachyury, and segmentation in *Hes7* mouse mutants does not occur properly (Bessho et al., 2003; Bessho et al., 2001; Faial et al., 2015). The direct regulation of Notch ligands and hairy enhancer of split transcription factors by Brachyury in both mouse and zebrafish shows that Brachyury has a conserved role in the segmentation process independent of its conserved role in NMP maintenance and mesoderm specification.

5.7   SIGNALING PATHWAYS COORDINATING MESODERM INDUCTION AND SEGMENTATION

As part of its role in axial extension, Brachyury regulates Wnt and retinoic acid signaling. These pathways, along with the FGF pathway, which is also involved in NMP maintenance and differentiation, simultaneously impact the somitogenesis process (Martin, 2016). The addition of new progenitor cells to the growing vertebrate body axis requires these cells to exit the tailbud and join the presomitic mesoderm. This is an active process that begins with an epithelial-to-mesenchymal transition (EMT) and subsequent directional migration into the mesodermal compartment (Figure 5.4A). The NMPs exist as an epithelium, where they undergo a canonical Wnt signaling-induced apical constriction and separation from the epithelium, and enter into a partial EMT state (Goto et al., 2017). Cells in the partial EMT state are migratory but lack directionality, and remain in the tailbud until canonical Wnt and FGF signaling induce the expression of the t-box transcription factors *Tbx6* (in mouse) or *tbx16* and *tbx6l* in zebrafish, as well as bHLH transcription factor *msgn1* in both mouse and zebrafish (Chapman and Papaioannou, 1998; Fior et al., 2012; Goto et al., 2017; Griffin et al., 1998; Ho and Kane, 1990; Kimmel et al., 1989; Manning and Kimelman, 2015; Morrow et al., 2017; Row et al., 2011; Yabe and Takada, 2012; Yoon and Wold, 2000). The induction of these t-box and bHLH transcription factors causes the cells to undergo a directional migration and exit into the mesodermal compartment, thus completing the EMT (Goto et al., 2017; Manning and Kimelman, 2015; Row et al., 2011).

Exogenously activating the Wnt pathway can accelerate the exit of NMP-derived mesoderm into the presomitic mesoderm through promotion of the EMT event (Bouldin et al., 2015; Goto et al., 2017). Wnt and FGF, which is also required for

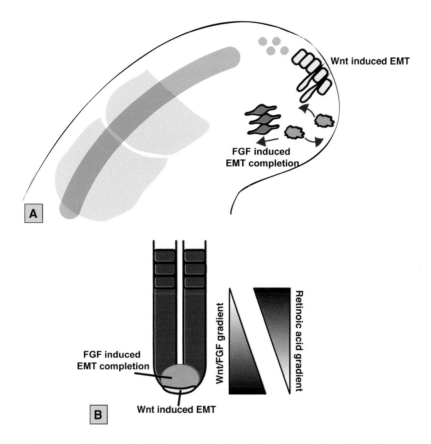

FIGURE 5.4   Secreted signals control the rate of mesoderm production and establish the position of segment formation. (A) NMPs undergo a two-step epithelial-to-mesenchymal transition as they become mesoderm, where Wnt signaling initiates the EMT (yellow), and FGF signaling promotes the completion of the EMT (orange). The process affects the rate at which new mesoderm joins the presomitic mesoderm. (B) The posteriorly localized Wnt and FGF signaling source creates a posterior-to-anterior gradient of signal across the presomitic mesoderm, which establishes the position of segment formation. A reciprocal anterior-to-posterior gradient of retinoic acid signaling represses FGF signaling, refining the FGF gradient across the presomitic mesoderm.

the EMT event, are the same signals that are involved in positioning the wavefront that establishes when a segment boundary forms. When FGF and Wnt signaling drop below a threshold level, segment boundary formation is initiated (Aulehla and Pourquie, 2010). The posteriorly localized expression of FGF and Wnt ligands creates a posterior-to-anterior gradient of signal through the presomitic mesoderm (Figure 5.4B). The dual activities of Wnt and FGF serve as a potential mechanism allowing somitogenesis to scale with presomitic mesoderm size. With higher Wnt and FGF, more cells will enter the presomitic mesoderm, creating a larger presomitic mesoderm. The higher amount of FGF and Wnt signal means that cells will have to be located farther away from the tailbud to reach the threshold level that induces border formation (Aulehla and Pourquie, 2010). Thus, border formation should scale with presomitic mesoderm size since the signals responsible for regulating NMP-derived mesoderm exit into the presomitic mesoderm, and therefore regulating PSM size, are the same signals establishing the wavefront position that dictates the location of border formation. A model has been developed to describe the scaling of somite size with the size of the presomitic mesoderm, which is termed a clock and scaled gradient model (Ishimatsu et al., 2018). This model explains a striking phenomenon, where size-reduced embryos generate the exact number of somites appropriate for that species despite the overall size of the embryo being reduced by nearly half (Cooke, 1975, 1981; Ishimatsu et al., 2018)

As previously mentioned, Brachyury maintains a low retinoic acid environment in the tailbud through direct regulation of *cyp26a1* expression (Martin and Kimelman, 2010). The signaling sink created by posteriorly localized *cyp26a1* expression and the location of the signal source in the newly formed somites creates a retinoic acid gradient across the presomitic mesoderm that is inversely correlated with the FGF and Wnt gradients (Diez del Corral et al., 2003; Shimozono et al., 2013). This RA gradient also helps position the somite boundary by negatively regulating the FGF gradient. In the mouse, RA signaling directly represses *Fgf8* expression, and FGF8 along with FGF4 establish the wavefront position (Kumar and Duester, 2014; Naiche et al., 2011). In the absence of RA signaling, FGF signaling increases, shifting the wavefront to a more anterior position and causing the formation of smaller somites (Diez del Corral et al., 2003). Interestingly, in zebrafish RA activates the expression of *fgf8* and *fgf17*, yet loss of RA signaling results in smaller somites just as in the mouse (Begemann et al., 2001; Hamade et al., 2006; Kawakami et al., 2005).

5.8 INDUCTION OF PARAXIAL/SOMITE FATE IN NMP-DERIVED MESODERM

Although the majority of the NMP-derived mesoderm will become paraxial mesoderm, a subset will become lateral mesoderm such as vascular endothelial cells. This decision is made locally by secreted signaling molecules that pattern the newly formed mesoderm (Figure 5.3). Clonal analysis of single-cell descendants in the mouse embryo showed that NMPs can give rise to lateral mesoderm, in addition to neural and paraxial mesoderm (Tzouanacou et al., 2009). An analysis of the role of canonical Wnt signaling in the zebrafish demonstrated that Wnt signaling not only plays an essential role in inducing mesoderm from NMPs, but also functions within the mesoderm progenitors to pattern them into paraxial and lateral subtypes (Martin and Kimelman, 2012). Continued activation of Wnt signaling leads to paraxial fate, whereas inhibition of Wnt signaling in cells just after they have become mesoderm causes them to adopt a vascular endothelial fate.

In addition to Wnt signaling, BMP and FGF signaling also play critical roles in patterning newly formed mesoderm. FGF signaling, like Wnt signaling, promotes paraxial fate, whereas BMP signaling induces lateral fates (Row et al., 2018). In the absence of FGF signaling, cells that would normally become paraxial mesoderm and somites instead become vascular endothelium. In the absence of BMP signaling, those NMP-derived mesodermal cells that would normally become vascular endothelium adopt a paraxial fate at the ventral midline (just under the notochord) where vasculogenesis normally occurs. These transfated cells also segment into somites with the same periodicity as the somites that normally form, and they differentiate into skeletal muscle. In the absence of both FGF and BMP signaling, newly derived mesoderm adopts an endothelial fate, suggesting that this is the default fate of NMP-derived mesoderm with respect to these two signals. BMP is required for endothelial fate to counteract the FGF and Wnt signals that are present initially during the mesoderm induction and EMT process. The patterning of the mesoderm downstream of FGF and BMP signaling is at the level of bHLH transcription factor activity. FGF signaling activates the expression of bHLH transcription factors *msgn1*, *myf5*, and *myod*, which in turn are required for the paraxial fate. In the absence of these three transcription factors, NMP-derived mesoderm becomes vascular endothelium. BMP signaling on the other hand induces the expression of the inhibitor of DNA binding (Id) genes, which are HLH proteins that act as natural dominant negative inhibitors of bHLH proteins. Thus, BMP signaling can antagonize the effects of FGF signaling by inhibiting the FGF-induced bHLH transcription factors (Row et al., 2018). This patterning mechanism is conserved in both zebrafish and mouse. In vitro–derived mouse NMPs that become mesoderm will give rise to paraxial fates in the presence of FGF signaling, or lateral endothelial fates in the presence of either BMP signaling or induced expression of the ID1 HLH protein (Row et al., 2018). Thus, it is a conserved vertebrate trait that NMP-derived mesoderm must be patterned properly into paraxial mesoderm by FGF and Wnt signaling, the same signals that act as the determination wavefront in the clock and wavefront mechanism of somitogenesis.

5.9 TERMINATION OF SOMITOGENESIS

In contrast to some annelids such as *Platynereis dumerilii*, which continue to add new posterior segments for the duration of their life (Fischer and Dorresteijn, 2004), all vertebrates eventually stop forming segments during embryogenesis. How this process comes to end is not well understood. Several mechanisms of body axis elongation and segment addition termination have been proposed, although a consensus has not been reached between different vertebrate species. A simple model is that the rate NMPs exit into the spinal cord and mesoderm exceeds the rate at which they divide, depleting the NMPs until they eventually run out (Kimelman, 2016). In the zebrafish, where tailbud NMPs rarely divide, this model seems plausible, although experiments in which additional NMPs are added to embryos to determine the effect on segment number and body length has not been performed.

In amniotes, where NMPs exhibit stem cell properties and divide more frequently than in zebrafish, other models have been proposed. During mouse development, the establishment of anterior–posterior identity along the axis is tied to dynamic changes in NMP gene expression as the axis elongates (Wymeersch et al., 2019). This is opposed to the neighboring progenitor cells of the notochord, which retain a consistent gene expression program throughout axial extension (Wymeersch et al., 2019). As NMPs mature during development, they progress through a complement of *Hox* transcription factor expression. *Hox* genes are well known for imparting axial identity to the body plan of animals, with anteriorly expressed *Hox* genes inducing anterior fates, and posterior *Hox* genes inducing posterior fates (Gaunt, 2018). This suggests that axial identity of the embryo is established, at least in part, by the transcriptionally dynamic NMP population that turns on *Hox* genes specific for a region of the axis and then contributes descendent cells to that region (Wymeersch et al., 2019). In this way, a clock mechanism within the NMPs controlling *Hox* gene expression may dictate the end of axial extension when the posterior terminal *Hox* genes are expressed. Indeed, posterior *Hox* gene expression in the chick and mouse is hypothesized to induce termination of axis extension, in part due to the posterior *Hox* gene–mediated repression of canonical Wnt and FGF signaling, which is required for continuous axis extension (Denans et al., 2015; Olivera-Martinez et al., 2012; Young et al., 2009).

The reduction in Wnt signaling as more posterior *Hox* genes are expressed causes a shortening of the presomitic mesoderm territory, which reduces the distance between the most recently formed somite and the NMPs (Denans et al., 2015). The signaling molecule retinoic acid is normally produced by the most recently formed somites, and is sufficient to inhibit the maintenance of NMPs and induce neural differentiation within them (Dobbs-McAuliffe et al., 2004; Martin and Kimelman, 2010; Niederreither et al., 1997; Olivera-Martinez et al., 2012). In the chick, as the source of RA gets closer to the NMPs and RA signaling becomes active in the tail-bud, *Brachyury* expression in the NMPs is repressed, leading to neural induction and the depletion of the NMPs (Olivera-Martinez et al., 2012). The retinoic acid signaling in the tailbud represses FGF signaling, which in turn causes a loss of *Brachyury* expression. This model does not apply to all vertebrates, however. In mouse and zebrafish, for example, loss of function of the major retinoic acid–producing enzyme *aldh1a2* does not affect the proper termination of axis elongation (Begemann et al., 2001; Cunningham et al., 2011).

5.10 PROGENITOR CELL BEHAVIORS THAT INFLUENCE THE SYNCHRONIZATION OF CYCLING GENE EXPRESSION

Mesodermal progenitors display several cell biological features that may be responsible for providing robustness to the segmentation process. Each individual presomitic cell functions as an autonomous oscillating unit, with cycling *her/Hes* gene expression that is part of the somite clock (Jiang et al., 2000). For proper segmentation to occur, neighboring presomitic cells must have synchronized oscillations. These oscillations are synchronized by Notch/Delta signaling, as *her/Hes* genes are targets of notch signaling (Jiang et al., 2000). The Notch receptor and Delta ligand are both transmembrane proteins, such that cells must be juxtaposed to each other for signaling to occur (called juxtacrine signaling) (Kopan and Ilagan, 2009). As previously mentioned, when cells enter the partial EMT state as they transition from NMP to mesodermal progenitor, they undergo a rapid mixing process with nondirectional movement (Goto et al., 2017; Lawton et al., 2013; Manning and Kimelman, 2015; Row et al., 2011). The reason for this random mixing is unknown, but it is hypothesized to rapidly synchronize gene expression oscillations in a group of cells through neighbor exchange-mediated Notch/Delta signaling (Uriu and Morelli, 2017). Disruption of Notch/Delta signaling causes asynchronous oscillations of clock gene expression and the subsequent failure of segment formation (Jiang et al., 2000). When Notch signaling is inhibited with small molecule drugs, somitogenesis fails to occur properly. Segmentation can recover after the drug is washed out and signaling is restored (Riedel-Kruse et al., 2007). In prior modeling simulations, recovery of segment formation after a transient Notch signaling inhibition did not occur until after tens to hundreds of oscillations, whereas in real embryos, segmentation recovers after just ten oscillations. When rapid random cell mixing of mesodermal progenitors is incorporated into the model, recovery of segmentation occurs with the same timeframe as real embryonic recovery (Uriu et al., 2010). Further modeling of cell mixing in the zebrafish tailbud using experimentally observed rates of movement reveals that the movement is fast enough to enhance the synchronization of neighboring cells and affect the coherence of oscillating gene expression (Uriu et al., 2017). This suggests that the rapid mixing of NMP-derived mesoderm as it leaves the tailbud is important in synchronizing clock gene oscillations in neighboring cells, allowing for proper somitogenesis (Figure 5.5A).

Another possible mechanism of clock oscillation synchrony relates to the cell cycle behavior of mesodermal progenitors. In zebrafish, post-gastrulation NMPs are arrested in the G2 phase of the cell cycle, and after being induced to become mesoderm, they undergo a synchronous mitotic event (Bouldin et al., 2014) (Figure 5.5B). A transgenic zebrafish line using the *her1* promoter driving the expression of a destabilized fluorescent protein, which is useful when rapid reporter turnover is necessary, allows for the visualization of endogenous segmentation clock oscillations in vivo (Delaune et al., 2012). Live imaging using this line revealed that oscillations of clock gene expression in daughter cells that have just finished dividing exhibit

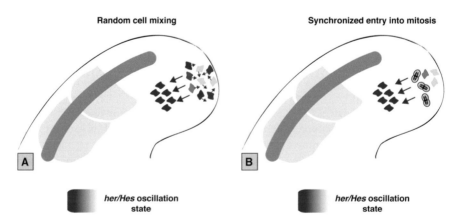

FIGURE 5.5 Progenitor cell behaviors help synchronize oscillating gene expression. (A) Mesodermal progenitors exhibit a period of random mixing, which helps synchronize the oscillatory expression of the *her/Hes* genes that are part of the segmentation clock. (B) Mesodermal progenitors have also been observed to undergo synchronized mitosis after exiting the G2 arrested NMP state. This may also help to synchronize the oscillatory expression of *her/Hes* genes between cells through Notch signaling dependent and independent mechanisms.

more highly synchronized clock oscillations relative to neighbor cells. This is true even in embryos where Notch signaling is disrupted, suggesting that this cell-cycle based entrainment is independent of Notch signaling (Delaune et al., 2012). Further effects of the cell cycle state relate to Notch signaling itself. When the Notch receptor is activated, the intracellular domain (NICD) of the transmembrane receptor is cleaved and enters the nucleus, where it functions as a transcription factor (Kopan and Ilagan, 2009). A recent report showed that the NICD is phosphorylated by the cyclin-dependent kinases CDK1 and CDK2, which facilitates SCF-mediated degradation of NICD (Carrieri et al., 2019). When cells are in the G1 phase of the cell cycle, where CDK1 and CDK2 activity are low, NICD will be stable which in turn will facilitate signaling. The dependence of NICD activity on cell cycle phase was shown to influence the rate of somitogenesis through control of clock oscillations (Carrieri et al., 2019). Thus, the enhanced stabilization of NICD during G1 of recently divided daughter cells may help entrain their oscillations. Additionally, this suggests that the prolonged G2 arrest of NMPs in zebrafish could prevent activation of the Notch pathway and premature cycling of the somite clock genes. In agreement with these cell cycle–related phenomena, experimental disruptions of the cell cycle impact the somitogenesis process. Zebrafish *emi1* mutants, which fail to progress from G2 to M phase throughout the embryo after gastrulation, have asynchronous *her1* expression and disorganized somite border formation (Zhang et al., 2008). Together these studies indicate that the synchronous division of mesodermal progenitors before they join the presomitic mesoderm may help to synchronize clock oscillations in the newly induced group of mesodermal progenitors.

5.11   NMPS AS A DEVELOPMENT MODULE AFFECTING EVOLUTIONARY CHANGE IN SEGMENT NUMBER AND BODY LENGTH

The variation in segment number and primary body axis length among different vertebrates suggests that NMP development has evolved, possibly contributing to the diversification of vertebrate body plans. There are several putative mechanisms by which changes to NMPs could affect the body plan, including the initial number of NMPs, the rate at which they are depleted, or the transcription of genes critical for the clock and wavefront patterning of the somites. Although it is difficult to experimentally demonstrate that changes in NMPs have occurred between species that directly impact differences in body plan, there is at least some evidence to suggest that molecular changes in NMPs lead to distinct body plan development. An example is the Brachyury gene, which as previously mentioned functions in an autoregulatory loop with canonical Wnt signaling to maintain the NMP population over the duration of body axis extension (Martin and Kimelman, 2008). In recent years, *Brachyury* mutations have been identified in naturally occurring tailless dogs and cats (Buckingham et al., 2013; Haworth et al., 2001; Hytonen et al., 2009). These data indicate that manipulating just the *Brachyury*/Wnt feedback loop is sufficient to change the body plan of animals, such that lowering the amount of functional Brachyury results in loss of tails and a shorter overall body length. This also suggests that evolutionary changes to *Brachyury* expression levels could modify body length, but there is no direct evidence to show that differential expression levels of *Brachyury* between species accounts for species-specific differences in body plan development.

Body length, as mentioned earlier, can also be controlled by the timing of *Hox* gene expression in the NMPs. Since the onset of terminal *Hox* gene expression correlates with the cessation of body axis extension, developmental changes to *Hox* gene expression could affect body length and segment number (Denans et al., 2015; Olivera-Martinez et al., 2012; Wymeersch et al., 2019). This was shown to be the case in a mouse *HoxB13* mutant. *HoxB13* is one of the terminal *Hox* genes expressed in NMPs as the most posterior part of the body is developing (Economides et al., 2003). *HoxB13* homozygous mutant mice have an increase body length and supernumerary segments, suggesting that changes to *Hox* gene regulation within NMPs could indeed affect the evolution of the vertebrate body.

Another important aspect of the vertebrate body plan is the trunk-to-tail ratio, or the number of trunk segments relative to tail segments. The trunk-to-tail transition point is generally accepted as the somite adjacent to the hindlimbs. Some animals, like snakes, have exceptionally long trunks compared to other vertebrates (Gomez et al., 2008). Although NMPs contribute to the somites along the entire body axis, they display different properties depending on whether they originate in the epiblast and contribute to the trunk, or within the tailbud where they contribute to the tail. NMPs exhibit a differential genetic requirement between the trunk and tail in mouse embryos. The trunk NMPs are dependent on *Oct4* transcription factor expression, whereas the tail NMPs depend on GDF11 signaling, which in turn represses the expression of the RNA binding protein Lin28 (Aires et al., 2019; Aires et al., 2016). When *Oct4* is artificially sustained in NMPs, there is striking lengthening of the trunk of the embryo, with up to six extra segments in the trunk compared to wild-type embryos (Aires et al., 2016). On the other hand, if *Lin28* is artificially sustained in NMPs, there is a lengthening of the tail, with up to five extra tail somites forming on average (Aires et al., 2019). These results indicate that modulations of the genetic networks regulating pre- or post-gastrulation NMPs can affect the evolution of the body plan through differential lengthening or shortening of the trunk vs. tail segment number.

Overall body length, the relative number of trunk vs. tail somites, and the total number of somites that form can impact adaptation to particular ecological niches. For instance, the large number of somites that form in snakes allows undulations associated with slithering on the ground, which is an important aspect of their locomotion and behavior. Since the molecular regulation of the clock and wavefront mechanism begins as mesoderm progenitors are established, changes in the expression of these components would affect the somite number and the ability to adapt to new environments. Thus, genetic changes that affect the induction, maintenance, and proliferation of NMPs, along with their subsequent segmentation after joining the mesoderm, can impact body plan evolution by affecting the length and/or number of segments. A comparison between clock and wavefront components has been made between species with varying segment number, including zebrafish, chicken, mouse, and the corn snake (Gomez et al., 2008). In these species, similar clock and wavefront genes are expressed, but in the corn snake, the rate of oscillations of clock genes relative to the overall rate of development is much faster than in the other species examined. The net result of this difference is the formation of many small somites in the corn snake. Together, the experimental evidence suggests that modulation of NMP maintenance and differentiation, along with clock and wavefront gene expression in NMP-derived mesoderm, is a major determinant of body plan evolution within the vertebrate clade.

5.12   DO NMPS EXIST OUTSIDE THE VERTEBRATE CLADE?

A major unresolved question in the field is whether NMPs are a vertebrate innovation. The vast majority of bilaterians undergo a posterior growth process during embryogenesis (Martin and Kimelman, 2009). Based on the diversity of extant organisms that undergo posterior growth, this mode of body plan formation is thought to have been present in the so-called urbilaterian (the ancestor to all bilaterians) (Martin and Kimelman, 2009). Those posteriorly growing metazoans that are also segmented exhibit a progressive posterior addition of new segments, similar to vertebrate somite addition. Many aspects of the posterior growth and the segmentation process are conserved between vertebrates and other bilaterians, such as the requirement for canonical Wnt signaling during posterior growth of both vertebrates and ecdysozoans (Martin and Kimelman, 2009).

Despite several commonalities between the posterior growth and segmentation process of vertebrates and invertebrates, a posteriorly localized cell type that gives rise to both neural and pre-segmentation mesoderm has yet to be formally identified (see Chapters 3 and 4). Some of the genes and signaling pathways play a conserved role during posterior elongation. In ecdysozoans, canonical Wnt signaling is localized to the segment addition zone, and disruption of the Wnt pathway leads to truncation of the body axis and a loss of posterior segments, similar to the effect of Wnt inhibition in vertebrates. Wnt signaling is also critical for the posterior elongation of the hemichordate *Saccoglossus kowalevskii*, and like vertebrates, functions in an autoregulatory loop with the Brachyury transcription factor (Fritzenwanker et al., 2019). Canonical Wnt signaling and Brachyury function in an autoregulatory loop in the sea urchin as well, suggesting that this regulatory relationship was present at the base of all deuterostomes (Sethi et al., 2012). Although canonical Wnt signaling is required for posterior elongation and segment addition in ecdysozoans, here it does not function in an autoregulatory loop with Brachyury. Loss of Brachyury function in *Tribolium* and crickets has no effect on posterior growth and segmentation (Berns et al., 2008; Shinmyo et al., 2006). Thus, the role of Wnt signaling during posterior growth is conserved across bilaterians, but Brachyury regulation of Wnt signaling appears to be a deuterostome innovation. Given that the essential role of Brachyury during zebrafish axis extension is to maintain Wnt signaling, and that Brachyury is dispensable during zebrafish axis extension as long as Wnt signaling is present (Martin and Kimelman, 2012), the absence of an important role for Brachyury during ecdysozoan axis extension could be due to another factor maintaining posterior Wnt signaling in this lineage.

The other factor that defines NMPs besides Brachyury expression is the expression of *Sox2*, which is a member of the SoxB subfamily of Sox transcription factors. Relatively little is known about the role of SOX2 in the context of NMPs, partially due to the early lethality of mouse *Sox2* mutants and lack of genetic knockouts in other models (Avilion et al., 2003), although conditional analysis suggests it plays a role in neural induction (Takemoto et al., 2011). Recent work has examined the role of SoxB transcription factors during axial elongation and segment addition in the spider *Parasteatoda tepidariorum* (Paese et al., 2018). Loss of SoxB function results in truncated embryos, with a loss of segments. Although this is a distinct phenotype from the mouse, it shows that SoxB factors play an important role in axial elongation outside of the vertebrate lineage. Thus, there appears to be important functions of both SoxB transcription factors and canonical Wnt signaling during axial elongation and segmentation in diverse organisms. Whether a germ-layer plastic progenitor homologous to the vertebrate NMPs exists outside of the vertebrate lineage remains to be determined.

5.13   CONCLUSIONS

Great progress has been made in understanding the source of new cells that contribute to the forming somites, as well as how the cell behavior and genetic network of these progenitor cells influence the segmentation process. The NMPs, which give rise to mesodermal progenitors that populate the somites, are maintained and induced to differentiate by some of the same signals that set the position of the newly formed somite boundary, thereby coordinating the NMP population and axial outgrowth with segmentation (Martin, 2016). Additionally, transcription factors such as Brachyury, among others, simultaneously directly regulate the transcription of target genes required for NMP maintenance and differentiation and the position and size of newly formed somites. And finally, as NMP-derived mesodermal cells join the presomitic mesoderm, they undergo behaviors such as random cell mixing and synchronized division that are thought to help coordinate groups of neighbor cells so that the segmentation process can occur normally.

Despite the advances that have been made, there are several key questions that remain unanswered about fundamental aspects of NMPs and the NMP-derived tissue segmentation process. These include (1) whether NMP-derived mesoderm gives rise to all cells of the somites, or if there are additional sources of cells; 2) what influence do NMPs have on somite number and somite size; and (3) how have changes to the induction, maintenance, and differentiation of NMPs over time affected the evolution of the vertebrate body plan. It will also be critical to determine whether an NMP-like cell type exists outside of the vertebrate lineage, or if this was a vertebrate innovation facilitating specific novelties of the vertebrate body. With the democratization of non-model systems due to the advent of CRISPR/Cas9-mediated genetic engineering and whole genome sequencing, answers to these questions should be feasible in the near future.

REFERENCES

Abu-Abed, S., P. Dolle, D. Metzger, B. Beckett, P. Chambon, and M. Petkovich. 2001. The retinoic acid-metabolizing enzyme, CYP26A1, is essential for normal hindbrain patterning, vertebral identity, and development of posterior structures. *Genes Dev.* 15: 226–240.

Ahn, D., K. H. You, and C. H. Kim. 2012. Evolution of the *tbx6/16* subfamily genes in vertebrates: Insights from zebrafish. *Mol. Biol. Evol.* 29: 3959–3983.

Aires, R., L. de Lemos, A. Novoa, A. D. Jurberg, B. Mascrez, D. Duboule, and M. Mallo. 2019. Tail bud progenitor activity relies on a network comprising *Gdf11*, *Lin28*, and *Hox13* genes. *Dev. Cell* 48: 383–395.e8.

Aires, R., A. D. Jurberg, F. Leal, A. Novoa, M. J. Cohn, and M. Mallo. 2016. Oct4 Is a Key regulator of vertebrate trunk length diversity. *Dev. Cell* 38: 262–274.

Amin, S., R. Neijts, S. Simmini, C. van Rooijen, S. C. Tan, L. Kester, A. van Oudenaarden, M. P. Creyghton, and J. Deschamps. 2016. Cdx and T Brachyury co-activate growth signaling in the embryonic axial progenitor niche. *Cell Rep.* 17: 3165–3177.

Attardi, A., T. Fulton, M. Florescu, G. Shah, L. Muresan, M. O. Lenz, C. Lancaster, J. Huisken, A. van Oudenaarden, and B. Steventon. 2018. Neuromesodermal progenitors are a conserved source of spinal cord with divergent growth dynamics. *Development* 145: dev166728.

Aulehla, A., and O. Pourquie. 2010. Signaling gradients during paraxial mesoderm development. *Cold Spring Harb. Perspect. Biol.* 2: a000869.

Avilion, A. A., S. K. Nicolis, L. H. Pevny, L. Perez, N. Vivian, and R. Lovell-Badge. 2003. Multipotent cell lineages in early mouse development depend on SOX2 function. *Genes Dev.* 17: 126–140.

Begemann, G., T. F. Schilling, G. J. Rauch, R. Geisler, and P. W. Ingham. 2001. The zebrafish neckless mutation reveals a requirement for *raldh2* in mesodermal signals that pattern the hindbrain. *Development* 128: 3081–3094.

Berenguer, M., J. J. Lancman, T. J. Cunningham, P. D. S. Dong, and G. Duester. 2018. Mouse but not zebrafish requires retinoic acid for control of neuromesodermal progenitors and body axis extension. *Dev. Biol.* 441: 127–131.

Berns, N., T. Kusch, R. Schroder, and R. Reuter. 2008. Expression, function and regulation of *Brachyenteron* in the short germband insect *Tribolium castaneum*. *Dev. Genes Evol.* 218: 169–179.

Bessho, Y., H. Hirata, Y. Masamizu, and R. Kageyama. 2003. Periodic repression by the bHLH factor *Hes7* is an essential mechanism for the somite segmentation clock. *Genes Dev.* 17: 1451–1456.

Bessho, Y., G. Miyoshi, R. Sakata, and R. Kageyama. 2001. *Hes7*: A bHLH-type repressor gene regulated by Notch and expressed in the presomitic mesoderm. *Genes Cells* 6: 175–185.

Bouldin, C. M., A. J. Manning, Y. H. Peng, G. H. Farr, K. L. Hung, A. Dong, and D. Kimelman. 2015. Wnt signaling and *tbx16* form a bistable switch to commit bipotential progenitors to mesoderm. *Development* 142: 2499–2507.

Bouldin, C. M., C. D. Snelson, G. H. Farr, and D. Kimelman. 2014. Restricted expression of *cdc25a* in the tailbud is essential for formation of the zebrafish posterior body. *Genes Dev.* 28: 384–395.

Brend, T., and S. A. Holley. 2009. Expression of the oscillating gene *her1* is directly regulated by hairy/enhancer of split, T-box, and suppressor of hairless proteins in the zebrafish segmentation clock. *Dev. Dyn.* 238: 2745–2759.

Buckingham, K. J., M. J. McMillin, M. M. Brassil, K. M. Shively, K. M. Magnaye, A. Cortes, A. S. Weinmann, L. A. Lyons, and M. J. Bamshad. 2013. Multiple mutant *T* alleles cause haploinsufficiency of Brachyury and short tails in Manx cats. *Mamm. Genome* 24: 400–408.

Cambray, N., and V. Wilson. 2002. Axial progenitors with extensive potency are localised to the mouse chordoneural hinge. *Development* 129: 4855–4866.

Cambray, N., and V. Wilson. 2007. Two distinct sources for a population of maturing axial progenitors. *Development* 134: 2829–2840.

Carrieri, F. A., P. J. Murray, D. Ditsova, M. A. Ferris, P. Davies, and J. K. Dale. 2019. CDK1 and CDK2 regulate NICD1 turnover and the periodicity of the segmentation clock. *EMBO Rep.* 20: e46436.

Chapman, D. L., and V. E. Papaioannou. 1998. Three neural tubes in mouse embryos with mutations in the T-box gene *Tbx6*. *Nature* 391: 695–697.

Chesley, P. 1935. Development of the short-tailed mutant in the house mouse. *J. Exp. Zool.* 70: 429–459.

Concepcion, D., and V. E. Papaioannou. 2014. Nature and extent of left/right axis defects in *T(Wis)/T(Wis)* mutant mouse embryos. *Dev. Dyn.* 243: 1046–1053.

Cooke, J. 1975. Control of somite number during morphogenesis of a vertebrate, *Xenopus laevis*. *Nature* 254: 196–199.

Cooke, J. 1981. Scale of body pattern adjusts to available cell number in amphibian embryos. *Nature* 290: 775–778.

Cooke, J., and E. C. Zeeman. 1976. Clock and wavefront model for control of number of repeated structures during animal morphogenesis. *J. Theor. Biol.* 58: 455–476.

Cunningham, T. J., and X. Zhao, G. Duester. 2011. Uncoupling of retinoic acid signaling from tailbud development before termination of body axis extension. *Genesis* 49: 776–783.

Delaune, E. A., P. Francois, N. P. Shih, and S. L. Amacher. 2012. Single-cell-resolution imaging of the impact of Notch signaling and mitosis on segmentation clock dynamics. *Dev. Cell* 23: 995–1005.

Denans, N., and T. Iimura, O. Pourquie. 2015. Hox genes control vertebrate body elongation by collinear Wnt repression. *Elife* 4: e04379.

Diez del Corral, R., I. Olivera-Martinez, A. Goriely, E. Gale, M. Maden, and K. Storey. 2003. Opposing FGF and retinoid pathways control ventral neural pattern, neuronal differentiation, and segmentation during body axis extension. *Neuron* 40: 65–79.

Dobbs-McAuliffe, B., Q. S. Zhao, and E. Linney. 2004. Feedback mechanisms regulate retinoic acid production and degradation in the zebrafish embryo. *Mech. Dev.* 121: 339–350.

Dobrovolskaia-Zavadskaia, N. 1927. Regarding the spontaneous mortification of the tail of a new-born mouse and the existence of a hereditary characteristic (factor). *Comptes Rendus Des Seances De La Societe De Biologie Et De Ses Filiales* 97: 114–116.

Duester, G. 2008. Retinoic acid synthesis and signaling during early organogenesis. *Cell* 134: 921–931.

Economides, K. D., and L. Zeltser, M. R. Capecchi. 2003. *Hoxb13* mutations cause overgrowth of caudal spinal cord and tail vertebrae. *Dev. Biol.* 256: 317–330.

Edri, S., P. Hayward, P. Baillie-Johnson, B. J. Steventon, and A. Martinez Arias. 2019a. An epiblast stem cell-derived multipotent progenitor population for axial extension. *Development* 146: dev168187.

Edri, S., P. Hayward, W. Jawaid, and A. Martinez Arias. 2019b. Neuro-mesodermal progenitors (NMPs): A comparative study between pluripotent stem cells and embryo-derived populations. *Development* 146: dev180190.

Emoto, Y., H. Wada, H. Okamoto, A. Kudo, and Y. Imai. 2005. Retinoic acid-metabolizing enzyme Cyp26a1 is essential for determining territories of hindbrain and spinal cord in zebrafish. *Dev. Biol.* 278: 415–427.

Evans, A. L., T. Faial, M. J. Gilchrist, T. Down, L. Vallier, R. A. Pedersen, F. C. Wardle, and J. C. Smith. 2012. Genomic targets of Brachyury (T) in differentiating mouse embryonic stem cells. *PLoS One* 7: e33346.

Faial, T., A. S. Bernardo, S. Mendjan, E. Diamanti, D. Ortmann, G. E. Gentsch, V. L. Mascetti, M. W. B. Trotter, J. C. Smith, and R. A. Pedersen. 2015. Brachyury and SMAD signalling collaboratively orchestrate distinct mesoderm and endoderm gene regulatory networks in differentiating human embryonic stem cells. *Development* 142: 2121–2135.

Fior, R., A. A. Maxwell, T. P. Ma, A. Vezzaro, C. B. Moens, S. L. Amacher, and J. Lewis, L. Saude. 2012. The differentiation and movement of presomitic mesoderm progenitor cells are controlled by Mesogenin 1. *Development* 139: 4656–4665.

Fischer, A., and A. Dorresteijn. 2004. The polychaete *Platynereis dumerilii* (Annelida): A laboratory animal with spiralian cleavage, lifelong segment proliferation and a mixed benthic/pelagic life cycle. *Bioessays* 26: 314–325.

Frank, D., and D. Sela-Donenfeld. 2019. Hindbrain induction and patterning during early vertebrate development. *Cell. Mol. Life Sci.* 76: 941–960.

Frisdal, A., and P. A. Trainor. 2014. Development and evolution of the pharyngeal apparatus. *Wiley Interdiscip. Rev. Dev. Biol.* 3: 403–418.

Fritzenwanker, J. H., K. R. Uhlinger, J. Gerhart, E. Silva, and C. J. Lowe. 2019. Untangling posterior growth and segmentation by analyzing mechanisms of axis elongation in hemichordates. *PNAS* 116: 8403–8408.

Garnett, A. T., T. M. Han, M. J. Gilchrist, J. C. Smith, M. B. Eisen, F. C. Wardle, and S. L. Amacher. 2009. Identification of direct T-box target genes in the developing zebrafish mesoderm. *Development* 136: 749–760.

Garriock, R. J., R. B. Chalamalasetty, M. W. Kennedy, L. C. Canizales, M. Lewandoski, and T. P. Yamaguchi. 2015. Lineage tracing of neuromesodermal progenitors reveals novel Wnt-dependent roles in trunk progenitor cell maintenance and differentiation. *Development* 142: 1628–1638.

Gaunt, S. J. 2018. Hox cluster genes and collinearities throughout the tree of animal life. *Int. J. Dev. Biol.* 62: 673–683.

Ghebranious, N., R. D. Blank, C. L. Raggio, J. Staubli, E. McPherson, L. Ivacic, K. Rasmussen, F. S. Jacobsen, T. Faciszewski, J. K. Burmester, R. M. Pauli, O. Boachie-Adjei, I. Glurich, and P. F. Giampietro. 2008. A missense *T* (*Brachyury*) mutation contributes to vertebral malformations. *J. Bone Miner. Res.* 23: 1576–1583.

Gomez, C., E. M. Ozbudak, J. Wunderlich, D. Baumann, J. Lewis, and O. Pourquie. 2008. Control of segment number in vertebrate embryos. *Nature* 454: 335–339.

Goto, H., S. C. Kimmey, R. H. Row, D. Q. Matus, and B. L. Martin. 2017. FGF and canonical Wnt signaling cooperate to induce paraxial mesoderm from tailbud neuromesodermal progenitors through regulation of a two-step epithelial to mesenchymal transition. *Development* 144: 1412–1424.

Gouti, M., J. Delile, D. Stamataki, F. J. Wymeersch, Y. Huang, J. Kleinjung, V. Wilson, and J. Briscoe. 2017. A gene regulatory network balances neural and mesoderm specification during vertebrate trunk development. *Dev. Cell* 41: 243–261.e7.

Gouti, M., A. Tsakiridis, F. J. Wymeersch, Y. Huang, J. Kleinjung, V. Wilson, and J. Briscoe. 2014. In vitro generation of neuromesodermal progenitors reveals distinct roles for wnt signalling in the specification of spinal cord and paraxial mesoderm identity. *PLoS Biol.* 12: e1001937.

Griffin, K. J. P., S. L. Amacher, C. B. Kimmel, and D. Kimelman. 1998. Molecular identification of *spadetail*: regulation of zebrafish trunk and tail mesoderm formation by T-box genes. *Development* 125: 3379–3388.

Gu, C., Y. Yoshida, J. Livet, D. V. Reimert, F. Mann, J. Merte, C. E. Henderson, T. M. Jessell, A. L. Kolodkin, and D. D. Ginty. 2005. Semaphorin 3E and plexin-D1 control vascular pattern independently of neuropilins. *Science* 307: 265–268.

Halpern, M. E., R. K. Ho, C. Walker, and C. B. Kimmel. 1993. Induction of muscle pioneers and floor plate is distinguished by the zebrafish no tail mutation. *Cell* 75: 99–111.

Hamade, A., M. Deries, G. Begemann, L. Bally-Cuif, C. Genet, F. Sabatier, A. Bonnieu, and X. Cousin. 2006. Retinoic acid activates myogenesis in vivo through Fgf8 signalling. *Dev. Biol.* 289: 127–140.

Haworth, K., W. Putt, B. Cattanach, M. Breen, M. Binns, F. Lingaas, and Y. H. Edwards. 2001. Canine homolog of the T-box transcription factor T; failure of the protein to bind to its DNA target leads to a short-tail phenotype. *Mamm. Genome* 12: 212–218.

Henrique, D., E. Abranches, L. Verrier, and K. G. Storey. 2015. Neuromesodermal progenitors and the making of the spinal cord. *Development* 142: 2864–2875.

Henry, C. A., M. K. Urban, K. K. Dill, J. P. Merlie, M. F. Page, C. B. Kimmel, and S. L. Amacher. 2002. Two linked hairy/Enhancer of split-related zebrafish genes, *her1* and *her7*, function together to refine alternating somite boundaries. *Development* 129: 3693–3704.

Herrmann, B. G., S. Labeit, A. Poustka, T. R. King, and H. Lehrach. 1990. Cloning of the *T*-gene required in mesoderm formation in the mouse. *Nature* 343: 617–622.

Ho, R. K., and D. A. Kane. 1990. Cell-autonomous action of zebrafish *spt-1* mutation in specific mesodermal precursors. *Nature* 348: 728–730.

Hofmann, M., K. Schuster-Gossler, M. Watabe-Rudolph, A. Aulehla, B. G. Herrmann, and A. Gossler. 2004. WNT signaling, in synergy with T/TBX6, controls Notch signaling by regulating *Dll1* expression in the presomitic mesoderm of mouse embryos. *Genes Dev.* 18: 2712–2717.

Holley, S. A., R. Geisler, and C. Nusslein-Volhard. 2000. Control of *her1* expression during zebrafish somitogenesis by a delta-dependent oscillator and an independent wave-front activity. *Genes Dev.* 14: 1678–1690.

Holley, S. A., D. Julich, G. J. Rauch, R. Geisler, and C. Nusslein-Volhard. 2002. *her1* and the notch pathway function within the oscillator mechanism that regulates zebrafish somitogenesis. *Development* 129: 1175–1183.

Horikawa, K., K. Ishimatsu, E. Yoshimoto, S. Kondo, andH. Takeda. 2006. Noise-resistant and synchronized oscillation of the segmentation clock. *Nature* 441: 719–723.

Hrabe de Angelis, M., J. McIntyre 2nd, and A. Gossler. 1997. Maintenance of somite borders in mice requires the *Delta* homologue *Dll1*. *Nature* 386: 717–721.

Hubaud, A., and O. Pourquie. 2014. Signalling dynamics in vertebrate segmentation. *Nat. Rev. Mol. Cell. Biol.* 15: 709–721.

Hytonen, M. K., A. Grall, B. Hedan, S. Dreano, S. J. Seguin, D. Delattre, A. Thomas, F. Galibert, L. Paulin, H. Lohi, K. Sainio, andC. Andre. 2009. Ancestral T-box mutation is present in many, but not all, short-tailed dog breeds. *J. Hered.* 100: 236–240.

Ishimatsu, K., T. W. Hiscock, Z. M. Collins, D. W. K. Sari, K. Lischer, D. L. Richmond, Y. Bessho, T. Matsui, and S. G. Megason. 2018. Size-reduced embryos reveal a gradient scaling-based mechanism for zebrafish somite formation. *Development* 145: dev161257.

Jahangiri, L., A. C. Nelson, and F. C. Wardle. 2012. A cis-regulatory module upstream of *deltaC* regulated by Ntla and Tbx16 drives expression in the tailbud, presomitic mesoderm and somites. *Dev. Biol.* 371: 110–120.

Jiang, Y. J., B. L. Aerne, L. Smithers, C. Haddon, D. Ish-Horowicz, and J. Lewis. 2000. Notch signalling and the synchronization of the somite segmentation clock. *Nature* 408: 475–479.

Julich, D., C. Hwee Lim, J. Round, C. Nicolaije, J. Schroeder, A. Davies, R. Geisler, J. Lewis, Y. J. Jiang, and S. A. Holley. 2005. *beamter/deltaC* and the role of Notch ligands in the zebrafish somite segmentation, hindbrain neurogenesis and hypochord differentiation. *Dev. Biol.* 286: 391–404.

Kawakami, Y., A. Raya, R. M. Raya, C. Rodriguez-Esteban, and J. C. I. Belmonte. 2005. Retinoic acid signalling links left-right asymmetric patterning and bilaterally symmetric somitogenesis in the zebrafish embryo. *Nature* 435: 165–171.

Keynes, R. 2018. Patterning spinal nerves and vertebral bones. *J. Anat.* 232: 534–539.

Kimelman, D. 2016. Tales of tails (and trunks): Forming the posterior body in vertebrate embryos. *Curr. Top. Dev. Biol.* 116: 517–536.

Kimelman, D., and B. L. Martin. 2012. Anterior-posterior patterning in early development: Three strategies. *Wiley Interdiscip. Rev. Dev. Biol.* 1: 253–266.

Kimmel, C. B., D. A. Kane, C. Walker, R. M. Warga, andM. B. Rothman. 1989. A mutation that changes cell movement and cell fate in the zebrafish embryo. *Nature* 337: 358–362.

Kinoshita, H., N. Ohgane, Y. Fujino, T. Yabe, H. Ovara, D. Yokota, A. Izuka, D. Kage, K. Yamasu, S. Takada, and A. Kawamura. 2018. Functional roles of the Ripply-mediated suppression of segmentation gene expression at the anterior presomitic mesoderm in zebrafish. *Mech. Dev.* 152: 21–31.

Kopan, R., and M. X. Ilagan. 2009. The canonical Notch signaling pathway: Unfolding the activation mechanism. *Cell* 137: 216–233.

Kumar, S., and G. Duester. 2014. Retinoic acid controls body axis extension by directly repressing *Fgf8* transcription. *Development* 141: 2972–2977.

Lawton, A. K., A. Nandi, M. J. Stulberg, N. Dray, M. W. Sneddon, W. Pontius, T. Emonet, and S. A. Holley. 2013. Regulated tissue fluidity steers zebrafish body elongation. *Development* 140: 573–582.

Lleras Forero, L., R. Narayanan, L. F. Huitema, M. VanBergen, A. Apschner, J. Peterson-Maduro, I. Logister, G. Valentin, L. G. Morelli, A. C. Oates, and S. Schulte-Merker. 2018. Segmentation of the zebrafish axial skeleton relies on notochord sheath cells and not on the segmentation clock. *Elife* 7: e33843.

Manning, A. J., and D. Kimelman. 2015. Tbx16 and Msgn1 are required to establish directional cell migration of zebrafish mesodermal progenitors. *Dev. Biol.* 406, 172–185.

Mara, A., and S. A. Holley. 2007. Oscillators and the emergence of tissue organization during zebrafish somitogenesis. *Trends Cell Biol.* 17: 593–599.

Mara, A., J. Schroeder, C. Chalouni, and Holley, S. A. 2007. Priming, initiation and synchronization of the segmentation clock by *deltaD* and *deltaC*. *Nat. Cell Biol.* 9: 523–530.

Maroto, M., R. A. Bone, and J. K. Dale. 2012. Somitogenesis. *Development* 139: 2453–2456.

Martin, B. L. 2016. Factors that coordinate mesoderm specification from neuromesodermal progenitors with segmentation during vertebrate axial extension. *Semin. Cell Dev. Biol.* 49: 59–67.

Martin, B. L., and D. Kimelman. 2008. Regulation of canonical Wnt signaling by Brachyury is essential for posterior mesoderm formation. *Dev. Cell* 15: 121–133.

Martin, B. L., and D. Kimelman. 2009. Wnt signaling and the evolution of embryonic posterior development. *Curr. Biol.* 19: R215–R219.

Martin, B. L., and D. Kimelman. 2010. Brachyury establishes the embryonic mesodermal progenitor niche. *Genes Dev.* 24: 2778–2783.

Martin, B. L., and D. Kimelman. 2012. Canonical Wnt signaling dynamically controls multiple stem cell fate decisions during vertebrate body formation. *Dev. Cell* 22: 223–232.

McGrew, M. J., A. Sherman, S. G. Lillico, F. M. Ellard, P. A. Radcliffe, H. J. Gilhooley, K. A. Mitrophanous, N. Cambray, V. Wilson, andH. Sang. 2008. Localised axial progenitor cell populations in the avian tail bud are not committed to a posterior Hox identity. *Development* 135: 2289–2299.

Morley, R. H., K. Lachani, D. Keefe, M. J. Gilchrist, P. Flicek, J. C. Smith, and F. C. Wardle. 2009. A gene regulatory network directed by zebrafish No tail accounts for its roles in mesoderm formation. *PNAS* 106: 3829–3834.

Morrow, Z. T., A. M. Maxwell, K. Hoshijima, J. C. Talbot, D. J. Grunwald, and S. L. Amacher. 2017. *tbx6l* and *tbx16* are redundantly required for posterior paraxial mesoderm formation during zebrafish embryogenesis. *Dev. Dyn.* 246: 759–769.

Naiche, L. A., N. Holder, and M. Lewandoski. 2011. FGF4 and FGF8 comprise the wavefront activity that controls somitogenesis. *PNAS* 108: 4018–4023.

Niederreither, K., P. McCaffery, U. C. Drager, P. Chambon, and P. Dolle. 1997. Restricted expression and retinoic acid-induced downregulation of the retinaldehyde dehydrogenase type 2 (RALDH-2) gene during mouse development. *Mech. Dev.* 62: 67–78.

Oates, A. C., and R. K. Ho. 2002. *Hairy/E(spl)-related (Her)* genes are central components of the segmentation oscillator and display redundancy with the Delta/Notch signaling pathway in the formation of anterior segmental boundaries in the zebrafish. *Development* 129: 2929–2946.

Oates, A. C., L. G. Morelli, and S. Ares. 2012. Patterning embryos with oscillations: structure, function and dynamics of the vertebrate segmentation clock. *Development* 139: 625–639.

Okubo, Y., T. Sugawara, N. Abe-Koduka, J. Kanno, A. Kimura, and Y. Saga. 2012. Lfng regulates the synchronized oscillation of the mouse segmentation clock via *trans*-repression of Notch signalling. *Nat. Commun.* 3: 1141.

Olivera-Martinez, I., H. Harada, P. A. Halley, and K. G. Storey. 2012. Loss of FGF-dependent mesoderm identity and rise of endogenous retinoid signalling determine cessation of body axis elongation. *PLOS Biology* 10: e1001415.

Ozbudak, E. M., and J. Lewis. 2008. Notch signalling synchronizes the zebrafish segmentation clock but is not needed to create somite boundaries. *PLoS Genet.* 4: e15.

Paese, C. L. B., A. Schoenauer, D. J. Leite, S. Russell, and A. P. McGregor. 2018. A SoxB gene acts as an anterior gap gene and regulates posterior segment addition in a spider. *Elife* 7: e37567.

Papaioannou, V. E. 2014. The T-box gene family: Emerging roles in development, stem cells and cancer. *Development* 141: 3819–3833.

Pourquie, O. 2011. Vertebrate segmentation: From cyclic gene networks to scoliosis. *Cell* 145: 650–663.

Pourquie, O. 2018. Somite formation in the chicken embryo. *Int. J. Dev. Biol.* 62: 57–62.

Riedel-Kruse, I. H., C. Muller, and A. C. Oates. 2007. Synchrony dynamics during initiation, failure, and rescue of the segmentation clock. *Science* 317: 1911–1915.

Rodrigo Albors, A., P. A. Halley, and K. G. Storey. 2018. Lineage tracing of axial progenitors using Nkx1-2CreER(T2) mice defines their trunk and tail contributions. *Development* 145: dev164319.

Row, R. H., J. L. Maitre, B. L. Martin, P. Stockinger, C. P. Heisenberg, and D. Kimelman. 2011. Completion of the epithelial to mesenchymal transition in zebrafish mesoderm requires Spadetail. *Dev. Biol.* 354: 102–110.

Row, R. H., A. Pegg, B. A. Kinney, G. H. 3rd Farr, L. Maves, S. Lowell, V. Wilson, B. L. Martin. 2018. BMP and FGF signaling interact to pattern mesoderm by controlling basic helix-loop-helix transcription factor activity. *Elife* 7: e31018.

Row, R. H., S. R. Tsotras, H. Goto, B. L. Martin. 2016. The zebrafish tailbud contains two independent populations of midline progenitor cells that maintain long-term germ layer plasticity and differentiate in response to local signaling cues. *Development* 143: 244–254.

Saga, Y. 2012. The mechanism of somite formation in mice. *Curr. Opin. Genet. Dev.* 22: 331–338.

Saga, Y., N. Hata, H. Koseki, and M. M. Taketo. 1997. *Mesp2*: A novel mouse gene expressed in the presegmented mesoderm and essential for segmentation initiation. *Genes Dev.* 11: 1827–1839.

Schultemerker, S., F. J. M. Vaneeden, M. E. Halpern, C. B. Kimmel, and C. Nussleinvolhard. 1994. *no tail (ntl)* is the zebrafish homolog of the mouse-*T* (*Brachyury*) gene. *Development* 120: 1009–1015.

Sethi, A. J., R. M. Wikramanayake, R. C. Angerer, R. C. Range, L. M. Angerer. 2012. Sequential signaling crosstalk regulates endomesoderm segregation in sea urchin embryos. *Science* 335: 590–593.

Shimozono, S., T. Iimura, T. Kitaguchi, S. Higashijima, and A. Miyawaki. 2013. Visualization of an endogenous retinoic acid gradient across embryonic development. *Nature* 496: 363–366.

Shinmyo, Y., T. Mito, T. Uda, T. Nakamura, K. Miyawaki, H. Ohuchi, and S. Noji. 2006. *brachyenteron* is necessary for morphogenesis of the posterior gut but not for antero-posterior axial elongation from the posterior growth zone in the intermediate-germ-band cricket *Gryllus bimaculatus*. *Development* 133: 4539–4547.

Takemoto, T., M. Uchikawa, M. Yoshida, D. M. Bell, R. Lovell-Badge, V. E. Papaioannou, and H. Kondoh. 2011. Tbx6-dependent *Sox2* regulation determines neural or mesodermal fate in axial stem cells. *Nature* 470: 394–398.

Tam, P. P., and S. S. Tan. 1992. The somitogenetic potential of cells in the primitive streak and the tail bud of the organogenesis-stage mouse embryo. *Development* 115: 703–715.

Taniguchi, Y., T. Kurth, S. Weiche, S. Reichelt, A. Tazaki, S. Perike, V. Kappert, and H. H. Epperlein. 2017. The posterior neural plate in axolotl gives rise to neural tube or turns anteriorly to form somites of the tail and posterior trunk. *Dev. Biol.* 422: 155–170.

Tsakiridis, A., and V. Wilson. 2015. Assessing the bipotency of in vitro-derived neuromesodermal progenitors. *F1000Res* 4: 100.

Turner, D. A., P. C. Hayward, P. Baillie-Johnson, P. Rue, R. Broome, F. Faunes, and A. Martinez Arias. 2014. Wnt/beta-catenin and FGF signalling direct the specification and maintenance of a neuromesodermal axial progenitor in ensembles of mouse embryonic stem cells. *Development* 141: 4243–4253.

Tzouanacou, E., A. Wegener, F. J. Wymeersch, V. Wilson, and J.-F. Nicolas. 2009. Redefining the progression of lineage segregations during mammalian embryogenesis by clonal analysis. *Dev. Cell* 17: 365–376.

Uriu, K., R. Bhavna, A. C. Oates, and L. G. Morelli. 2017. A framework for quantification and physical modeling of cell mixing applied to oscillator synchronization in vertebrate somitogenesis. *Biol. Open* 6: 1235–1244.

Uriu, K., and L. G. Morelli. 2017. Determining the impact of cell mixing on signaling during development. *Dev. Growth Differ.* 59: 351–368.

Uriu, K., Y. Morishita, and Y. Iwasa. 2010. Random cell movement promotes synchronization of the segmentation clock. *PNAS* 107: 4979–4984.

vanEeden, F. J. M., M. Granato, U. Schach, M. Brand, M. FurutaniSeiki, P. Haffter, M. Hammerschmidt, C. P. Heisenberg, Y. J. Jiang, D. A. Kane, R. N. Kelsh, M. C. Mullins, J. Odenthal, R. M. Warga, M. L. Allende, E. S. Weinberg, and C. NussleinVolhard. 1996. Mutations affecting somite formation and patterning in the zebrafish, Danio rerio. *Development* 123: 153–164.

Wilkinson, D. G. 2018. Establishing sharp and homogeneous segments in the hindbrain. *F1000Res* 7: 1268.

Wymeersch, F. J., Y. Huang, G. Blin, N. Cambray, R. Wilkie, F. C. Wong, and V. Wilson. 2016. Position-dependent plasticity of distinct progenitor types in the primitive streak. *Elife* 5: e10042.

Wymeersch, F. J., S. Skylaki, Y. Huang, J. A. Watson, C. Economou, C. Marek-Johnston, S. R. Tomlinson, and V. Wilson. 2019. Transcriptionally dynamic progenitor populations organised around a stable niche drive axial patterning. *Development* 146: dev168161.

Yabe, T., K. Hoshijima, T. Yamamoto, and S. Takada. 2016. Quadruple zebrafish mutant reveals different roles of Mesp genes in somite segmentation between mouse and zebrafish. *Development* 143: 2842–2852.

Yabe, T., and S. Takada. 2012. Mesogenin causes embryonic mesoderm progenitors to differentiate during development of zebrafish tail somites. *Dev. Biol.* 370: 213–222.

Yamaguchi, T. P., S. Takada, Y. Yoshikawa, N. Y. Wu, and A. P. McMahon. 1999. *T* (Brachyury) is a direct target of Wnt3a during paraxial mesoderm specification. *Genes Dev.* 13: 3185–3190.

Yoon, J. K., and B. Wold. 2000. The bHLH regulator *pMesogenin1* is required for maturation and segmentation of paraxial mesoderm. *Genes Dev.* 14: 3204–3214.

Young, T., J. E. Rowland, C. van de Ven, M. Bialecka, A. Novoa, M. Carapuco, J. van Nes, W. de Graaff, I. Duluc, J. N. Freund, F. Beck, M. Mallo, and J. Deschamps. 2009. *Cdx* and *Hox* genes differentially regulate posterior axial growth in mammalian embryos. *Dev. Cell* 17: 516–526.

Zhang, L., C. Kendrick, D. Julich, and S. A. Holley. 2008. Cell cycle progression is required for zebrafish somite morphogenesis but not segmentation clock function. *Development* 135: 2065–2070.

6 Teloblasts in Crustaceans

Gerhard Scholtz

CONTENTS

6.1 INTRODUCTION

The great diversity of adult crustacean forms, body organizations, and life styles finds its correspondence in a comparable variety of ontogenetic pathways. Crustaceans are unmatched by other arthropod groups concerning their manifold cleavage and gastrulation patterns, segmentation processes, and larval types and development. This fascinating ontogenetic diversity is even greater if recent views on crustacean phylogeny are taken into account. According to these views, crustaceans are paraphyletic with Remipedia, either alone or together with Cephalocarida, as sister taxon to Hexapoda (Regier et al. 2010; von Reumont et al. 2012; Schwentner et al. 2017). Hence, insects are deeply nested within a group that was traditionally called Crustacea but now has been named Pancrustacea or Tetraconata (Zrzavý and Štys 1997; Dohle 2001).

Large cells that undergo a series of asymmetric divisions in one direction can be found in several taxa of Pancrustacea/Tetraconata including hexapods. These characteristic stem cells occur in different proliferation processes such as neurogenesis, then they are called neuroblasts, or the longitudinal growth of germ bands, then they are called teloblasts. In any case, neuroblasts and teloblasts are involved in the quick

125

generation of competent cell material. Neuroblasts have been found in several crustacean groups, such as malacostracans, branchiopods, and copepods, and in hexapods (e.g., Brenneis et al. 2013; Hartenstein and Stollewerk 2015; Hein and Scholtz 2018). Neuroblasts are neuronal precursors and proliferate a set of other neuronal precursors (ganglion mother cells) that divide into neurons.

In contrast to neuroblasts, which might be an apomorphy of Pancrustacea/Tetraconata (Ungerer and Scholtz 2008), the occurrence of teloblasts is restricted to malacostracan crustaceans (Dohle et al. 2004; Fischer et al. 2010). Malacostracan teloblasts are situated in the pre-anal region of the germ band and generate the ectodermal (ectoteloblasts) and mesodermal (mesoteloblasts) cellular material for segmentation. Hence, teloblasts are a special case of a growth zone.

Ecto- and mesoteloblasts have been first described in annelids, in particular in clitellates (oligochaetes and hirudineans) (see Anderson 1973; Dohle 1999), where they have a similar pre-anal position and function like those of malacostracans but show lower numbers (Shankland and Savage 1997).

The frequent occurrence of large stem cells with asymmetric divisions in different developmental processes and various taxa suggests that this cell type evolved several times independently. It seems an effective way for the proliferation of many cells in a longitudinal arrangement. However, the molecular mechanisms of asymmetric cell divisions are not yet completely understood (see Knoblich 2010).

6.2 FORMATION OF THE GERM DISC

Either during gastrulation or after it has been completed, superficial cells assemble to form the germ rudiment, which represents the starting point for the formation of the adult body organization (Figure 6.1). As has been shown in amphipods, this process can involve dramatic cell migration (Scholtz and Wolff 2002). In crustaceans with yolky eggs, this phase is characterized by the germ disc that marks the future ventral side of the embryo (Figure 6.1A). The germ disc comprises outer ectodermal cells and inner layers or groups of mesodermal, endodermal, and primordial germ cells (e.g. Samter 1900; Manton 1928; Weygoldt 1958; Browne et al. 2005). These cells are free of yolk and densely packed. In some cases, vitellophages are situated in the yolk mass, which digest the yolk (Scholtz and Wolff 2002). The remaining area of the egg surface is called the extraembryonic region, because it plays no formative role. In this region, the cells are large, contain yolk, and are loosely arranged (Figure 6.1A, B). Probably some of these cells also act as vitellophages. With the formation of the germ disc, the anteroposterior axis of the future body becomes recognizable (Figure 6.1A). The germ disc elongates and turns into the germ band. Characteristic landmarks are the anterior and posterior invaginations of the stomodaeum (ectodermal mouth and foregut) and the proctodaeum (ectodermal anus and hindgut) (Figure 6.1B). At the germ band's anterior margin, the head lobes (Scheitelplatten of Cladocera; see Samter 1900; Kühnemund 1929) form, which are the anlagen of the lateral eyes and some anterior brain parts (Manton 1928; Kühnemund 1929; Weygoldt 1958; Scholl 1963; Dohle 1972; Browne et al. 2005; Mittmann et al. 2014) (Figure 6.1B). At the posterior end, the pre-anal growth zone is differentiated (Ooishi 1959, 1960; Dohle 1970; Scholtz 1992) (Figures 6.1B

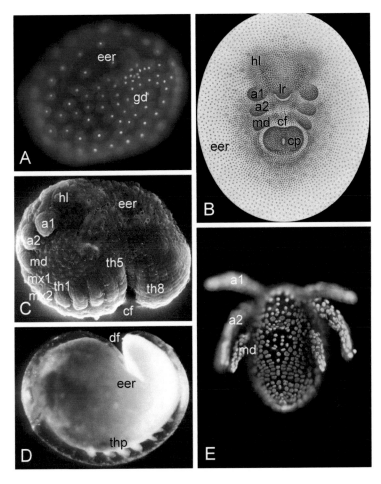

FIGURE 6.1 Formation of the germ disc and the transformation into the germ band. (A) The early germ disc of the amphipod *Cryptorchestia garbinii*. The future anterior end of the embryo is on top. Blastoderm cells getting rid of their yolk migrate together forming the germ disc (gd), whereas loosely arranged cells form the extraembryonic region (eer). (B) Ventral view (anterior on top) of the early germ band showing an egg nauplius of the crayfish *Astacus astacus* with the advanced anlagen of the eyes (head lobes, hl), the first and second antennae and the mandibles (a1, a2, md) compared with more posterior segmental structures. At the anterior, there is the anlage of the labrum (lr) covering the forming stomodaeum, in the posterior there is the caudal furrow (cf) and the forming caudal papilla (cp) with the proctodaeum (modified after Reichenbach 1886). (C) SEM of an advanced embryo of the amphipod *Gammarus pulex* (lateral view, anterior to the left, dorsal on top). Here, an egg nauplius is lacking because the naupliar region shows no advanced development. The naupliar appendages are in a similar stage as in the crayfish in B, but in contrast to this the fifth thoracic segment (th5) shows early limb buds. (D) Lateral view of the embryo of an isopod (lateral view, anterior to the left, dorsal on top). The germ band shows no caudal furrow but stretches along the ventral side. In contrast to the situation shown in B and C, a dorsal furrow (df) is formed. Some thoracopods (thp) are visible. (E) The nauplius larva of the northern krill *Meganyctiphanes norvegica*. This is a reduced non-feeding nauplius with few cells (nuclear fluorescent stain bisbenzimide).

and 6.2A–E). This produces the ectodermal and mesodermal cellular material that elongates the germ band and eventually gives rise to the ectodermal segmental structures such as the intersegmental furrows, ganglia, limbs, and the mesodermal musculature. During this process, the endodermal midgut precursor cells form the midgut and the midgut glands, if present (Reichenbach 1886; Fioroni 1970). With the elongation of the germ band and the differentiation of the segmental structures, the germ band extends laterally and the extraembryonic region gets smaller until the lateral halves of the embryo meet at the dorsal side (Figure 6.1C, D). This dorsal closure is accompanied by the digestion of the yolk mass, which at the hatchling stage is reduced to a large degree or entirely gone (Scholl 1963).

However, these stages are frequently overlapping and not strictly separated. Furthermore, a proper germ disc and the subsequent germ band are absent in species with small, almost yolk-free eggs that undergo total or mixed cleavage such as some copepods, cladocerans, and malacostracans (e.g., Kühn 1913; Fuchs 1914). In these cases, the germ anlage comprises the entire embryo including the dorsal side (Figure 6.1E).

6.3 GERM BAND

6.3.1 SHAPE

After gastrulation and during the differentiation of the germ band, the embryos assume a characteristic shape that differs between taxa. The plesiomorphic condition within malacostracans is characterized by the formation of a transverse ventral groove, the caudal furrow (Figure 6.1B, C). With advanced development, this groove deepens and subdivides the embryo in an anterior yolky region and a posterior ventrally folded mostly yolk-free caudal papilla (Scholtz 2000) (Figures 6.1B and 6.2A). The latter starts as a dome-shaped bud in the posterior region of the embryo (Figure 6.1B). Concomitant with the budding of cells in the posterior growth zone, the caudal papilla elongates and becomes ventrally flexed (Figure 6.2A). Since the growth zone encircles the caudal papilla (Figure 6.2A, C, D), the ventral and dorsal parts of the embryo and the subsequent segments are formed. With hatching, the caudal papilla stretches backward and the egg envelopes rupture. The groove between the anterior region and the caudal papilla does not correspond to the boundary between thorax and pleon. Rather the caudal papilla comprises the anlagen of a varying number of thoracic, all pleonic segments, and the telson (Scholtz 2000) (Figure 6.2A). A caudal papilla of this sort has been found in embryos of Leptostraca, Stomatopoda, Decapoda, Syncarida, and Thermosbaenacea (Scholtz 2000). In the latter, the caudal papilla contains some yolk but the growth zone still encircles the papilla (Zilch 1974). Among peracarids, Mysidacea possess a caudal papilla, but the yolk content is relatively high and the growth of the germ band is restricted to the ventral side (Scholtz 1984). Amphipoda show a caudal groove, but the posterior body is almost as wide as the anterior part, and as in mysidaceans only the ventral side constitutes the germ band (Weygoldt 1958; Scholtz 1990; Ito et al. 2011) (Figure 6.1C). In contrast to this, the embryos of Isopoda, Tanaidacea and Cumacea, Spelaeogriphacea, and Mictacea show neither a caudal groove nor a caudal papilla (Scholtz 2000;

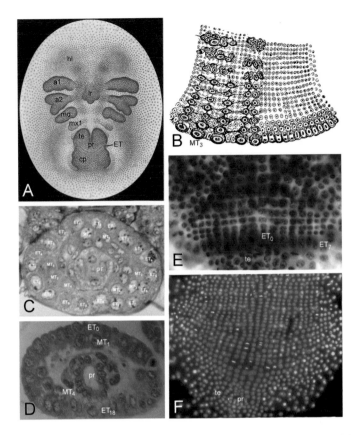

FIGURE 6.2   Teloblast patterns. (A) The germ band with a ventrally folded caudal papilla of *Astacus astacus* (modified after Reichenbach 1886). The head lobes (hl), the labrum (lr), the first and second antennae (a1, a2), the mandibles (md), the first maxillae (mx1), and the caudal papilla (cp) show an advanced development, when compared with Figure 6.1B. The proctodaeum (pr) is situated in the area of the telson (te). The ectoteloblasts (ET) are clearly recognizable based on their relatively large size; there are more than 19 in number (see D). Reichenbach's publication is the first record of this type of stem cells. (B) The posterior region of the germ band of the isopod *Cymothoa* sp. (anterior on top) (modified after Patten 1890). This is the first record of mesoteloblasts (MT). The mesoteloblasts and their derivatives of the right animal's side are highlighted (yellow). Note the regular row arrangement (compare with Figures 6.5C and 6.6) and the cytoplasmic longitudinal and horizontal connections of the mesoteloblasts and their derivatives. Furthermore, there is a certain distance between the median mesoteloblast and the more lateral mesoteloblasts, exemplified by the third mesoteloblast from the midline (MT_3) (compare with C and D). (C) A cross section through the caudal papilla in the area of teloblasts of *Homarus americanus* (modified after Dohle et al. 2004). An outer circle of 19 ectoteloblasts (ET) and an inner circle of 8 mesoteloblasts (MT) around the proctodaeum are visible. This is the ancestral condition within Malacostraca. (D) A similar section in the freshwater crayfish *Cherax destructor*. About 40 ectoteloblasts (ET) are combined with 8 mesoteloblasts (MT) both situated around the proctodaeum. (E) The growth zone of the peracarid *Neomysis integer*. The ectoteloblasts (ET) form a transverse row of about 15 cells anterior to the telson ectoderm (te). (F) The absence of ectoteloblasts in the germ band of the amphipod *Cryptorchestia garbinii*. The transverse ectoderm rows are formed by cell rearrangements in front of the telson ectoderm with the proctodaeum.

Olesen et al. 2014) (Figure 6.1D). The germ band grows out in one plane and only later a dorsal furrow forms that straightens with hatching. Accordingly, this condition has been considered as an apomorphy for the Mancoidea within the Peracarida (see Richter and Scholtz 2001).

6.3.2 Growth

Like hexapods, myriapods, and chelicerates, non-malacostracan crustaceans do not show specific cell types in their growth zone. Furthermore, the cells in the germ band do not display a recognizable stereotyped pattern of cell division and arrangement. This is different in malacostracans. They show individually recognizable cells and reproducible cell division patterns during the formation and differentiation of the germ band up to early segmentation. Parts of this idiosyncratic developmental process were detected in the 19th century. Reichenbach (1886) was the first to describe the special stem cells of the growth zone of malacostracans, the ectodermal teloblasts, and Patten (1890) reported similar cells in the mesoderm, the mesoteloblasts (Figure 6.2A, B). Bergh (1893, 1894) and McMurrich (1895) were the first to resolve aspects of the regular stereotyped cleavage pattern in the malacostracan germ band. Since then, the stereotyped formation and differentiation of the malacostracan germ band (see Figures 6.2–6.7) has been intensely studied, leading to the most detailed knowledge of the early segmentation process in crustaceans (Dohle et al. 2004).

The growth zone of most Malacostraca is characterized by transversely arranged relatively large stem cells, the teloblasts (Figures 6.2–6.7). As mentioned earlier, teloblasts occur in the ectoderm (ectoteloblasts) and mesoderm (mesoteloblasts). They divide asymmetrically producing smaller cells in an anterior direction (Figures 6.2 and 6.5–6.7). This generates the cellular material for the subsequent segmentation of the germ band. A ring of 19 ectoteloblasts (one unpaired ventral median ectoteloblast and nine paired ectoteloblasts on either side of the midline) and an inner ring of 8 mesoteloblasts (four on either side, a median mesoteloblast is absent) that surround the caudal papilla anterior to the telson anlage are found in Leptostraca, Stomatopoda, most Decapoda, Euphausiacea, Anaspidacea, and Thermosbaenacea (Scholtz 2000) (Figures 6.2C and 6.3A). This distribution allows for the conclusions that this pattern is the original condition for Malacostraca (Scholtz 2000). However, within malacostracans some evolutionary changes have taken place. In the lineage to freshwater crayfishes, the number of ectoteloblasts increased to about 40 and the ring arrangement persisted (Scholtz 1993; Scholtz et al. 2009) (Figure 6.2D). In contrast to this, the teloblast rings have been evolutionarily transformed to transverse rows with a variable number (15–23) of ectoteloblasts in Peracarida (Figure 6.2E). The Amphipoda are a notable exception, because they lost ectoteloblasts entirely (Bergh 1894; Scholtz and Wolff 2002) (Figure 6.2F). In contrast to the situation in ectoteloblasts, the mesoteloblasts remain conservative, and no exception of the number 8 has been found (Figures 6.2 and 6.4). Even amphipods that evolutionarily lost ectoteloblasts are equipped with 8 mesoteloblasts (Figure 6.4). The only evolutionary change in mesoteloblasts occurs in peracarids, which show a transverse row of mesoteloblasts instead of the plesiomorphic ring arrangement (Scholtz 2000) (see Figures 6.2B, 6.5C, and 6.7).

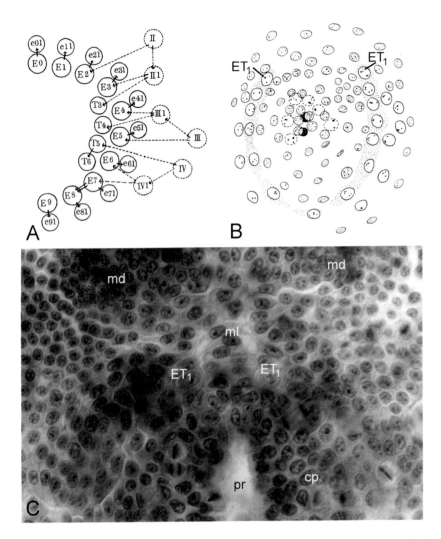

FIGURE 6.3 Differentiation of ectoteloblasts. (A) The complex cell division pattern during the differentiation of ectoteloblasts in one body half in some decapod crustaceans (modified after Ooishi 1959, with permission from Wiley & Sons). The Roman numerals label ectotelo-blast precursors; the ectoteloblasts are designated with an uppercase E and their descendants with lowercase e. (B) The *in situ* formation of ectoteloblasts in the early germ disc of the cumacean *Diastylis rathkei* (modified after Dohle 1970, with permission from Springer). The ectoteloblasts form two half rings that migrate around the gastrulation center (immigrated cell are shown as dark and dotted nuclei) and meet anteriorly in the middle. One of the cells between the two first ectoteloblasts (ET_1) will be transformed into the median, unpaired ectoteloblast. (C) The formation of ectoteloblasts (ET) in the early germ disc of the decapod *Cherax destructor*. As in the cumacean (B) and in contrast to other decapods, the ectotelo-blasts of crayfish differentiate *in situ*. This happens during the formation of the caudal papilla (cp) and the proctodaeum (pr) at a stage where the naupliar appendages begin to differentiate, as is exemplified by the mandibular buds (md) (compare with Figure 6.1B). One of the mid-line cells (ml) will become the median unpaired ectoteloblast (see Figure 6.2D, ET_0).

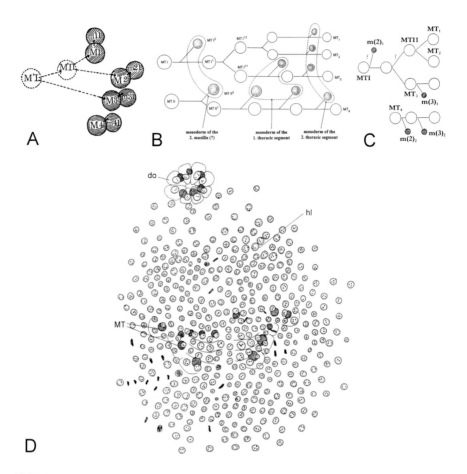

A B C

D

FIGURE 6.4   Differentiation of mesoteloblasts. (A) The cell division pattern of the differentiation of mesoteloblasts in one body half of some decapods (modified after Ooishi 1959, with
permission from Wiley & Sons). M labels mesoteloblasts (Arabic numerals) and mesoteloblast precursors (Roman numerals). The mesoteloblast descendants are labeled with a lowercase m. Corresponding patterns in (B) the cumacean *Diastylis rathkei* (modified after Dohle
et al. 2004) and (C) the amphipod *Gammarus pulex* (modified after Scholtz 1990). In both
cases mesoteloblasts (Arabic numerals) and their precursors (Roman numerals) are labeled
MT, the descendants are designated with lowercase m. (D) The germ disc of the amphipod
Gammarus pulex at the transition to a germ band showing the subectodermal position of the
mesoteloblasts/precursors) (MT) (modified after Scholtz 1990). Posterior to the dorsal organ
(do) are the head lobes (hl; compare with Figures 6.1 and 6.2). The first transverse ectoderm
rows are forming (lines).

Ectoteloblasts are generated in several different ways. Ooishi (1959, 1960)
described a complex stereotyped cell division pattern for a decapod shrimp, a her-
mit, and a brachyuran crab (Figure 6.3). In these species, the ectoteloblasts are
differentiated in a stepwise manner and the most ventral ectoteloblasts begin their
characteristic asymmetric divisions before the more dorsal ectoteloblasts have been

differentiated. The ectoteloblasts in a freshwater crayfish and a cumacean have been reported to form *in situ* without any special lineage but nevertheless in a ventral–dorsal sequence (Dohle 1970; Scholtz 1992). By contrast, the differentiation of meso-teloblasts follows a cell lineage in decapods, cumaceans, and amphipods (Ooishi 1959, 1960; Dohle 1970; Scholtz 1990; Price and Patel 2008; Hunnekuhl and Wolff 2012) (Figure 6.4).

With their asymmetric divisions, the teloblasts generate regular transverse rows or rings of cells that form a regular gridlike pattern (Hejnol et al. 2006). The divisions of the teloblasts follow more or less a mediolateral wave of mitoses on either side of the midline (Figure 6.5B). The unpaired median ectoteloblast generates a midline column of unpaired cells that matches the symmetry axis of the body in the post-naupliar region (Figure 6.5B). Each ectodermal transverse row or ring of ectoteloblast derivatives forms a genealogical unit (Figures 6.5 and 6.6). In the ectoderm, the cells of each of these rows divide twice following a mediolateral wave of mitoses with an anteroposterior spindle orientation (Dohle et al. 2004; Scholtz and Wolff 2013) (Figures 6.5 and 6.6). Each row generates four descendant rows, which are still arranged in a regular grid (Figures 6.5 and 6.6). This process elongates the germ band further. Hence, germ band elongation is a two-step process: first the generation of ectoteloblast rows (founders of the genealogical units), second the two waves of divisions in longitudinal direction of each of these rows (Scholtz and Wolff 2013) (Figures 6.5B, C and 6.6). After that the differential cleavages begin. These mitoses have varying spindle orientations and are sometimes asymmetric but nonetheless still follow a stereotyped pattern (Figures 6.5 and 6.6). With these differential cleavages, the germ band becomes three-dimensional and the segmental structures such as intersegmental furrows, limb buds, and ganglion primordia are formed (Dohle et al. 2004) (Figures 6.5, 6.6, and 6.11).

When the gridlike arrangement of the ectoderm cells was detected, it was thought that each of the ectoteloblast transverse rows or rings gave rise to a morphological adult segment (Bergh 1893; McMurrich 1895; Manton 1928). However, more detailed analyses have shown that the segmental boundaries run transversely and slightly obliquely through the progeny of each ectoteloblast row (Dohle 1972, 1976; Scholtz 1984; Dohle and Scholtz 1988; Scholtz and Dohle 1996). Hence, every morphological segment in the ectoderm is formed by the progeny of two genealogical units and likewise segmental ganglia and legs are composite structures (Dohle et al 2004) (Figures 6.5, 6.6, and 6.11). Interestingly, amphipods show the same regular pattern of the ectoderm cells in the germ band despite the absence of ectoteloblasts (Scholtz 1990; Dohle et al. 2004). In this case, the rows form by an arrangement of previously scattered ectodermal cells in an anteroposterior sequence (Figures 6.2F, 6.5C, D, and 6.6).

It appears that the midline cells play an organizing role during the process of row generation and segment differentiation. This is indicated by the early and accelerated formation of the midline propagating in the anteroposterior direction and the subsequent generation of ectoderm rows from the midline toward lateral (Scholtz 1990). Experiments in which midline cells of the amphipod *Parhyale hawaiensis* have been ablated with laser beams and the expression of the *single-minded* (*sim*) gene has been suppressed strongly corroborated this view. Both approaches led to

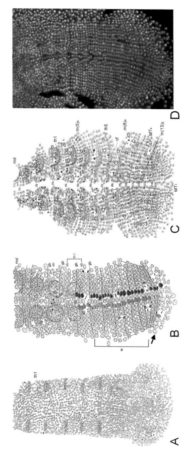

FIGURE 6.5 The germ bands of a cladoceran and three malacostracans. The first thoracic segment (th1) is labeled for comparison. Parts A–C are camera lucida drawings (A is modified after Gerberding 1997; B and C are modified after Scholtz 1984, 1990). Part D is a fluorescent staining of cell nuclei with Bisbenzimide (Hoechst Blue). (A) The post-naupliar germ band of the cladoceran *Leptodora kindtii*. There are many small cells with an irregular arrangement, and ectoteloblasts are lacking. (B) Germ band of *Neomysis integer*. There are relatively few cells, which are regularly arranged in a gridlike pattern of longitudinal columns (including a midline) and transverse rows. The large ectoteloblasts (arrow) give rise to regular rows that follow a determined sequence of division (yellow: undivided rows; blue: rows in the phase of the first wave of division; green: rows in the phase of the second wave of division; orange: beginning differential cleavages and morphological segmentation; red: the offspring of one ectoteloblast along the anteroposterior axis of the germ band). ie, intercalary elongation; sb, segmental border; gb, genealogical boundary (for explanation see text). In malacostracans, the telson is formed posterior to the teloblasts. (C) The advanced germ band of the amphipod *Gammarus pulex*. In this case, the regular gridlike pattern of ectoderm cells forms without ectoteloblasts. This is an apomorphy of Amphipoda. The midline is omitted; the mesoderm is drawn in bold lines. The eight mesoteloblasts (MT) at the end of the germ band and the transverse rows of mesoteloblast derivatives are visible (numbers in brackets). Some individual cells of the mesoderm are labeled. (D) The post-naupliar germ band of the amphipod *Cryptorchestia garbinii* showing the regular cell arrangement. The stage is slightly younger than that in (C).

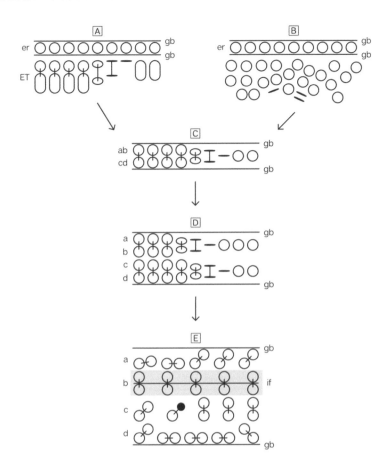

FIGURE 6.6 Schematic summary of row formation and segmentation in the post-naupliar germ band of malacostracan crustaceans. Only the animal's left side is shown and the midline is on the left side. Transverse lines indicate the genealogical borders (gb) between the ectoderm rows. The transverse ectoderm rows are formed either by ectoteloblasts (ET), a condition found in the posterior part of most peracarids and decapods examined (A), or by scattered blastoderm cells (B), a condition found in anterior rows of most peracarids and decapods and in the entire post-naupliar germ band in amphipods. After formation, each row (except the anteriormost rows that show a somewhat different pattern) undergoes two medio-lateral mitotic waves with only longitudinally oriented spindles and equal mitoses, resulting in four transverse descendant rows named a, b, c, d (C, D). Thereafter, the differential cleavages begin. These show a stereotyped pattern of mitoses with regard to size and position of the division products. (E) A simplified schematic pattern of the first differential cleavage up to the fifth cells from the midline. Some characteristics of the individual mitoses differ among the studied species, a phenomenon not shown here (for comparison see Dohle et al. 2004). With the differential cleavages, segmentation begins. The segmental boundary (shaded area) marked by the intersegmental furrow (if) does not match the genealogical border (gb). The intersegmental furrows run transversely and slightly obliquely through the descendants of one ectoderm row in the area of descendant rows a and b (compare with Figure 6.11). Thus, the descendants of each ectoderm row contribute to two segments.

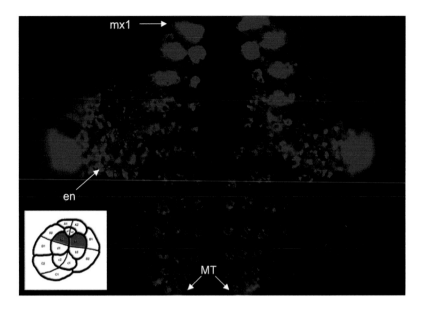

FIGURE 6.7   Mesoteloblast and their descendants (modified after Wolff and Scholtz 2002, with permission from Elsevier). The result of the *in vivo* labeling of two blastomeres of the 16-cell stage (insert: red cells) of *Cryptorchestia garbinii* with the fluorescent dye DiI. The eight mesoteloblast (MT), their descendant transverse rows (beginning at the first maxillary segment [mx1]) and the endodermal midgut-gland anlagen (en) are stained. The cells of the more anterior mesoderm rows begin their first wave of division with slightly oblique spindle orientations.

a dorsalization of ventral areas of the germ band (Vargas-Vila et al. 2010). The *sim* gene is known from *Drosophila* to play a role in the differentiation of midline and ventral cell fates (see Vargas-Vila et al. 2010). Apparently, a mediolateral gradient of *sim* is also responsible for the ventrodorsal differentiation of segmental structures in the amphipod *Parhyale hawaiensis* (Vargas-Vila et al. 2010). In malacostracans with ectoteloblasts, the role of the midline must begin with the differentiation of the median ectoteloblast cell. The latter produces most of the midline cells in an anterior direction (Dohle et al. 2004).

The mesoderm rows show a much larger distance between each other and undergo a different sequence of mitoses. Furthermore, there is no unpaired midline population as in the ectoderm. Nevertheless, the mesoderm cells also reveal a stereotyped arrangement and cell division pattern (Dohle 1972; Scholtz 1990; Price and Patel 2008; Hunnekuhl and Wolff 2012) (Figures 6.2B and 6.7). In contrast to the ectoderm rows, each mesoteloblast row generates the mesodermal equipment of one morphological segment. Hence, in the mesoderm the genealogical units and the segments match. There is only one curious exception, which has been described in two amphipod species. This involves the innermost-paired mesoderm cells of the first thoracic segment. These are generated in the second maxillary segment and migrate posteriorly, joining the mesoteloblasts descendants of the first thoracic segment to form the full complement of four mesoderm cells per side (Scholtz 1990; Price and

Patel 2008). The segmental mesoteloblast descendant cells of each side form the ventral body musculature (innermost mesoteloblast derivatives), the muscles of the limb (the two median mesoteloblast derivatives), and the dorsal heart (the most lateral mesoteloblast derivatives) (Hunnekuhl and Wolff 2012).

The teloblasts do not generate the material for all segments. The cells of the prospective naupliar region are formed before the teloblasts are differentiated. Ectodermal and mesodermal cells seem irregularly distributed and a stereotyped cell lineage cannot be identified (Dohle et al. 2004; Scholtz and Wolff 2013). In addition, the ectodermal cells giving rise to the segments of the first and second maxillae and the anterior part of the first thoracic segment do not descend from the ectoteloblasts. Nevertheless, they are arranged in transverse rows like the ectoteloblast derivatives and they show a similar further cell division pattern, with the notable exception of the anterior part of the first maxilla, which shows a slightly different cell arrangement and division pattern (Dohle et al. 2004; Wolff and Scholtz 2006). After the formation of 12 rows in the cumacean *Diastylis rathkei* or 13 ectoderm rows in the decapod *Cherax destructor*, the ectoteloblasts quit their characteristic asymmetric divisions. Two more rows are formed with somewhat more irregular divisions, which produce the cells for the transition between the last segment and the telson (Dohle 1970; Scholtz 1992).

The post-naupliar mesoderm behaves differently. Descendants of mesoteloblast precursors and differentiated mesoteloblasts form the mesodermal equipment of the post-naupliar segments beginning with the second maxilla (Hunnekuhl and Wolff 2012). Yet, as in the ectoteloblasts, the final divisions of the mesoteloblasts are either inverted with respect to the smaller daughter cells or symmetrical (Dohle 1970; Scholtz 1990). Whether derivatives of mesoteloblast contribute to the telson mesoderm is not clear.

6.4 EVOLUTION OF THE STEREOTYPED DIVISION PATTERN

As mentioned earlier, the regular gridlike arrangement of post-naupliar ectoderm and mesoderm cells and the elaborate stereotyped cell division pattern is restricted to malacostracan crustaceans. A closer look at some branchiopod, cirripede, and copepod species did not reveal a comparable pattern. Neither teloblasts nor the gridlike cell arrangement have been identified (Gerberding 1997; Dohle et al. 2004; Ponomarenko 2014; Hein and Scholtz 2018) (Figures 6.5A and 6.8). Furthermore, the figures of Stegner and Richter (2015) on the development of a cephalocarid species do not indicate the presence of regular cell arrangements and teloblasts. In addition, claims that cirripedes and anostracans possess teloblasts (Anderson 1967, 1969) could not be substantiated by more recent investigations (Benesch 1969; Dohle et al. 2004; Ponomarenko 2014). However, in some non-malacostracan crustaceans (Anostraca, Copepoda) an unpaired midline cell population has been recognized (e.g., Dohle et al. 2004; Hein et al. 2019). Moreover, the images of nauplii of parasitic copepods suggest that some sort of regular cell arrangement occurs (McClendon 1907). In particular, in the light of recent phylogenetic hypotheses of Tetraconata/Pancrustacea (Regier et al. 2010; Schwentner et al. 2017), in which Copepoda are either the sister group of Malacostraca or the sister group of Thecostraca plus Malacostraca, this deserves a reinvestigation to reveal whether it shows similarities to the pattern found in malacostracans.

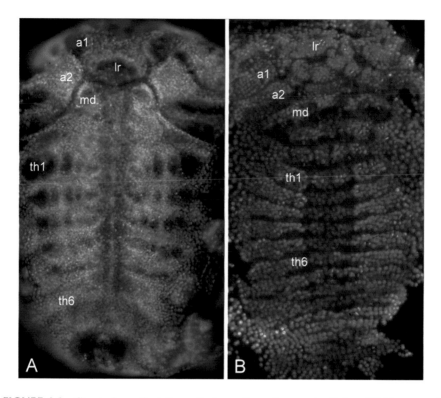

FIGURE 6.8   Comparison of relative cell size, cell numbers, and cell fate. (A) The ventral aspect of an embryo of the cladoceran *Leptodora kindtii*. (B) The ventral aspect of the germ band of the amphipod *Cryptorchestia garbinii*. In both embryos (nuclei stained with the fluorescent dye: Bisbenzimide, Hoechst Blue) the labrum (lr), the antennae (a1, a2), the mandibles (md), and corresponding thoracic segments (th1, th2) are labeled. It is obvious that in the non-malacostracan species (a) many more and relatively small cells form the corresponding structures compared with the malacostracan (B).

An invariant cell division pattern in the germ band has not been found in Chelicerata, Myriapoda, and Hexapoda (Anderson 1973). Thus, it appeared evolutionarily somewhere within the Pancrustacea/Tetraconata, most likely in the lineage leading to Malacostraca (Fischer et al. 2010). Furthermore, there are indications that some aspects of this elaborated pattern evolved only within malacostracans, namely in the Caridoida (Richter and Scholtz 2001). The leptostracans and stomatopods differentiate teloblasts with characteristic behavior (Manton 1934; Shiino 1942; Fischer et al. 2010). Yet, both groups reveal irregularities of the cell division pattern during later stages compared with those of decapods and peracarids (Fischer et al. 2010) (Figure 6.9).

The reasons for the evolution of the invariant cell division patterns of malacostracans are unknown. However, a possible explanation might be that malacostracans show a relatively low number of cells in their germ bands when compared to other crustaceans and arthropods (Figure 6.8). With a low number of cells, a stereotyped

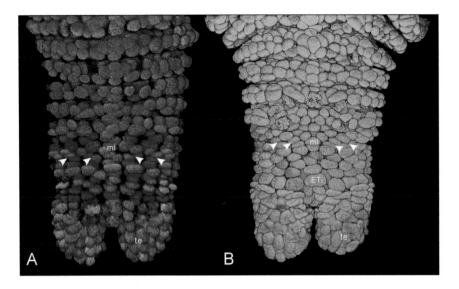

FIGURE 6.9 Row formation in the ectoderm of Leptostraca and Stomatopoda (surface rendering of confocal stacks; modified after Fischer et al. 2010, with permission from Elsevier). (A) Ventral side of the caudal papilla of the leptostracan *Nebalia* sp. (fluorescent dye: bisbenzimide, Hoechst Blue). An area with cells arranged in transverse rows and longitudinal columns in front of the telson ectoderm (te). The ectoteloblasts already finished their characteristic asymmetric divisions and are no longer recognizable. (B) Caudal papilla and ventral germ band of the stomatopod *Gonodactylaceus falcatus* (fluorescent dye: Sytox Green). In both species, the sister cells of the first wave of division of the cell rows show an oblique orientation (arrowheads). This is different from the situation in the remaining malacostracans (Caridoida sensu Richter and Scholtz 2001) (see Figures 6.5 and 6.6). Hence, the pattern found in Leptostraca and Stomatopoda can be considered as ancestral among Malacostraca.

cell division is necessary to put the right cell in the right place at the right time to generate a specific determination of cell fate, as, for instance, a neuronal precursor or the tip of a limb bud. If there are numerous cells to form a limb or a ganglion anlage, an exact position or number of individualized cells is not required (Schnabel 1997).

6.5 ANOMALY OF CELL SHAPES AND BEHAVIOR

The shape and behavior of the ectoderm cells in the malacostracan germ band are also interesting from a cytological perspective. The normal appearance of cells in a cell layer is a hexagonal shape (e.g., Irvine and Wieschaus 1994: Stollewerk et al. 2001). This shape allows the greatest density of cells, comparable to the arrangement of wax and paper combs in social hymenopterans or the facets of most arthropod compound eyes. Accordingly, the cells in the germ band of most arthropod embryos are hexagonal. In contrast to this, the cells of the malacostracan germ band are more or less squared and arranged in regular rows (Figure 6.10). The reason for this is unknown. In those species that differentiate ectoteloblasts, the explanation can be seen in the regular gridlike arrangement of the ectoteloblasts, which is passed on to

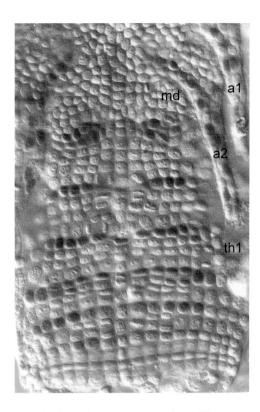

FIGURE 6.10   Cell shape in the malacostracan germ band. The post-naupliar germ band of the mysidacean *Neomysis integer* depicting rectangular cell shapes, exemplified in the first thoracic (th1) and more posterior segments. This stands in contrast to the naupliar region with the first and second antennae (a1, a2) and the mandibles (md) that do not show the regular cell arrangement as the post-naupliar region and that represents the ancestral condition of germ bands of arthropods in general (compare with Figures 6.5A and 6.8A). The dark cells show expression of the Engrailed protein (compare with Figures 6.11 and 6.12)

their progeny. However, this assumption is falsified by the row formation process in amphipods. Here cell rows are formed by migrating cells, which nevertheless form squares.

The netlike cytoplasmic connections between the derivatives of the mesoteloblasts are another unusual feature (Patten 1890; Scholtz 1990) (Figure 6.2). As yet, it is not clear whether these connections are indicative of a syncytial post-naupliar mesoderm. Nonetheless, this appears likely and a syncytium might be sensible if the cells are as distantly arranged as the mesodermal segment precursors. Currently, this is only a speculation.

Consecutive division planes in animal cells are mostly oriented perpendicularly to each other (Strome 1993). The movements of the centrosomes serve as explanation for this. Prior to each mitosis, the centrosome duplicates, and the two daughter centrosomes separate to opposite sides of the nucleus. Here they serve as the microtubule-organizing centers of the mitotic spindle. The division plane is perpendicular to the orientation of the spindles and each daughter cell gains one centrosome. Due to their position, they divide at right angles to the previous centrosome division. Hence, deviations from this alternating 90° pattern require different or additional movements of the centrosomes. The cells of the regular ectoderm rows in the germ band of malacostracans divide twice in the same direction with spindles oriented parallel to the longitudinal axis of the embryo. During the subsequent differential cleavages several of these cells continue with this spindle direction, whereas most others show all various spindle orientations, which are nevertheless stereotyped (see earlier). The mechanisms for this unusual cell behavior are not clear. Extrinsic signals or intrinsic cues as well as the cell shape are possible explanations (Gillies and Cabernard 2011). However, this requires further studies.

6.6 HIERARCHY OF GERM LAYERS

There is a long debate as to whether the mesoderm determines the fate of the ectoderm or vice versa. Experimental approaches in insects revealed that the ectoderm induces segmental fates to the mesoderm cells (Bock 1942; Haget 1953). Since gastrulation in pterygote insects follows a largely different path than that of crustaceans (Anderson 1973), one could expect to find different hierarchic relationships between germ layers. Indeed, comparative data of malacostracans might at first sight argue for a leading role of the mesoderm in the differentiation and segment formation of the germ band (Scholtz 1990). Whereas ectoteloblasts underwent some evolutionary changes with respect to differentiation, arrangement, and number (up to total loss), mesoteloblasts behave conservatively in this respect. There are always eight mesoteloblasts. Moreover, as far as is known all mesoteloblasts are formed via a corresponding stereotyped cell division pattern (Dohle et al. 2004; Hunnekuhl and Wolff 2012) (Figure 6.4). However, a more detailed comparison of the segmentation in the ectoderm and mesoderm layers reveals that the differentiation of the ectoderm precedes that of the mesoderm. Hence, the ectoderm might have an impact on the mesoderm rather than the other way round (Scholtz 1990). This inference has been convincingly confirmed by experimental ablations of ectodermal cells, which led to a distorted pattern of the mesoderm, including absence of segmental structures and gene expression (Hannibal et al. 2012). In contrast to this, the early formation and differentiation of the mesoteloblasts and their descendant rows are independent of ectodermal influences (Hannibal et al. 2012).

6.7 SEGMENTATION

In most crustacean embryos, segment formation follows a more or less anteroposterior sequence with more anterior segments showing an advanced development. In these cases, the naupliar region comprising the eye region, the segments of the first and second antennae and of the mandibles develops first, more or less simultaneously. There is a distinct developmental gap between the naupliar segments and the post-naupliar region (Figure 6.1B). This is true for species that hatch as nauplius larva and for most of those with a later hatchling stage such as a zoea larva or with a direct development. In the latter two cases, one speaks of an egg nauplius (Scholtz

2000, but see Jirikowski et al. 2013) (Figure 6.1B). An egg nauplius is not formed in amphipod, tanaidacean, cumacean, and isopod embryos. In these cases the gradient between the naupliar and post-naupliar regions is smooth and a distinct gap between the development of the naupliar and the post-naupliar segments cannot be observed (Scholtz 2000) (Figure 6.1C,D).

6.8   SHORT AND LONG GERM DEVELOPMENT

The stepwise addition of segments in relation to an extension of the germ band is called short germ development (Krause 1939; Scholtz 1992; Patel et al. 1994). In species with direct development or another mode of advanced hatching, the short germ mode of development can be explained by the embryonization of the nauplius and metanauplius stages, including the anamorphic larval development (Figures 6.1 and 6.2). However, some crustacean embryos even adapted a kind of long germ development (Scholtz 1992). In these cases, the germ disc is turned into an extended germ band, largely based on cell rearrangement. Anteroposterior segmentation is delayed with respect to germ band formation and shows a very flat gradient, i.e., the segments are formed more or less simultaneously over the entire germ band. The long germ development is correlated with direct development. Hence, it occurs only in species with an epimorphic developmental mode in which all segments are present at the hatching stage. Examples are cladocerans such a *Daphnia* (Schwartz 1973; Mittmann et al. 2014) and to a certain extent amphipods (Scholtz 1992) (Figure 6.5C, D).

Based on comparative and experimental data it has been suggested that two processes can be discriminated that are involved in germ band differentiation. One is formation and elongation of the germ band in an anterior direction. The other is the anteroposterior propagation of the subdivision of the germ band into serially repeated units (Dohle 1972; Scholtz 1992; Williams et al. 2012; Scholtz and Wolff 2013). Hence, the growth zone does not generate segments but just the competent cellular material, which is eventually segmented. This view is consistent with the idea that short and long germ development do not necessarily imply fundamentally different mechanisms but can be the result of a heterochronic shift between germ band elongation and subsequent segment formation (Scholtz 1992; Scholtz and Wolff 2013).

6.9   SEGMENT MORPHOGENESIS

The first signs of forming segments are the invaginations of the intersegmental furrows, the limb buds, and the early ganglion anlagen (Figure 6.11). Again, the germ bands of malacostracans show the most detailed resolution of these processes. Intersegmental furrows form as transverse, slightly obliquely oriented invaginations. As mentioned earlier, they appear within the descendants of each ectoderm row (Dohle et al. 2004). The findings at the cellular level of malacostracan segmentation have been corroborated at the molecular level. Namely, the expression of segment polarity genes such as *engrailed* and *wingless* concurs with the morphological results on the formation of segmental boundaries. The *engrailed* gene is

FIGURE 6.11   Segment formation and expression of the segment polarity gene, *engrailed* in *Cryptorchestia garbinii* (modified after Dohle et al. 2004). (A) Differential interference contrast image of the whole mount of a germ band showing a thoracic segment during formation. The expression of *engrailed* is made visible with an antibody (brown nuclei). Each transverse stripe including the midline (m) marks the posterior margin of a segment. Posterior to the *engrailed* expression the intersegmental furrows form. The stereotyped arrangement of ectoderm cells is recognizable. (B) Drawing of the preparation of A with an analysis of the mitoses of the differential cleavages and the clonal composition. Lines connect sister cells. (C) SEM image of three segments of the embryonic thorax (th) with false-color blue staining of the cells that express *engrailed*. This technique demonstrates the morphogenesis of the segmental furrows and the limb buds. Lines connect sister cells. The labels mark individually identified cells. ml, midline. Compare with Figures 6.10 and 6.12.

responsible for the establishment and maintenance of a posterior segmental cell fate in *Drosophila melanogaster* and other arthropods such as spiders, millipedes, and crustaceans (e.g. Patel et al. 1988; Hidalgo 1998; Damen 2002; Hughes and Kaufman 2002) (Figures 6.10–6.12). Accordingly, it is expressed in transverse stripes in the posterior region of forming segments. This has also been shown for a number of crustacean representatives such as decapods, cirripedes, copepods, isopods, amphipods, mysids, ostracods, and branchiopods (e.g., Patel et al. 1988; Manzanares et al. 1996; Scholtz and Dohle 1996; Abzhanov and Kaufman 2004; Deutsch et al. 2004; Wolff 2009; Ikuta 2018; Hein et al. 2019). In all these species, *engrailed* is expressed at the posterior margin of forming segments (Figure 6.11). Yet, the knowledge of the cell lineage in the germ band of malacostracans allows for an unmatched cellular resolution of *engrailed* expression. It has been shown that *engrailed* is expressed in the anteriormost cells of the ectodermal genealogical units after the second round of mediolateral divisions just in front of the forming intersegmental furrows (Scholtz and Dohle 1996) (Figures 6.10 and 6.11). Hence, the result of the lineage analyses demonstrating that the anterior cells of a genealogical unit contribute to the posterior region of a morphological segment has been corroborated by the *engrailed* marker. Moreover, this result shows that the genealogical units of malacostracans can be compared to the insect parasegments, i.e., fundamental initial developmental units that subdivide the anteroposterior body axis into repeated structures that are offset compared to the morphological segments (Lawrence 1992).

6.10   CELL LINEAGE AND CELL FATE

It has been suggested that stereotyped cell division patterns during embryonic development indicate cell autonomous differentiation (e.g., Stent 1985; Weisblat and Shankland 1985). According to this view, cells gain the instructions for their further

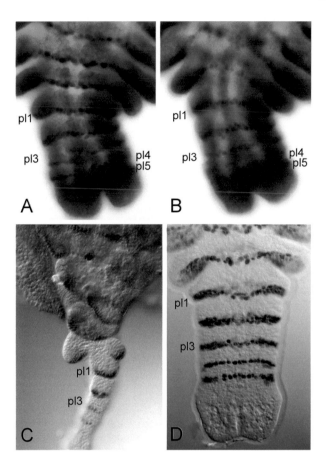

FIGURE 6.12   Cell ablation experiments in the decapod *Cherax destructor*. The dark brown cells express *engrailed*. The loss of cellular material in early germ bands by injection of thin glass electrodes led to a number of different malformations. (A and B) Ventral and dorsal aspects of an embryonic pleon showing spiral segmentation (helicomery). The *engrailed* stripe of the third pleonic segment anlage (pl3) is medially interrupted and somewhat shifted. At the dorsal level the stripe is intact. On the animal's right side, it is connected to the fourth pleonic stripe (pl4), which continues at the dorsal side and is ventrally confluent with the fifth pleonic segment anlage (pl5). Thus, a spiral with two loops is formed. (C) A more dramatic loss of cellular material resulting in a very thin pleon, which nevertheless shows a more or less regular *engrailed* expression. The posterior thoracic segments are completely distorted and fused with a greatly irregular *engrailed* pattern. (D) A normally developed pleon in a corresponding stage for comparison.

fate based on the lineage that led to their existence. Thus, the teloblasts are thought to imprint a specific fate on their descendant cells. This view has been put forward based on experimental studies on clitellate worms (Shankland 1991; Shankland and Savage 1997). Like malacostracans but convergently (see Scholtz 2002), Clitellata differentiate ecto- and mesoteloblasts that give rise to columns of cells with a specific cell fate during the formation of segmental structures (Storey 1989; Shankland and Savage 1997). However, the comparative data of malacostracan germ bands are ambivalent. The fact that the cells of non-ectoteloblastic origin anterior to the first ectoteloblast show a corresponding division pattern and adopt the same fate as the latter suggests that it is more the region rather than the origin that influences cell differentiation (Scholtz and Dohle 1996). By contrast, cell ablation and perturbation experiments in *Cherax destructor* embryos shortly after the formation of the telo-blast rings using fine glass electrodes and the application of the *engrailed* antibody suggest that at least in teloblast descendants, the fate seems determined by their origin. These experiments resulted in extremely malformed germ bands with differ-ent patterns. Two examples are shown here (Figure 6.12). One embryo possesses a very thin posterior region. At the transition between the unaffected and the distorted region *engrailed* expression is clustered and does not show a regular segmental pat-tern. By contrast, the expression pattern in the narrow posterior region is quite regu-lar, comparable to the situation in normal embryos of that stage (Figure 6.12C, D). The other embryo shows a case of spiral segmentation or helicomery (see Morgan 1895; Leśniewska et al. 2009; Scholtz 2017) (Figure 6.12A, B). This phenomenon occurs if in the area of a segment anlage one half is shifted out of phase and thus connected with an adjacent segmental half, creating one or more spiral loops around the body (see Chapter 1). These results indicate that *engrailed* expression is based on cell lineage. The cells maintain *engrailed* expression despite being out of register compared to the wild-type condition and lying in close contact to other engrailed positive cells. The same explanation holds true for the spiral segments. The displace-ment of cells does not alter their fate as *engrailed* positive cells. The expression pattern in the narrow posterior end can be explained by a low number of remaining ectoteloblasts, which nevertheless produce a normal pattern of descendants.

Yet, this conclusion concerns only the normal differentiation processes during the embryonic development of recent species. If the course of evolution is considered, the causal connection of ontogenetic stages can be dramatically altered (Scholtz and Dohle 1996; Scholtz 2008). One example is the loss of ectoteloblasts in amphipods, in which ectoteloblasts are not a necessary prerequisite for the regular row arrange-ment and the stereotyped cleavage pattern in the malacostracan germ band. In other words, the tight relation between teloblasts and the fate of their descendants has been evolutionarily decoupled.

6.11 HOMOLOGY ISSUES

The observed evolutionary alterations of the ectoteloblast patterns raise some inter-esting questions concerning homology. The very detailed correspondence between the row arrangement and the subsequent division patterns of ectodermal cells in the germ band in amphipods and other malacostracans strongly suggests homol-ogy (Dohle and Scholtz 1988; Scholtz and Dohle 1996; Dohle et al. 2004). Yet, this homologous pattern is generated by different processes. In the ancestral case, ecto-teloblasts form the rows by specific asymmetric mitoses in a certain sequence. By contrast, in the lineage leading to amphipods, these ectoteloblasts are lost and the migration of scattered ectodermal cells leads to row formation (see Figures 6.5 and 6.6). This means that homologous structures are not necessarily formed via the same developmental pathway. There are numerous examples of this phenomenon (see Scholtz 2005). Yet, because it is counterintuitive, many discussions of homology stress similar ontogeny as necessary or at least indicative for inferring homology (see Scholtz 2005 for discussion). Moreover, different developmental pathways are often used to negate assumptions of homology of similar structures or to show their convergence based on this argument.

The alteration of the ancestral ring of 19 ectoteloblasts to about 40 in the freshwater crayfish bears on the problem of serial homology (see Scholtz 1993). Are the ectoteloblast rings as such homologous despite the different numbers of cells? Alternatively, are just the inner (ventral) 19 ectoteloblasts of the crayfish homologous to the 19 ectoteloblasts of other malacostracans? But what about the 20 more dorsal ectoteloblasts of crayfish? These form the dorsal side of the posterior thoracic and the pleonic segments. By contrast, in other malacostracans these dorsal parts are formed by the dorsal ectoteloblasts within the ring of 19 ectoteloblast cells (see Figure 6.2). No one questions the homology of the tergites of a lobster (19 ectoteloblasts) and a crayfish (about 40 ectoteloblasts). In any case, the cell division pattern and the fate of the ventral ectoderm cells in crayfish correspond to those of other malacostracans. This means that, again, the question of the homology of teloblasts has to be seen as independent of the question of the homology of the cell division pattern in the germ band.

Yet, this example touches the more general problem of the homology of serial structures, since similar problems occur with the number of cervical vertebrae among tetrapods, the number of segments in arthropod tagmata, or the number of vertebrate teeth. In all these examples there is a defined entity or frame (neck, a tagma such as thorax, mouth dentition) that is considered homologous but the number of serial elements that constitute this entity varies (see Goodrich 1913; Scholtz 1993; Müller and Wagner 1996; Böhmer et al. 2018).

6.12   PERSPECTIVES

Malacostracan crustaceans show a number of specific apomorphic features regarding germ band differentiation and segment formation. These include the formation of teloblasts as a special expression of a growth zone in the strict sense and the stereotyped cell lineage of the teloblasts' progeny. This cell lineage allows a very detailed resolution of cell fate during germ band growth and segment formation. Furthermore, a combination of comparative and experimental studies indicates the role of cell lineages for cell determination and the interaction of germ layers. On the other hand, the reasons how and why the stereotyped cell division patterns evolved in the malacostracan stem lineage and why the teloblasts underwent some evolutionary changes concerning their number and arrangement up to a total loss of ectoteloblasts in amphipods circumstances are still obscure. In any case, the observed changes imply interesting problems related to the interface of ontogeny and homology. To address these developmental and evolutionary questions in even greater detail, studies combining new imaging techniques, molecular approaches, and comparative approaches are desirable.

ACKNOWLEDGMENTS

The research summarized in this review has been supported over the years by grants from the Deutsche Forschungsgemeinschaft and from the Einstein Stiftung. This contribution is a largely modified part of a more comprehensive treatment of crustacean embryology that will be published in the series "The Natural History of the Crustacea."

REFERENCES

Abzhanov, A., and T. C. Kaufman. 2004. Hox genes and tagmatization of higher Crustacea (Malacostraca). In *Evolutionary Developmental Biology of Crustacea*, Ed. G. Scholtz, 43–74. Lisse: A.A. Balkema.

Anderson, D. T. 1967. Larval development and segment formation in the branchiopod crustaceans *Limnadia stanleyana* King (Conchostraca) and *Artemia salina* (L.) (Anostraca). *Austr. J. Zool.* 15: 47–91.

Anderson, D. T. 1969. On the embryology of the cirripede crustacean *Tetraclita rosea* (Krauss), *Tetraclita purpurascens* (Wood), *Chthamalus antennatus* (Darwin) and *Chamaesipho columna* (Spengler) and some considerations of crustacean phylogenetic relationships. *Phil. Trans. R. Soc. London* B 256: 183–235.

Anderson, D. T. 1973. *Embryology and Phylogeny in Annelids and Arthropods*. Oxford: Pergamon Press.

Benesch, R. 1969. Zur Ontogenie und Morphologie von *Artemia salina* L. *Zool. Jahrb. Anat.* 86: 307–458.

Bergh, R. S. 1893. Beiträge zur Embryologie der Crustaceen. I. Zur Bildungsgeschichte des Keimstreifens von *Mysis*. *Zool. Jahrb. Anat.* 6: 491–528.

Bergh, R. S. 1894. Beiträge zur Embryologie der Crustaceen. II. Die Drehung des Keimstreifens und die Stellung des Dorsalorgans bei *Gammarus pulex*. *Zool. Jahrb. Anat.* 7: 235–248.

Bock, E. 1942. Wechselbeziehungen zwischen den Keimblättern bei der Organbildung von *Chrysopa perla* (L.) I. Die Entwicklung des Ektoderms in mesodermdefekten Keimteilen. *Wilhelm Roux's Arch. Entwickl.-Mech.* 141: 159–279.

Böhmer, C., E. Amson, P. Arnold, A. H. van Heteren, and J. A. Nyakatura. 2018. Homeotic transformations reflect departure from the mammalian 'rule of seven' cervical vertebrae in sloths: Inferences on the Hox code and morphological modularity of the mammalian neck. *BMC Evol. Biol.* 18: 84.

Brenneis, G., A. Stollewerk, and G. Scholtz. 2013. Embryonic neurogenesis in *Pseudopallene* sp. (Arthropoda, Pycnogonida) includes two subsequent phases with similarities to different arthropod groups. *EvoDevo* 4: 32.

Browne, W. E., A. L. Price, M. Gerberding, and N. H. Patel. 2005. Stages of embryonic development in the amphipod crustacean, *Parhyale hawaiensis*. *Genesis* 42: 124–149.

Damen, W. G. M. 2002. Parasegmental organization of spider embryos implies that the parasegment is an evolutionary conserved entity in arthropod embryogenesis. *Development* 129: 1239–1250.

Deutsch, J. S., E. Mouchel-Viel, H. È. Quéinnec, and J.-M. Gibert. 2004. Genes, segments, and tagmata in cirripedes. In *Evolutionary Developmental Biology of Crustacea*, Ed. G. Scholtz, 19–42. Lisse: A.A. Balkema.

Dohle, W. 1970. Die Bildung und Differenzierung des postnauplialen Keimstreifs von *Diastylis rathkei* (Crustacea, Cumacea) I. Die Bildung der Teloblasten und ihrer Derivate. *Zeitschr. Morph. Okol.* 67: 307–392.

Dohle, W. 1972. Über die Bildung und Differenzierung des postnauplialen Keimstreifs von *Leptochelia* spec. (Crustacea, Tanaidacea). *Zool. Jahrb. Anat.* 89: 503–566.

Dohle, W. 1976. Die Bildung und Differenzierung des postnauplialen Keimstreifs von *Diastylis rathkei* (Crustacea, Cumacea) II. Die Differenzierung und Musterbildung des Ektoderms. *Zoomorphologie* 84: 235–277.

Dohle, W. 1999. The ancestral cleavage pattern of the clitellates and its phylogenetic deviations. *Hydrobiologia* 402: 267–283.

Dohle, W. 2001. Are the insects terrestrial crustaceans? A discussion of some new facts and arguments and the proposal of the proper name "Tetraconata" for the monophyletic unit Crustacea + Hexapoda. *Ann. Soc. Ent. France* 37: 85–103.

Dohle, W., M. Gerberding, A. Hejnol, and G. Scholtz. 2004. Cell lineage, segment differentiation, and gene expression in crustaceans. In *Evolutionary Developmental Biology of Crustacea*, Ed. G. Scholtz, 95–133. Lisse: A.A. Balkema.

Dohle, W., and G. Scholtz. 1988. Clonal analysis of the crustacean segment - the discordance between genealogical and segmental borders. *Development* 104: 147–160.

Fioroni, P. 1970. Die organogenetische und transitorische Rolle der Vitellophagen in der Darmentwicklung von *Galathea* (Crustacea, Anomura). *Zeitschr. Morph. Okol.* 67: 263–306.

Fischer, A., T. Pabst, and G. Scholtz. 2010. Germ band differentiation in the stomatopod *Gonodactylaceus falcatus* and the origin of the stereotyped cell division pattern in Malacostraca (Crustacea). *Arthropod Struct. Dev.* 39: 411–422.

Fuchs, K. 1914. Die Keimbahnentwicklung von *Cyclops viridis* Jurine. *Zool. Jahrb. Anat.* 38: 103–156.

Gerberding, M. 1997. Germ band formation and early neurogenesis of *Leptodora kindtii* (Cladocera): First evidence for neuroblasts in the entomostracan crustaceans. *Invert. Repr. Dev.* 32: 63–73.

Gillies, T. E., and C. Cabernard. 2011. Cell division orientation in animals. *Curr. Biol.* 21: R599–R609.

Goodrich, E. S. 1913. Metameric segmentation and homology. *Quart. J. Microsc. Sci.* 59: 227.

Haget, A. 1953. Analyse expérimentale des facteurs de la morphogenèse embryonnaire chez le coléoptère *Leptinotarsa*. *Bull. Biol. Fr. Belg.* 87: 123–217.

Hannibal, R. L., A. L. Price, and N. H. Patel. 2012. The functional relationship between ectodermal and mesodermal segmentation in the crustacean, *Parhyale hawaiensis*. *Dev. Biol.* 361: 427–438.

Hartenstein, V., and A. Stollewerk. 2015. The evolution of early neurogenesis. *Dev. Cell* 32: 390–407.

Hein, H., and G. Scholtz. 2018. Larval neurogenesis in the copepod *Tigriopus californicus* (Tetraconata, Multicrustacea). *Dev. Genes Evol.* 228: 119–129.

Hein, H., S. Smyth, X. Altamirano, and G. Scholtz. 2019. Segmentation and limb formation during naupliar development of *Tigriopus californicus* (Copepoda, Harpacticoida). *Arthropod Struct. Dev.* 50: 43–52.

Hejnol, A., R. Schnabel, and G. Scholtz. 2006. A 4D-microscopic analysis of the germ band in the isopod crustacean *Porcellio scaber* (Malacostraca, Peracarida) – developmental and phylogenetic implications. *Dev. Genes Evol.* 216: 755–767.

Hidalgo, A. 1998. Growth and patterning from the *engrailed* interface. *Int. J. Dev. Biol.* 42: 317–324.

Hughes, C. L., and T. C. Kaufman. 2002. Exploring myriapod segmentation: The expression patterns of *even-skipped*, *engrailed*, and *wingless* in a centipede. *Dev. Biol.* 247: 47–61.

Hunnekuhl, V., and C. Wolff. 2012. Reconstruction of cell lineage and spatiotemporal pattern formation of the mesoderm in the amphipod crustacean *Orchestia cavimana*. *Dev. Dyn.* 241: 697–717.

Ikuta, K. 2018. Expression of two *engrailed* genes in the embryo of *Vargula hilgendorfii* (Müller, 1890) (Ostracoda: Myodocopida) *J. Crust. Biol.* 38: 23–26.

Irvine, K. D., and E. Wieschaus. 1994. Cell intercalation during *Drosophila* germband extension and its regulation by pair-rule segmentation genes. *Development* 120: 827–841.

Ito, A., M. N. Aoki, K. Yahata, and H. Wada. 2011. Embryonic development and expression analysis of *Distal-less* in *Caprella scaura* (Crustacea, Amphipoda, Caprellida). *Biol. Bull.* 221: 206–214.

Jirikowski, G. J., S. Richter, and C. Wolff. 2013. Myogenesis of Malacostraca - the "egg-nauplius" concept revisited. *Front. Zool.* 10: 76.

Knoblich, J. A. 2010. Asymmetric cell division: Recent developments and their implications for tumour biology. *Nature Rev. Mol. Cell Biol.* 11: 849–860.

Krause, G. 1939. Die Eitypen der Insekten. *Biol. Zentr.* 59: 495–536.

Kühn, A. 1913. Die Sonderung der Keimesbezirke in der Entwicklung der Sommereier von *Polyphemus pediculus* De Geer. *Zool. Jahrb. Anat.* 35: 243–340.

Kühnemund, E. 1929. Die Entwicklung der Scheitelplatte von *Polyphemus pediculus* De Geer von der Gastrula bis zur Differenzierung der aus ihr hervorgehenden Organe. *Zool. Jahrb. Anat.* 50: 385–432.

Lawrence, P. A. 1992. *The Making of a Fly.* Oxford: Blackwell Scientific Publications.

Leśniewska, M., L. Bonato, A. Minelli, and G. Fusco. 2009. Trunk anomalies in the centipede *Stigmatogaster subterranea* provide insight into late-embryonic segmentation. *Arthropod Struct. Dev.* 38: 417–426.

Manton, S. M. 1928. On the embryology of a mysid crustacean *Hemimysis lamornae*. *Phil. Trans. R. Soc. London B* 216: 363–463.

Manton, S. M. 1934. On the embryology of the crustacean *Nebalia bipes*. *Phil. Trans. R. Soc. London B* 498: 163–238.

Manzanares, M., T. A. Williams, R. Marco, and R. Garesse. 1996. Segmentation in the crustacean *Artemia*: *Engrailed* expression studied with an antibody raised against the *Artemia* protein. *Roux's Arch. Dev. Biol.* 205: 424–431.

McClendon, J. F. 1907. On the development of parasitic copepods. II. *Biol. Bull.* 12: 57–88.

McMurrich, J. P. 1895. Embryology of the isopod Crustacea. *J. Morphol.* 11: 63–154.

Mittmann, B., P. Ungerer, M. Klann, A. Stollewerk, and C. Wolff. 2014. Development and staging of the water flea *Daphnia magna* (Straus, 1820; Cladocera, Daphniidae) based on morphological landmarks. *EvoDevo* 5: 12.

Morgan, T. H. 1895. A study of metamerism. *Quart. J. Microsc. Sci.* 37: 395–476.

Müller, G. B., and G. P. Wagner. 1996. Homology, Hox genes, and developmental integration. *Am. Zool.* 36: 4–13.

Olesen, J., T. Boesgaard, T. M. Iliffe, and L. Watling. 2014. Thermosbaenacea, Spelaeogriphacea, and Mictacea. In *Atlas of Crustacean Larvae*, Eds. J. W. Martin, J. Olesen, and J. T. Høeg, 195–198. Baltimore: Johns Hopkins University Press.

Ooishi, S. 1959. Studies on the teloblasts in the decapod embryo: I. Origin of teloblasts in *Heptacarpus rectirostris* Stimpson. *Embryologia* 4: 283–309.

Ooishi, S. 1960. Studies on the teloblasts in the decapod embryo: II. Origin of teloblasts in *Pagurus samuelis* Stimpson and *Hemigrapsus sanguineus* De Haan. *Embryologia* 5: 270–282.

Patel, N. H., T. B. Kornberg, and C. S. Goodman. 1988. Expression of *engrailed* during segmentation of grasshopper and crayfish. *Development* 107: 201–212.

Patel, N. H., B. G. Condron, and K. Zinn. 1994. Pair-rule expression patterns of *even-skipped* are found in both short and long-germ beetles. *Nature* 367: 429–434.

Patten, W. 1890. On the origin of vertebrates from arachnids. *Quart. J. Micr. Sci.* 31: 317–378.

Ponomarenko, E. A. 2014. *The Embryonic Development of* Elminius modestus *Darwin*, 1854 (*Thecostraca: Cirripedia*). Doctoral Dissertation, Humboldt-Universität zu Berlin. Berlin, Germany.

Price, A. L., and N. H. Patel. 2008. Investigating divergent mechanisms of mesoderm development in arthropods: The expression of *Ph-twist* and *Ph-mef*2 in *Parhyale hawaiensis*. *J. Exp. Zool. B: Mol. Dev. Evol.* 310: 24–40.

Regier, J. C., J. W. Shultz, A. Zwick, A. Hussey, B. Ball, R. Wetzer, J. W. Martin, and C. W. Cunningham. 2010. Arthropod relationships revealed by phylogenomic analysis of nuclear protein-coding sequences. *Nature* 463: 1079–1083.

Reichenbach, H. 1886. Studien zur Entwicklungsgeschichte des Flusskrebses. *Abh. Senckenb. Naturf. Ges.* 14: 1–137.

Richter, S., and G. Scholtz. 2001. Phylogenetic analysis of the Malacostraca (Crustacea). *J. Zool. Syst. Evol. Res.* 39: 113–136.

Samter, M. 1900. Studien zur Entwicklungsgeschichte der *Leptodora hyalina* Lillj. *Z. Wiss. Zool.* 68: 169–260.

Schnabel, R. 1997. Why does a nematode have an invariant lineage? *Sem. Cell Dev. Biol.* 8: 341–349.

Scholl, G. 1963. Embryologische Untersuchungen an Tanaidaceen (*Heterotanais oerstedi* Kröyer). *Zool. Jahrb. Anat.* 80: 500–554.

Scholtz, G. 1984. Untersuchungen zur Bildung und Differenzierung des postnauplialen Keimstreifs von *Neomysis integer* Leach (Crustacea, Malacostraca, Peracarida). *Zool. Jahrb. Anat.* 112: 295–349.

Scholtz, G. 1990. The formation, differentiation and segmentation of the post-naupliar germ band of the amphipod *Gammarus pulex* L. (Crustacea, Malacostraca, Peracarida). *Proc. R. Soc. London B* 239: 163–211.

Scholtz, G. 1992. Cell lineage studies in the crayfish *Cherax destructor* (Crustacea, Decapoda): Germ band formation, segmentation, and early neurogenesis. *Roux's Arch. Dev. Biol.* 202: 36–48.

Scholtz, G. 1993. Teloblasts in decapod embryos: An embryonic character reveals the mono-phyletic origin of freshwater crayfishes (Crustacea, Decapoda). *Zool. Anz.* 230: 45–54.

Scholtz, G. 2000. Evolution of the nauplius stage in malacostracan crustaceans. *J. Zool. Syst. Evol. Res.* 38: 175–187.

Scholtz, G. 2002. The Articulata hypothesis - or what is a segment? *Org. Divers. Evol.* 2: 197–215.

Scholtz, G. 2005. Homology and ontogeny: Pattern and process in comparative developmental biology. *Theory Biosci.* 124: 121–143.

Scholtz, G. 2008. On comparisons and causes in evolutionary developmental biology. In *Evolving Pathways: Key Themes in Evolutionary Developmental Biology*, Eds. A. Minelli, and G. Fusco, 144–159. Cambridge: Cambridge University Press.

Scholtz, G. 2017. Segmentierung. Ein zoologisches Konzept von Serialität. In *Serie und Serialität. Konzepte und Analysen in Gestaltung und Wissenschaft*, Ed. G. Scholtz, 139–166. Berlin: Reimer.

Scholtz, G., and W. Dohle. 1996. Cell lineage and cell fate in crustacean embryos - a comparative approach. *Int. J. Dev. Biol.* 40: 211–220.

Scholtz, G., and C. Wolff. 2002. Cleavage pattern, gastrulation, and germ disc formation of the amphipod crustacean *Orchestia cavimana*. *Contr. Zool.* 71: 9–28.

Scholtz, G., and C. Wolff. 2013. Arthropod embryology: Cleavage and germ band development. In *Arthropod Biology and Evolution*, Eds. A. Minelli, G. Boxshall, and G. Fusco, 63–90. Heidelberg: Springer.

Scholtz, G., A. Abzhanov, F. Alwes, C. Biffis, and J. Pint. 2009. Development, genes, and decapod evolution. In *Decapod Crustacean Phylogenetics*, Eds. J. W. Martin, K. A. Crandall, and D. L. Felder, 31–46. Boca Raton: CRC Press.

Schwartz, V. 1973. *Vergleichende Entwicklungsgeschichte der Tiere*. Stuttgart: Thieme.

Schwentner, M., D. J. Combosch, J. Pakes Nelson, and G. Giribet. 2017. A phylogenomic solution to the origin of insects by resolving crustacean-hexapod relationships. *Curr. Biol.* 27: 1818–1824.

Shankland, M. 1991. Leech segmentation: Cell lineage and the formation of complex body patterns. *Dev. Biol.* 144: 221–231.

Shankland, M., and R. M. Savage. 1997. Annelids, the segmented worms. In *Embryology: Constructing the Organism*, Ed. S. F. Gilbert, and A. M. Raunio, 219–235. Sunderland: Sinauer Associates.

Shiino, S. M. 1942. Studies on the embryology of *Squilla oratoria* de Haan. *Mem. Coll. Sci., Kyoto Imp. Univ.* 17: 77–174.

Stegner, M. E. J., and S. Richter. 2015. Development of the nervous system in Cephalocarida (Crustacea): Early neuronal differentiation and successive patterning. *Zoomorphology* 134: 183–209.

Stent, G. S. 1985. The role of cell lineage in development. *Phil. Trans. R. Soc. London B* 312: 3–19.

Stollewerk, A., M. Weller, and D. Tautz. 2001. Neurogenesis in the spider *Cupiennius salei*. *Development* 128: 2673–2688.

Storey, K. G. 1989. Cell lineage and pattern formation in the earth-worm embryo. *Development* 107: 519–532.

Strome, S. 1993. Determination of cleavage planes. *Cell* 72: 3–6.

Ungerer, P., and G. Scholtz. 2008. Filling the gap between neuroblasts and identified neurons in crustaceans adds new support for Tetraconata. *Proc. R. Soc. B* 275: 369–376.

Vargas-Vila, M. A., R. L. Hannibal, R. J. Parchem, P. Z. Liu, and N. H. Patel. 2010. A prominent requirement for *single-minded* and the ventral midline in patterning the dorsoventral axis of the crustacean *Parhyale hawaiensis*. *Development* 137: 3469–3476.

von Reumont, B. M., R. A. Jenner, M. A. Wills, E. Dell'Ampio, G. Pass, I. Ebersberger, B. Meyer, S. Koenemann, T. M. Iliffe, A. Stamatakis, O. Niehuis, K. Meusemann, and B. Misof. 2012. Pancrustacean phylogeny in the light of new phylogenomic data: Support for Remipedia as the possible sister group of Hexapoda. *Mol. Biol. Evol.* 29: 1031–1045.

Weisblat, D. A., and M. Shankland. 1985. Cell lineage and segmentation in the leech. *Phil. Trans. R. Soc. London B* 312: 39–56.

Weygoldt, P. 1958. Die Embryonalentwicklung des Amphipoden *Gammarus pulex pulex* L. *Zool. Jahrb. Anat.* 77: 51–110.

Williams, T. A., B. Blachuta, T. A. Hegna, and L. M. Nagy. 2012. Decoupling elongation and segmentation: Notch involvement in anostracan crustacean segmentation. *Evol. Dev.* 14: 372–382.

Wolff, C. 2009. The embryonic development of the malacostracan crustacean *Porcellio scaber* (Isopoda, Oniscidea). *Dev. Genes Evol.* 219: 545–564.

Wolff, C., and G. Scholtz. 2002. Cell lineage, axis formation, and the origin of germ layers in the amphipod crustacean *Orchestia cavimana. Dev. Biol.* 250: 44–58.

Wolff, C., and G. Scholtz. 2006. Cell lineage analysis of the mandibular segment of the amphipod *Orchestia cavimana* reveals that the crustacean paragnaths are sternal outgrowths and not limbs. *Front. Zool.* 3: 19.

Zilch, R. 1974. Die Embryonalentwicklung von *Thermosbaena mirabilis* Monod (Crustacea, Malacostraca, Pancarida). *Zool. Jahrb. Anat.* 93: 462–576.

Zrzavý, J., and P. Štys. 1997. The basic body plan of arthropods: Insights from evolutionary morphology and developmental biology. *J. Evol. Biol.* 10: 353–367.

7 Segmentation in Leeches

David A. Weisblat and Christopher J. Winchell

CONTENTS

7.1 INTRODUCTION

As discussed in Chapter 4, the mixed distribution of non-metameric, metameric, and segmented body plans across what molecular phylogenetics defines as the three superphyla of bilaterally symmetric animals (Deuterostomia, Ecdysozoa, Lophotrochozoa/Spiralia) evidences a much greater degree of evolutionary plasticity in developmental processes than was thought to be the case when phylogenetic trees were constructed using morphological comparisons, as exemplified in the Articulata hypothesis. To more fully understand the evident evolutionary plasticity of developmental pathways leading to the gain or loss of segmentation and metamerism, it is necessary to compare axial growth and patterning among diverse models, including both segmented and unsegmented taxa representing different branches of the phylogenetic tree.

In the 1960s and 1970s, the nervous system of the medicinal leech *Hirudo* emerged as a powerful system in which to study how nervous systems function at the level of individually identified cells. This work was initiated by Stephen Kuffler and John Nicholls (Kuffler and Potter 1964; Nicholls and Baylor 1968), building on Retzius' neuroanatomical descriptions from the 19th century, and was greatly expanded by John Nicholls, Gunther Stent, and their many disciples (Muller et al. 1981). Modern studies of leech development stem largely from the decision by Gunther Stent in the mid-1970s to use these animals as models for studying neural development of leeches. He was guided in this decision contemporaneously by Roy Sawyer (Sawyer 1986; Weisblat et al. 1978) and Juan Fernández (Fernández and Stent 1980), and also by Whitman's pioneering 19th century work on embryonic cell lineages (1878, 1887). In the decades since Stent's decision, *Helobdella* has become among the best known models for studying spiralian development; in this chapter, we summarize our current understanding of just one aspect of *Helobdella* development, i.e., segmentation.

In this chapter, we present glossiphoniid leeches, more specifically species from the genus *Helobdella*, as useful models for studying segmentation among annelids, within the relatively understudied superphylum Lophotrochozoa/Spiralia. Breeding populations of *H. austinensis* (Kutschera et al. 2013), the most commonly used species at this point, are readily maintained in lab culture and produce year-round supplies of relatively large and hardy embryos that undergo stereotyped and unequal cleavages. Thus, as will be described in detail later, the early embryo contains large individually identified, experimentally accessible, lineage-restricted stem cells from which segmental mesoderm and ectoderm arise in anteroposterior progression (Weisblat and Kuo 2009, 2014). In contrast, the midgut, which also exhibits metameric organization, arises in parallel from a complex merger of multiple embryonic lineages and involves the initial formation and later cellularization of a syncytial yolk cell (SYC).

In addition to their experimental tractability, leeches are of interest because genomic and phylogenetic analyses of recent years have highlighted them as a group whose genome is dramatically more extensively rearranged than almost any other known animal group, relative to that inferred for the bilaterian ancestor, as judged by the loss of macrosynteny (Simakov et al. 2013). If evo-devo has a central dogma, it is that genomic changes give rise to changes in developmental mechanisms and processes, and that changes in development underlie evolutionary changes in phenotype (e.g., morphology). Thus, we can envision species evolving by proceeding stochastically through a very highly dimensional "evo-devo space" made up of all the possible combinations of genome, development, and phenotype. For each species, its evolutionary trajectory through evo-devo space is a constrained random walk, in which each step is subject to the constraints that the species remain reproductively viable at each step, and that there are different probabilities associated with different types of genomic change.

From this, it would follow that increasing the range of possible changes that are "allowed" for the evolving genome in leeches and their allies would increase their ability to explore the surrounding evo-devo space. Thus, while much of evo-devo research is essentially *retrospective*, for instance, comparing segmentation mechanisms across taxa with the goal of understanding how segmentation proceeded in the last common segmented ancestor at each point where this feature originated, leeches may help us to expand our appreciation of the range of possible developmental mechanisms consistent with a segmented body plan.

Traditionally, the phylum Annelida was thought to comprise three classes of (exclusively) segmented worms: polychaetes, oligochaetes, and leeches. Molecular phylogenies reveal that polychaetes are paraphyletic with respect to clitellates (comprising oligochaetes, leeches, and their allies), and that oligochaetes are paraphyletic with respect to leeches. In addition, the phylum Annelida now includes unsegmented taxa that were previously classified as separate phyla (McHugh 1997; Struck et al. 2007, 2011). This chapter summarizes the cellular processes involved in leech segmentation, which appears to be representative of the process in clitellate annelids, a monophyletic group comprising the oligochaetes and euhirudinids (the true leeches), plus two transitional groups, acanthobdellids and branchiobdellids.

7.2 SEGMENTATION IN THE *HELOBDELLA* BODY PLAN

The leech body plan (Figure 7.1) contains 32 segments, distributed along three tagmata: the head of the animal, including the anterior sucker, which includes four rostral segments, designated R1–R4, respectively, plus non-segmental, prostomial structures; the midbody region contains 21 segments, designated M1–M21; and finally the caudal region, associated with the rear sucker contains seven segments, designated C1–C7. The leech body is paradigmatically segmented, featuring spatially coherent metameric structures in derivatives of all three germ layers: ectoderm, mesoderm, and endoderm (Sawyer 1986). Metameric mesodermal derivatives include muscles, nephridia, and even a subset of mesodermally derived neurons. Metameric ectodermal derivatives include the circumferential body wall divisions (annuli) for which the phylum is named, peripheral sensory structures called sensillae, some types of pigment cells, and, most prominently, the ganglia of the ventral nerve cord. Anatomical studies from the 19th century and a large body of neurobiological studies beginning in the 1960s reveals that neurons in the segmental ganglia are uniquely identifiable (in terms of their physiology and morphology) and largely conserved from segment to segment, even across species (Retzius 1891; Muller et al. 1981). Finally, metameric organization of endoderm-derived structures is evidenced by the iterated lobes of the crop (aka cecae in anterior midgut) and of the intestine (posterior midgut). Structures that are not segmented in *Helobdella* or most leeches include the dorsal anterior ganglion of the nerve cord, the foregut (proboscis and esophagus), salivary glands, ovary, and hindgut (rectum and anus).

7.3 BOUNDARY-DRIVEN VERSUS LINEAGE-DRIVEN SEGMENTATION

Most organisms used to study segmentation are drawn from either arthropods (superphylum Ecdysozoa) or vertebrates (superphylum Deuterostomia). In these

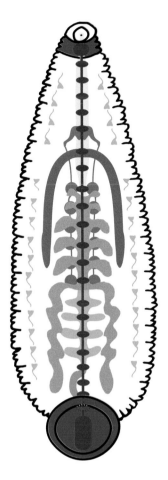

FIGURE 7.1 Segmentation of mesoderm, ectoderm, and endoderm in the *Helobdella* body plan. This schematic shows a ventral view with anterior up. The rostral and caudal tagmata contain four and seven fused segments (dark blue), respectively, with the latter forming the rear sucker. The midbody tagma consists of 21 separate segments. Segmental muscles are illustrated in Figure 7.4 and are not shown here. Ectodermal segmentation is indicated by: 3 epidermal annuli in each segment (small black bulges, flanked by deeper constrictions of the body wall to mark segment boundaries); bilaterally symmetric ganglia of the ventral nerve cord, linked by connective nerves (small red ovals and red lines, respectively; due to segment fusion, the rostral and caudal tagmata possess larger ganglia (large red ovals) than midbody segments; the nerve cord in midbody segment 21 (dashed oval and line) is obscured by the posterior sucker; segmental components of the peripheral nervous system are not shown. Mesodermal segmentation is indicated by the male genital organs and multiple paired testisacs (orange), which develop in midbody segments 5 and 8–13, respectively; the female reproductive system (purple), originating in midbody segment 6, with ovaries extending posteriorly; the excretory system of paired metanephridia (green) in midbody segments 2–5 and 8–18 (these two sets have reversed orientation relative to one another, filtering coelomic fluid anteriorly and posteriorly, respectively). Endodermal segmentation is indicated by ceca of the midgut, which form distinct segmental evaginations in the crop (taupe) and the intestine (light blue). Non-segmented gut structures include the proboscis and esophagus (yellow) and the rectum (pink).

models, segmentation entails the formation of boundaries that divide fields of (usu-ally) mesodermal or ectodermal cells into metameres. As detailed in Chapter 2, these metameres may arise sequentially (as is the case with vertebrates and short germ insects) or simultaneously (as in long germ insects) along the anterior–pos-terior (A–P) axis, and may correspond to segmental, parasegmental, or double-segmental primordia. The actual patterns of cell division leading to the formation of morphological pattern elements within the units are often variable (the vari-able and irregularly shaped clones of cells within *Drosophila* compartments, for example), and cell clones arising within one metamere are restricted from crossing boundaries separating them from adjacent metameres. This is analogous to creat-ing a repeating pattern in a row of bushes by trimming them so that their branches do not intermingle, without regard for the branching patterns of the individual bushes. We refer to these processes as boundary-driven segmentation mechanisms (Figure 7.2A).

Segmentation by boundary-driven mechanisms is so widespread among the com-monly studied animal models that one may be forgiven in assuming that this is the

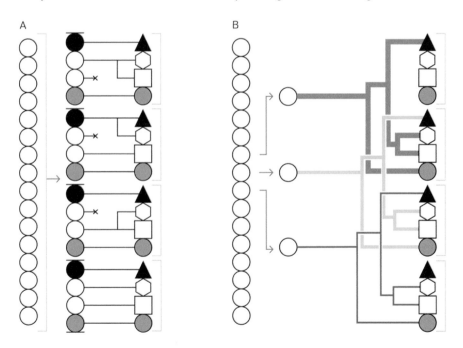

FIGURE 7.2 Boundary-driven and lineage-driven segmentation. In boundary-driven seg-mentation (left), as exemplified in arthropods and vertebrates, segmental boundaries are imposed on initially unspecified regions of cells (syncytial blastoderm or presomitic meso-derm). Segmentally iterated structures arise within the segmental boundaries despite variable patterns of cell division, in part due to restrictions on the ability of cells to cross the bound-aries. In lineage-driven segmentation (right), as exemplified in leeches and other clitellate annelids, stereotyped patterns of cell division among longitudinally arrayed precursors cells lead to segmentally repeating structures even if their clones interdigitate across morphologi-cal segment boundaries.

only way to generate metameres and segments. In fact, however, another process, which we call lineage-driven segmentation, has been described for clitellate annelids and also for malacostracan crustaceans (see Chapter 6). Returning to the analogy of the row of bushes, if the bushes are carefully pruned to achieve the same branching pattern for each bush, then the row of bushes will present a repeating pattern even if the branches of adjacent bushes intermingle (Figure 7.2B). In *Helobdella*, the metameric structures associated with segmental mesoderm and ectoderm arise in a lineage-driven process as will be described later.

7.4 AN AXIAL POSTERIOR GROWTH ZONE (PGZ) ORIGINATES FROM THE D QUADRANT IN *HELOBDELLA* AND OTHER CLITELLATE ANNELIDS

Embryos of annelids, mollusks, and most other lophotrochozoan taxa undergo a conserved pattern of early cell divisions called spiral cleavage, in which the third cleavage is obliquely oriented and unequal. This results in eight-cell embryos composed of quartets of animal micromeres and vegetal macromeres, respectively, that are offset from one another by roughly 45 degrees around the animal–vegetal axis. In many annelids (Dohle 1999), the first and second mitoses are unequal, so that the four-cell stage already contains a uniquely determined D quadrant blastomere, indicating a heterochronic shift of the D quadrant specification process relative to the presumed equal cleaving spiralian ancestor (Freeman and Lundelius 1992). The D quadrant ultimately gives rise to much of the mesodermal and ectodermal tissues. In Clitellata, the unequal first and second cleavages segregate yolk-deficient domains of cytoplasm to the D macromere. In leeches, the yolk-deficient cytoplasm is enriched in mitochondria, maternal mRNAs, and ribosomes (Fernández and Olea 1982; Fernández et al. 1990, 1998; Holton et al. 1989, 1994) and it is this material, rather than the difference in cell size that is critical for specifying the D quadrant in *Helobdella* (Astrow et al. 1987; Nelson and Weisblat 1991, 1992).

C.O. Whitman's pioneering cell lineage studies in the 19th century accurately established the outline and many of the details of the early development of glossiphoniid leeches (Whitman 1878; see also Sandig and Dohle 1988; Bissen and Weisblat 1989). In *Helobdella* and other clitellate embryos (Figure 7.3), the D macromere undergoes an arcane and yet remarkably well conserved series of mostly unequal divisions, generating a mix of smaller, yolk-free cells (*micromeres*) and five bilateral pairs of larger, yolk-rich cells (M, N, O/P, O/P, and Q *teloblasts*; Figure 7.3). The D macromere generates a total of 16 micromeres, and the other three quadrants each contribute three more. These 25 micromeres gives rise to exclusively non-segmental tissues, chiefly the proboscis, the dorsal anterior ganglion, and the squamous epithelium of a temporary integument that spread to cover the embryo during gastrulation. In contrast, as described later, the teloblasts function as individually identifiable, lineage-restricted stem cells. They constitute a posterior growth zone (PGZ) from which segmental mesoderm and ectoderm arise in anteroposterior progression.

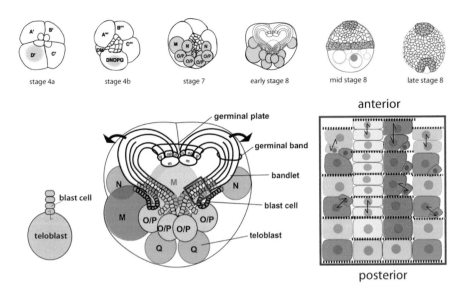

FIGURE 7.3 Formation of the *Helobdella* posterior growth zone (PGZ). Top: Selected stages of *Helobdella* development. Bottom left: Each teloblast undergoes repeated, highly unequal divisions, forming a column of segmental founder cells (blast cells) for that lineage. Bottom center: Detailed depiction of an early stage 8 embryo, showing that the five ipsilateral columns of teloblast-derived bandlets merge into a parallel array (germinal band) at the surface of the embryo in the vicinity of the animal pole, with the four ectodermal bandlets lying over the mesodermal bandlet (not visible here) on each side of the embryo. Teloblasts and their proximal blast cells constitute the PGZ. For clarity, the micromere-derived epithelium that covers the germinal bands and the region behind them is not illustrated for early stage 8, but is depicted by small irregular contours for mid and late stage 8. Bottom right: Schematic representation of stereotyped blast cell divisions and other events in the early germinal band (red boxes). Grandparental N and Q lineages are depicted by alternating dark and light cells in the blue (n) and green (q) bandlets. Fate specification in the O/P equivalence group is represented by the transition from orange/yellow (equipotent o/p) cells at the posterior end of the germinal band to distinct yellow (P lineage) and orange (O lineage) cells more anteriorly. Horizontal dotted lines indicate that individual segmental complements in the O and P lineages are not yet in register with those in the N and Q lineages. See text for details.

7.5 THE LEECH PGZ PROVIDES A HIGHLY SIMPLIFIED AND EXPERIMENTALLY ACCESSIBLE EXAMPLE OF AXIAL GROWTH AND PATTERNING

During segmentation, each teloblast undergoes a series of several dozen highly unequal stem cell divisions (Figure 7.3). The smaller daughter cell at each division is called a blast cell and is designated with the lowercase letter corresponding to the teloblast of origin. The blast cells exhibit significantly prolonged cell cycle durations (12–44 hours to their first mitosis, depending on the lineage), while the teloblasts divide with cell cycle times of 75 to 90 minutes (depending on the temperature and the species). Moreover, in each lineage, each newly born blast cell maintains contact with the blast cell formed in the previous division. Thus, the progeny of each telo-blast arise in a remarkably well-ordered column of cells (called a bandlet), within which the blast cells are strictly arranged according to birth order (Figure 7.3). The five bandlets on each side of the embryo coalesce ipsilaterally, forming an array of parallel bandlets called the germinal band (Figure 7.3). The germinal bands and the region between them are covered by a micromere-derived squamous epithelium.

As teloblast divisions add cells to the posterior ends of the germinal bands, the germinal bands lengthen and move over the surface of the embryo, eventually coalescing to form a bilaterally symmetric sheet of cells called the germinal plate (Figures 7.3 and 7.4). Concomitantly, the micromere-derived epithelium spreads to cover the surface of the embryo, providing it with a temporary protective covering. The germinal plate forms in anteroposterior progression (i.e., starting with the cells distal to the teloblasts) along the ventral midline of the embryo (Figures 7.3 and 7.4).

In the *Helobdella* embryo, the germinal plate initially wraps most of the way around the spherical, yolk-rich embryo, so that its anterior and posterior end are close to one another. During subsequent stages of development, cell proliferation and differentiation of segmental tissues within the germinal plate cause it to gradu-ally expand dorsolaterally and straighten out, eventually meeting along the dorsal midline of the embryo and completing the formation of the tubular leech body plan. During this process, the yolky core elongates and is converted to midgut.

The bandlets within the left and right germinal bands are arrayed in a stereotyped manner—the four ectodermal bandlets lie superficial to the mesodermal bandlet, and in alphabetic order (now designated n, o, p, and q) so that the n bandlets are at the leading edge of the germinal bands, and come into direct apposition along the ventral midline as the germinal bands coalesce to form the germinal plate (Figures 7.3 and 7.4). As cells proliferate and differentiate during later stages of development, the lateral edges of the germinal plate move dorsally around the yolk cells, eventually zippering together along the dorsal midline to form the tubular body of the adult leech. This broad understanding of the development of glossiphoniid leeches was provided by C.O. Whitman in the 19th century (Whitman 1878). Roughly one cen-tury later, the development and application of cell lineage tracing by microinjection of precursor cells with marker enzymes or fluorophores (Weisblat et al. 1978, 1980; Gimlich and Braun 1985) opened the door to a more detailed understanding of leech development in general and segmentation in particular.

Injecting individual teloblasts with lineage tracers revealed that segmental meso-derm and ectoderm arise as the composite of five segmentally repeating, bilaterally symmetric patterns of cells. As will be described in Section 7.8, these segmentally iterated patterns are *not* simply blast cell clones, and are therefore designated as M, N, O, P, and Q "kinship groups," respectively, defined as all the cells within one segment that arise from the corresponding teloblast (the relationship between the ambiguously named O/P teloblasts and the distinct O and P kinship groups will be detailed in Section 7.9).

Axial growth from the posterior end of the embryo, accompanied by patterning in an anteroposterior progression are common traits of embryos in all three of the bilaterian superphyla. Thus, it seems likely that this feature may have been present in the bilaterian ancestor. In most such embryos, however, as cells exit the posterior zone to generate segments or other axially patterned tissues, they are replaced by a poorly characterized and perhaps variable mixture of cell proliferation, including stem cell processes, and cell movements, i.e., influx of cells from other parts of the embryos (see Chapter 2). Uncertainty as to whether the posterior zone is replenished by growth (cell division) or cell movements has led to this region being called a segment addition zone (Janssen et al. 2010).

By contrast, in leeches and certain crustacean embryos (see Chapter 6), the more precise term of PGZ is appropriate because the segmental founder cells arise entirely via stem cell divisions of a small number of relatively large and individually identifiable cells. The presence of teloblastic PGZs in leeches and crustaceans was at one point taken as evidence for the Articulata hypothesis, i.e., a shared origin of their segmentation processes in a common ancestor of annelids and arthropods (Giribet 2003). In light of current molecular phylogenies, and with a better understanding of the limited similarities and significant differences between leeches and crustaceans PGZs, we now view these similarities as further evidence for evolutionary plasticity of development.

7.6 EVOLUTIONARY ANTIQUITY OF THE M TELOBLASTS

As described earlier (Figure 7.3), the PGZ in *Helobdella* and most clitellates comprises exactly ten teloblasts, one pair of mesoteloblasts (M), and four pairs of ectoteloblasts (designated N, O/P, O/P, and Q in *Helobdella*). The M teloblasts are of particular interest evolutionarily, in that they arise from the zygote by an embryonic cell lineage that is conserved across a wide diversity of spiralian taxa, reflecting a pattern of cell divisions and cell fate assignments that was already established as long as 600 million years ago (Lambert 2008; Peterson et al. 2008). Specifically, the M teloblasts arise from the bilaterally symmetric division of a precursor born at the equivalent cell division (sixth embryonic cleavage within the D quadrant lineage). This precursor is named cell 4d in the standard nomenclature used for describing spiralian development; the corresponding cell is designated blastomere DM″ in a specific nomenclature used for *Helobdella*. Notwithstanding the evolutionary antiquity and extensive conservation of the association between the identity of cell 4d and its fate as precursor of bilateral mesoderm, scientists have discovered a fascinating instance of evolutionary plasticity here as well. In the polychaete annelid *Capitella teleta*, the left and right longitudinal bands of mesoderm arise from third quartet micromeres in the D and C quadrants, respectively, rather than from the left and right progeny of the 4d cell as in most spiralians (Meyer et al. 2010).

In spiralian embryos generally, the bilateral pair of M teloblasts undergo a stereotyped series of unequal cell divisions that typically contribute individually identifiable sets of mesodermal and germline precursor cells to their respective embryos in a fixed sequence. Despite the apparent homology among spiralian M teloblasts, the conspicuous functioning of paired teloblasts as stem cells in posterior growth is known only from clitellate annelids. Thus, in *Helobdella* and other clitellates, these unequal divisions are more highly asymmetric in size, more numerous, and more regular than in other spiralian taxa. The first born m blast cells in the left and right mesodermal bandlets arise in contact with one another and this contact is maintained for some time as the distal ends of the bandlets move to the surface of the embryo. As the ectodermal bandlets arise, they seem to move along the underlying m bandlets on each side, thus bringing the distal ends of the left and right germinal bands into contact at the future anterior end of the nascent germinal plate.

7.7 KINSHIP GROUPS ARE WELL CONSERVED WITHIN EACH LINEAGE, BUT HETEROGENEOUS IN TERMS OF CELL-TYPE COMPOSITION AND SPATIAL DISTRIBUTION

Among the four ectodermal lineages (N, O, P, and Q), the N kinship group (defined in Section 7.5) is primarily composed of neurons in the ventral ganglia, while the Q kinship group constitutes primarily dorsal epidermis, while the O and P kinship groups contribute primarily ventrolateral and dorsolateral ectoderm (Figure 7.4), respectively. There is extensive mediolateral intercalation among the kinship groups, however. Even the dorsalmost (Q) kinship group contributes a small set of neurons to the ventral ganglion (Weisblat et al. 1984). These cells arise by a small set of precursors that separate from the main Q lineage and migrate medially within the germinal plate (Torrence and Stuart 1986). For the giant glossiphoniid leech species, *Haementeria ghilianii*, it has been possible to combine lineage tracing with neurophysiological characterization of individual neurons in the segmental ganglia of late-stage embryos. This approach has confirmed that individually identified neurons arise from specific kinship groups (Kramer and Weisblat 1985), and these lineage relationships appear to be conserved at least across glossiphoniid species. Moreover, the distinct patterns of teloblast-derived ectodermal kinship groups can also be recognized in oligochaetes (Goto et al. 1999a, 1999b; Arai et al. 2001; Storey 1989a, 1989b), indicating that many of the cellular details of segmentation via a teloblastic growth are conserved from an ancestral clitellate annelid.

7.8 KINSHIP GROUPS ARE NOT CLONES

What is the spatial relationship between a teloblast's kinship group (i.e., all the descendants of that teloblast that occur within one morphologically defined segment) and the blast cell clones in that teloblast lineage (i.e., all the descendants of individual blast cells)? To characterize the segmental contributions of individual blast cells, the blast cells themselves can be injected with lineage tracers in *Helobdella* or related species (M. Leviten, unpublished observations; Gleizer and Stent 1993; Shankland 1987a; Shain et al. 1998; Kuo and Shankland 2004a), albeit only with considerable difficulty. An alternative method for following the fates of individual blast cells is to inject the same teloblast during successive cell cycles with two different tracers, thereby labeling one individual blast cell uniquely (Zackson 1982; Gline et al. 2011; Kuo and Shankland 2004a). A third way to examine the distribution of individual blast cell clones is to inject a teloblast with tracer after it has already produced some blast cells, and then examine the distribution of progeny at the boundary between the last unlabeled and first labeled blast cells in the lineage (Weisblat and Shankland 1985).

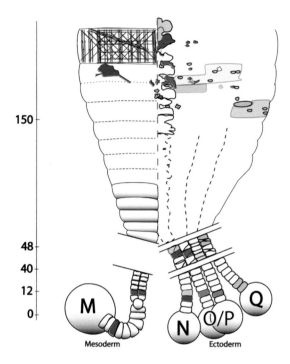

FIGURE 7.4   Stereotyped spatial distribution of blast cell clones in parental and grandparental stem cell lineages. Schematic representation of the *Helobdella* PGZ (lower portion of drawing) and germinal plate (upper portion), depicting the anteroposterior progression of segmentation in terms of individual blast cell clones. For simplicity, only the mesodermal (M) lineage is shown in the left side of the figure and only the four ectodermal (N, O, P, Q) lineages are shown in the right side of the figure; the ventral midline is indicated by the dotted line at the center of the drawing, while the edges correspond to the dorsal midline. In the mesodermal (M teloblast) lineage (pink), individual blast cells initiate their stereotyped pattern of divisions prior to entering the germinal band. Each cell makes one segment's worth of definitive progeny, including body wall muscles, nephridia, and a few ganglionic neurons, but the final clone of cells arising from a single m blast cell is distributed over three segments. The O and P lineages arise from initially equipotent O/P teloblasts and blast cells, which assume distinct O (orange) and P (yellow) fates within the germinal bands (see Figure 7.3). Individual o and p blast cell clones make one segment's worth of definitive progeny for ventrolateral and dorsolateral ectoderm, respectively, including specific sets of ganglionic neurons, epidermis, and peripheral neurons, but the final clones of cells are distributed over two morphological segments. The N (blue) and Q (green) lineages also contribute specific sets of ganglionic neurons, epidermis, and peripheral neurons to ventral and dorsal ectoderm, respectively, but in each of these lineages, two blast cell clones are required to make a single segment's worth of definitive progeny.

These overlapping and complementary approaches have revealed that in contrast with arthropod and vertebrate models, leech segmentation is a lineage-driven process and not a boundary-driven one. Specifically, for the M, O, and P lineages, serially homologous blast cell clones within the same lineage interdigitate longitudinally. Thus, the differentiated cell types (definitive progeny) that make up one blast cell's mature clone are distributed across more than one morphologically defined segment, and the set of cells in each segment that comprise the M, O, or P kinship group arise from more than one blast cell clone (Figures 7.2 and 7.4).

For example, the O-lineage neurons on the left side of each segmental ganglion arise from two different o bandlet blast cells (Figure 7.4) (Weisblat and Shankland 1985). For the mesodermal (M) lineage, the situation is even more extreme—the progeny of individual m blast cell clones are distributed longitudinally over three consecutive segments (Figure 7.4). Overall, it is difficult if not impossible to reconcile these results with the compartment model of segmentation that has been applied to arthropod taxa, but they are fully consistent with the notion of lineage-driven segmentation.

Obviously this conclusion breaks down at the anterior and posterior ends of the animal, where there is no "next" segment to which cells can migrate and no "next blast cell" from which they migrate. The full details of how blast cell lineages are altered in these regions remains to be determined, but careful application of lineage tracing techniques has provided partial answers. For example, in the early mesodermal (M) lineage, the first six m blast cells (designated em1 through em6, respectively) contribute progeny to a variety of non-segmental tissues (Gline et al. 2011). Of these six cells, em1 through em4 make no contributions at all to segmental mesoderm; em5 and em6, however, each generate a clone that is a hybrid of segmental and non-segmental progeny. Specifically, they contribute progeny to the first two mesodermal segments, segments R1 and R2, that in midbody segments would be produced by anterior m blast cells.

7.9   THE O–P EQUIVALENCE GROUP

Another interesting result from a long series of experiments in *Helobdella* is that the distinct O and P lineages arise from what are initially equipotent blast cells, produced by two O/P teloblasts on each side of the embryo (Weisblat and Blair 1984; Zackson 1984; Shankland and Weisblat 1984; Ho and Weisblat 1987; Shankland 1987a, 1987b). The default condition for these equipotent o/p blast cells is to assume the ventrolateral (O lineage) fate, which is normally seen in the bandlet lying next to the ventral, n bandlet. In normal development, the other column of o/p blast cells is induced to follow the dorsolateral (P lineage) fate by signaling from the adjacent q blast cells (Huang and Weisblat 1996). In at least one *Helobdella* species, the P lineage can also be specified by a redundant signal from the M lineage (Kuo and Shankland 2004a). The Q-derived dorsalization of the P lineage involves BMP-mediated signaling, in common with dorsalization processes in other systems (Kuo and Weisblat 2011). The process in *Helobdella* differs significantly from that in commonly studied vertebrate and insect models, however: (1) BMP5-8 rather than BMP2/4 is the critical signaling ligand; (2) Gremlin, rather than Chordin, acts as the antagonist, and is expressed dorsolaterally rather than ventrally; (3) the signaling appears to be mediated by cell–cell contacts, rather than diffusion of the ligand (Kuo and Weisblat 2011; Tao et al. in preparation).

The anterior components of the O and P lineages also deviate from their midbody counterparts, but in an entirely different manner than seen for the M lineage. Intriguingly, the pair of equipotent O/P teloblasts on each side of the embryo arises as sister cells from the roughly equal division of an OP proteloblast. But before its equal division, the OP proteloblast itself undergoes four highly asymmetric teloblastic divisions (Sandig and Dohle 1988; Bissen and Weisblat 1989). The op blast cells contribute four segment's worth of lateral ectoderm to segments R1 through R4. Lineage analysis has confirmed that these op blast cell clones contribute segmental sets of epidermal cells plus central and peripheral neurons that are roughly equivalent to the canonical O and P kinship group progeny of the midbody segments (Kuo and Shankland 2004b). Curiously, however, the initial divisions of the op blast cells do not immediately create progenitors of canonical O and P kinship group cells; even when the op clone comprises 4 cells, the anteriormost daughter cell in the clone (cell op.aa) is fated to contribute a mix of nominal O and P progeny. These observations support previous work showing that highly specific cellular phenotypes can arise by alternative pathways (Shankland 1987b). In addition, there is no evidence that BMP signaling plays a role in cell fate specification with the lateral ectoderm of the four rostral segments (Kuo et al. 2012).

7.10 GRANDPARENTAL STEM CELL LINEAGES

Analyzing the segmental contributions of individual blast cells in the N and Q lineages, as described earlier for the M, O, and P lineages, also yielded surprising results (Figure 7.4). Namely, in these lineages, the respective blast cells adopt two different fates in exact alternation. In addition to generating distinct complements of final, differentiated progeny to the respective N and Q kinship groups, the different blast cell types exhibit characteristic differences in the timing and asymmetry of their subsequent mitoses (Zackson 1984; Bissen and Weisblat 1987, 1989; Zhang and Weisblat 2005). These distinct blast cell identities are designated f and s, based on differences in the timing of their initial mitosis. Cells nf and qf have shorter cell cycle times than cells ns and qs, respectively.

One possible explanation for the fact that n (and q) blast cells arise as two distinct cell types in exact alternation is that the N (and Q) teloblasts from which these cells arise switch back and forth with each division, in terms of the two different cell types produced in these lineages. Thus, if we define canonical stem cell divisions as being "parental" in nature because one daughter cell at each division recapitulates the fate of the parent cells, then the N and Q teloblasts are undergoing "grandparental divisions." The other possible explanation is that the differences between f and s blast cell types are imposed only after the cells arise from the N and Q teloblasts, through precise patterning processes of one sort or another. Unfortunately, currently available evidence (summarized later) is not conclusive in favor of either the intrinsic or extrinsic models for specifying the f and s fates in the N and Q lineages. Therefore, we designate these intriguing phenomena as grandparental stem cell lineages, rather than as grandparental stem cells per se.

Given the differences between f and s blast cells in cell cycle duration, it was originally thought that differences in the duration of their G1 phase might yield information about when the f and s cells become different, and this possibility was addressed using BrdUTP incorporation to label S phase nuclei in carefully timed experiments (Bissen and Weisblat 1989). This approach revealed that blast cells enter

S phase immediately upon birth from the teloblast, and that the differences in cell cycle duration reflect differences in duration of the G2 phase. However, dye-coupling analysis revealed differences in the persistence of cytoplasmic bridges between the parent teloblast and nascent nf and ns blast cells, which suggested the possibility of more subtle differences in the cell cycles by which N teloblasts give rise to nf and ns blast cells (Bissen and Weisblat 1987). In another approach, photoablating individual blast cells within the n bandlet revealed that the flanking cells within the bandlet have assumed distinct nf or ns identities prior to their first mitoses (Bissen and Weisblat 1987). Unfortunately, experimental constraints imposed by the geometry of the embryos made it impossible to carry out these experiments until the blast cell had entered the germinal bands, which is many hours after their birth from the parent teloblast. Intriguingly, such ablation experiments have also revealed differences in cellular dynamics between nf and ns blast cells. As will be discussed further elsewhere (Section 7.13), it had been shown that killing one blast cell in a germinal bandlet has the potential to delay the forward progress of cells in that bandlet that are behind the lesioned cell (Shankland 1984). When this experiment was carried out in the N teloblast–derived bandlet, it was discovered that the effect of the lesion depends in part on the identity (nf or ns) of the cell that was killed (Shain et al. 2000). Specifically, when an nf cell is killed, the two ns cells flanking the lesion often maintain their positions and even close the gap in the bandlet within the germinal band, so that the only deficit in the resultant nerve cord is the clone that would otherwise have arisen from the ablated cell. In contrast, when an ns blast cell is killed, the cells behind the lesion invariably slip backward by one or more segments relative to the ipsilateral bandlets. Thus, these experiments suggest a positive affinity between ns cells that is lacking between nf cells.

Two different molecular approaches offer hope for eventually elucidating the mechanisms by which f and s blast cell fates are assigned in the grandparental stem cell lineages. One set of experiments suggests that differentially upregulated expression and activation of a CDC42 member of the small Rho GTPase family is associated with the ns cell fate (Zhang et al. 2009). Given the connections between CDC42 activity and filopodia formation (Nobes and Hall 1995), we speculate that the observed differences between nf and ns cells in CDC42 activity may also underlie the differences observed by Shain et al. (2000) in the behavior of nf and ns cells following the ablation of neighboring blast cells.

More recently, it has been discovered that some components of the WNT signaling pathway are differentially expressed between pre-mitotic f and s blast cells and within their early clones (Cho, Yoo et al. in preparation); the timing of these expression differences suggests that it is more likely to be a consequence of the initial specification of f and s fates, but perhaps characterizing the cell-type specific enhancers of genes will allow us to work our way upstream toward the initial specification event.

Studies in oligochaetes suggest that the mix of parental and grandparental teloblast lineages may be a general feature of clitellate annelid development (Goto et al. 1999a, 1999b; Arai et al. 2001; Storey 1989a). Apart from certain lineage mutations in *Caenorhabditis elegans* that exhibit a similar cell fate alternation (Chalfie et al. 1981), we are not aware of any other examples outside this group.

7.11 EXPRESSION PATTERNS OF *DROSOPHILA* SEGMENTATION GENE HOMOLOGS

"Modern" studies of segmentation in *Helobdella* were initiated in the 1970s, prior to the reorganization of animal phylogenies based on molecular sequence comparisons. At that time, it was commonly assumed that annelids and arthropods arose from a segmented common ancestor. In light of this assumption, which is now very much in doubt, the dramatic discoveries elucidating the molecular genetic basis of segmentation in *Drosophila* led to speculation as to how the various types of "segmentation genes" might function in what was obviously the very different cellular process of segmentation that was being worked out for leeches.

Within the prevailing belief that annelids and arthropods share a segmented common ancestor, for example, the discovery of the grandparental N and Q lineages in *Helobdella* resonated with the anterior and posterior compartments of *Drosophila* segments. This led to the hypothesis that *engrailed* or other segment polarity genes might be specifying the differences between the f and s blast cells in the N and Q lineages. Initial immunohistochemical studies characterizing the expression of an Engrailed homolog in segmentally iterated transverse stripes in the *Helobdella* germinal plate seemed to support this hypothesis (Wedeen and Weisblat 1991; Lans et al. 1993; Ramírez et al. 1995), but the late onset of Engrailed expression was somewhat disconcerting, as was the eventual realization that Engrailed-expressing cells appear to play no role in regulating segment polarity in leeches (Seaver and Shankland 2001) and that the stripes of Engrailed-expressing cells only appear *after* morphological events of segmentation were well underway (Shain et al. 1998). In other work, the leech *hedgehog* gene was characterized, but turned out not to be expressed in a segmentally iterated pattern within the germinal plate during the early process of segmentation; more as in vertebrates, the leech *hedgehog* gene plays an important role in gut formation (Kang et al. 2003). Thus, despite considerable effort, there is no evidence as yet that the segment polarity genes identified in *Drosophila* play homologous roles in *Helobdella* segmentation.

An alternative possibility for establishing homology at the cellular and molecular levels between *Drosophila* segmentation genes and the f and s cells of the grandparental teloblast lineages in *Helobdella* was via homologs of the pair-rule genes. To investigate this possibility, homologs of *even-skipped* (*eve*) and *hairy/enhancer-of-split* (*hes*) were characterized (Song et al. 2002, 2004). Curiously, both genes proved to be expressed at the earliest zygotic stages. This expression was maintained in apparently all the teloblasts and early blast cells of the PGZ, but no differences in expression between the f and s cells of the N or Q lineages were observed.

Even more curiously, transcripts for both *hes* and *eve* genes were associated with the chromatin of dividing cells (Song et al. 2002, 2004); in the case of the leech *hes* homolog, it was shown that transcripts accumulated during mitosis and nuclear accumulation of translated protein began as cells exited mitosis. The significance of this antiphasic, cell cycle-driven expression remains to be determined. The injection of antisense morpholinos targeting *eve* expression led to a partial disruption of germinal band morphology and of neurogenesis in the injected lineage, but both these genes are ripe for reinvestigation using genome editing approaches that were unavailable at the time.

Yet another attempt to draw parallels between the segmentation processes in leeches and flies involved the characterization of a leech *nanos* gene (Pilon and Weisblat 1997). In this undertaking, the rationale was that since *nanos* expression marks the posterior pole of the early *Drosophila* embryo, it might either be expressed in the teloblasts of the PGZ throughout its function, since they constitute the posterior pole of the growing embryo, or perhaps just during their last divisions, which generate founder cells for the posteriormost segments. Rather than either of these possibilities, however, the leech *nanos* is an abundant and broadly expressed maternal transcript that declines markedly during cleavage, is essentially absent from the PGZ, and then is reexpressed broadly within the germinal bands and germinal plate, before finally becoming restricted to presumptive germline precursors (Kang et al. 2002). During cleavage, expression of Nanos protein appears to turn on during the two-cell stage and peak at stage 4b. At this stage, both transcript and protein levels are higher in DNOPQ than in sister cell DM (Pilon and Weisblat 1997).

In summary, while the leech homologs of *Drosophila* segmentation genes all show interesting and/or suggestive patterns of expression, there is no obvious point of homology with the critical roles of related genes in *Drosophila* segmentation. In retrospect, these results are not surprising given the vast phylogenetic separation between annelids and arthropods and the real possibility that segmentation arose independently in Ecdysozoa and Lophotrochozoa. Given the complex signaling requirements associated with evolution of segmentation in *any* taxon, we do not find it surprising that the transcription factor pair-rule genes *eve* and *hes* are expressed in the PGZ, along with other widely used developmental regulators in the Notch, WNT, and BMP signaling pathways (Rivera et al. 2005, Rivera and Weisblat 2009; Cho et al. 2010; Yoo et al. in preparation; Kuo and Weisblat 2011), some of which regulate segmentation of vertebrates and/or arthropods.

7.12 SPATIOTEMPORAL REGISTRATION OF BLAST CELL CLONES

The juxtaposition of parental (M, O, and P) lineages and grandparental (N and Q) lineages within the germinal bands of the *Helobdella* embryo presents a pair of fascinating conundrums regarding the spatial and temporal registration of blast cells that are destined to contribute to a given segment.

First, given that all the blast cell progeny are about the same size, one segment's worth of cells in the n and q bandlets takes up twice as much space as one segment's worth of cells in the m, o, or p bandlets as the cells enter the germinal band Figures 7.3 and 7.4). This initial spatial discrepancy is corrected as cells begin dividing, because the individual m, o, and p blast cell clones elongate to match the combined length of the neighboring f and s blast cell clones in the n and q bandlets. This positional correction is manifested by the continual movement of n and q bandlets relative to the m, o, and p bandlets at the posterior ends of the germinal bands (Weisblat and Shankland 1985).

A second puzzle arises from the observation that all the teloblasts divide at a rate that is roughly constant among teloblasts and throughout their production of segmental founder cells (Wordeman 1983). In *Helobdella austinensis*; the telo-blast cell cycle time is roughly 90 minutes at 23°C (Yoo and Bylsma, in prepara-tion). All four subsets of teloblasts (M, N, O/P, and Q) begin making segmental founder cells at about the same time, at roughly 30 hours after zygote deposition (hrs AZD; leech embryos are fertilized internally but arrest at meiosis I until they are deposited into cocoons, so zygote deposition marks the onset of postfertiliza-tion development). Thus, the M and O/P teloblasts have made their entire comple-ment of 32 segmental founder cells at about 78–80 hrs AZD. At this time, the N and Q teloblasts have also generated about 32 segmental founder cells, but these correspond to only 16 segments, and the N and Q teloblasts do not finish making 64 segmental cells each for another 2 days (148–150 hrs AZD; Yoo and Bylsma in preparation).

Put another way, leech embryos exhibit a progressive and segment-specific dis-parity in the age of consegmental blast cell clones—the blast cells from all segments that contribute to segment R1 are all born within a few hours of each other, but the n and q blast cells contributing to posterior segments are born more than a day later than the m, o, and p blast cells contributing to the same segments (Lans et al. 1993). Moreover, so far as has been determined for *Helobdella*, each of the seven types of blast cell clones contributing progeny to a segment (m, nf, ns, o, p, qf, qs) develops according to an autonomous clonal clock rather than synchronizing their develop-ment to an external segmental clock (Lans et al. 1993).

The temporal discrepancy between parental and grandparental lineages in *Helobdella* also raises questions concerning segmentation in oligochaetes, the paraphyletic annelid taxon from which leeches evolved (Erséus and Källersjö 2004; Struck et al. 2011). So far as is known, the process by which segmental mesoderm and ectoderm arise during embryogenesis is well conserved among clitellate annelids, including the production of homologous, segmentally repeat-ing M, N, O, P, and Q kinship groups from lineage-restricted, teloblastic stem cells and the same combination of parental and grandparental lineages as in *Helobdella*. An intriguing difference, however, is that whereas segmentation in leeches is a determinate process, many oligochaetes exhibit indeterminate growth and segmentation. That is, leeches produce their fixed complement of exactly 32 segments during embryogenesis, and with no post-embryonic seg-ment formation. In contrast, many oligochaetes add somewhat indeterminate numbers of homonymous segments post-embryonically from a posterior growth zone (Zattara and Bely 2013; Bely et al. 2014). If oligochaetes maintain the same rules as leeches for producing segmental founder cells, with an increasing dis-crepancy between the birthdate of the N or Q versus M, O and P founder cell clones in posterior segments, it would suggest that posterior segments of mature specimens should be missing their N and Q kinship groups for substantial peri-ods of time after the M, O, and P-derived cells have completely differentiated. Alternatively, it could be that post-embryonic segmentation in oligochaetes entails the emergence of new, more homogeneous stem cells, and/or that cell cycle times of post-embryonic segmentation stem cells are regulated differently than in the embryonic teloblasts.

7.13   SEGMENT IDENTITIES AND THE PARADOXICAL SPECIFICATION OF REGIONAL DIFFERENCES ALONG THE ANTERIOR–POSTERIOR AXIS

The extent to which segmentation in leeches is viewed as being homonymous versus heteronomous or tagmatized is partly a matter of definition and partly a matter of how closely one looks. For the purposes of this discussion, the essential point is that there are reproducible regional and segment-specific differences along the anterior–posterior (A–P) axis. Most obvious is the organization of the four rostral segments and seven caudal segments (segments R1–R4 and C1–C7, respectively) into specialized anterior and posterior suckers, innervated by fused segmental ganglia (Figure 7.1). The four fused anterior ganglia connect via circumesophageal connectives to a non-segmental, dorsal anterior ganglion, comprising progeny descended from the first micromere quartet, which arises at third cleavage (Weisblat et al. 1984). In addition, and notwithstanding their rather uniform external morphologies, the midbody segments (segments M1-M21) also exhibit a variety of segment-specific features such as: specialization of segments 5 and 6 for male and female reproductive functions (Sawyer 1986), respectively; segment-specific distribution and orientation of metanephridia (Weisblat and Shankland 1985; Martindale and Shankland 1988); segment-specific distribution of primordial germ cells and definitive testisacs (Sawyer 1986; Kang et al. 2002); and segment-specific differences in the occurrence and distribution of various neuronal phenotypes (Stuart et al. 1987; Shankland and Martindale 1989; Martindale and Shankland 1990; Shankland et al. 1991).

There is also ample evidence for the specification of segmental differences within the endodermal derivatives of the midgut (Figure 7.1). The crop (anterior midgut) comprises six segmentally iterated, bilaterally paired outpocketings called crop ceca that serve in food storage; the posterior pair of crop ceca are particularly large and extend rearward through most of the midbody. Crop ceca are particularly important for the blood-feeding leeches, which may feed only rarely and yet can take in several times their own weight at one feeding. The intestine (posterior midgut) features smaller and more uniformly sized ceca.

These morphological markers indicate that segments acquire specific identities along the A–P axis in leeches. Molecular markers corresponding to segment-specific identities come largely from analyses of the expression of the leech Hox genes, but this work has led to fascinating new puzzles as well.

Initial studies of leech Hox genes were carried out without benefit of having a sequenced genome or transcriptome. A number of putative Hox orthologs were identified, however, and their expression patterns were interpreted as showing the expected succession of A–P boundaries consistent with the canonical Hox clusters of *Drosophila* and vertebrates, all of which seems in line with the conclusions that (1) Hox genes function to specify segment or regional identities along the A–P axis, and; (2) their expression is driven by spatiotemporal colinearity responding to their sequence within the evolutionarily conserved Hox clusters (Kourakis et al. 1997). However, a significant body of work has revealed that the situation is more complicated than it first appeared.

First, a variety of ablation experiments indicate that ectodermal blast cells have assumed specific identities at least by the time they enter the germinal bands. For example, as described earlier, blast cell ablation experiments were used to confirm the commitments of blast cells to their f and s identities in the grandparental N lineage (Bissen and Weisblat 1987; Shain et al. 2000). More definitively, precise combinations of lineage tracing and immunostaining were used to identify segment-specific patterns of presumptive peptidergic neurons within the N lineage (Shankland and Martindale 1989). Then, blast cell ablations were carried out to induce the rearward slippage of o blast cells as previously demonstrated (Shankland 1984; Martindale and Shankland 1990). The experimentally induced slippage was timed to reposition o blast cells into segments posterior to those that would normally contain the identifiable peptidergic neurons—the repositioned o blast cells still gave rise to the kinship group appropriate to the segment for which it was originally destined rather than adapting to its new position. This result indicates that the blast cell had already acquired a segmental identity by the time it had entered the germinal band, well prior to its first mitosis. Similar approaches yielded similar results for blast cells in the mesodermal (M) lineages of a different glossiphoniid leech species (Gleizer and Stent 1993). An important advance was that in this work, it was possible to ablate individual m blast cells soon after they had been produced by the parent M teloblasts when they were only a few hours in the lineage.

Thus, the results indicated that these m blast cells had acquired their segment-specific fates at or within a couple of hours after their birth from the parent stem cell. And yet, most of the leech Hox genes that have been examined are not expressed until much later, when the blast cells have undergone many subsequent divisions and the differentiation of segments is well underway within the germinal plate (Kourakis et al. 1997; Nardelli-Haefliger et al. 1994; Cho et al. in preparation).

Moreover, notwithstanding the general result that Hox gene expression boundaries fall within particular segments across multiple teloblast lineages, blast cell ablation experiments followed by immunostaining revealed that these boundaries of Hox gene expression are established by the cells of each lineage independently of interactions with other lineages, as are the other segment-specific markers (Nardelli-Haefliger et al. 1994).

Yet another aspect of this puzzle emerged when whole genome sequencing was carried out on three lophotrochozoan species (Simakov et al. 2013). The genomes of a mollusk (*Lottia*) and a polychaete annelid (*Capitella*) each contain 11 mutually orthologous Hox genes. The molluscan Hox genes appear in a single genomic scaffold and the polychaete genes appear in two or possibly three scaffolds, but lie in the same syntenic order as the molluscan genes, in accord with expectations based on vertebrates and *Drosophila*. In stark contrast, the *Helobdella* genome contains 18 or 19 Hox genes with multiple duplications (two paralogs each of *labial*, *deformed*, and *lox4*; three of *post2*; and five of *sex combs reduced*) and loss of *proboscipedia* and *post1*. The organization of the cluster is also severely disrupted. Many genes appear as singletons, flanked by non-Hox genes, and the longest single cluster contains only four genes, none of which are neighbors in the inferred ancestral lophotrochozoan Hox cluster (Simakov et al. 2013).

Thus, since the last shared ancestor of *Capitella* and *Helobdella*, there has been an extensive disruption of the annelid Hox cluster in the evolutionary lineage leading to leeches. This disruption is emblematic of a comparably extensive rearrangement of the entire leech genome relative to other sequenced taxa; available evidence suggests that such rearrangements are characteristic of clitellate annelids as a group.

In summary then, it appears that the segmental founder cells in leeches acquire specific segmental identities before they undergo their first mitoses and possibly as they are born from the parent stem cells/teloblasts. In addition, these segmental identities are assigned independently within each lineage and before the onset of most Hox gene expression, by mechanisms that remain to be determined. And yet, despite the extensive disruption of the ancestral Hox cluster organization, these genes are expressed in patterns that are fairly consistent with those inferred for the ancestral annelid Hox genes.

7.14 THE "LEECH PERSPECTIVE" ON THE COUNTING PROBLEM IN SEGMENTATION

Comparisons of indeterminate and determinate growth in clitellate annelids highlight two related and frustratingly unanswered questions of general interest in segmentation. First, how do embryos that *do* exhibit determinate growth count out the appropriate number of segments for their species. A related second question, that might be expected to shed light on the first, is how precisely they count out their segments. Many oligochaetes make well over 100 segments, and there appear to be species-specific differences in the number of adult segments. The exact number of segments varies among individuals, however, in a roughly normal distribution around some value. For example, newly hatched *Lumbricus terrestris* were reported as having 130–160 segments, with 150 as the most common value (Evans 1946), equivalent to a natural variation of around 10% from the mean value. In leeches, as described earlier, the number of segments is much smaller (32) and is also more tightly controlled. We are aware of no naturally occurring variants of this value, whereas a 10% variation would be expected to produce animals ranging from 29 to 35 segments or so.

In the 1970s, when annelids and arthropods, as segmented protostomes, were considered as likely sister taxa, hypotheses were based in part on efforts to homologize the segmentation process in *Helobdella* with what was known about *Drosophila*. Thus, one model to explain the precise control of segment number in leeches was that the teloblast nuclei might undergo five rounds of syncytial nuclear proliferation (in M and O/P teloblasts, or six in N and Q teloblasts) after which the established set of 32 or 64 segmental founder cells would be counted out by successive cellularization events. This idea was ruled out when the (then) new technique of staining nuclear DNA with fluorescent Hoechst dyes revealed that the teloblasts are mononucleate and produce blast cells by standard mitoses (Zackson 1982).

Another significant outcome from this work was the discovery that the teloblasts in each lineage undergo additional mitoses after their full complement of segmental founder cells has been produced—moreover, the number of these so-called supernumerary blast cells varies from teloblast to teloblast within and among embryos

(Zackson 1982; Desjeux and Price 1999; Yoo et al. unpublished). Thus, while the total number of segments is tightly controlled, the numbers of blast cells produced is not. One interpretation of these observations is that variability in the number of stem cell divisions is an ancestral feature in clitellate annelids, and that the leech clade has superimposed a distinct secondary patterning process that gives precise control over the number of segments produced. The molecular bases of both of these inferred processes remain to be determined.

It should also be noted that the specific values cited for segment numbers in leech taxa, e.g., 15 for branchiobdellids, 29 for acanthobdellids, and 32 for true leeches, are based on necessarily subjective interpretations of morphological analyses (Purschke et al. 1993). Such analyses are particularly problematic for the head region; in leeches, for example, the number of segments is also reported as being 33 or 34 based on different interpretations of head morphology (Verdonschot 2015; Sawyer 1986). The value of 32 segments used in this chapter is based on the count of serially homologous neural metamers (4 in the head, 21 in the midbody, and 7 in the tail) (Muller et al. 1981). This value was reinforced by results from cell lineage tracing that showed that only 32 serially homologous segmental primordia arise from the teloblasts of the PGZ and that other head structures have a distinct embryonic origin.

7.15   GENESIS OF ENDODERM FROM A SYNCYTIAL YOLK CELL

It was originally assumed that the supernumerary blast cells described earlier were fated to die after they failed to enter the germinal bands and germinal plate, from which segments arise (Shankland 1984), and that the segmented midgut structures arose exclusively from the three non-D quadrants of the early embryo as proposed by C.O. Whitman in the 19th century, after they had completed producing micromere production, perhaps in combination with the teloblasts after they had ceased dividing (Sawyer 1986; Nardelli-Haefliger and Shankland 1993; Shankland and Savage 1997). As with other aspects of leech segmentation, these assumptions are not entirely correct, and the cellular origins of the segmented midgut are quite interesting.

Generally speaking, the two goals of gut formation for any embryo are (1) to provide a mechanism for digesting the egg's yolk as fuel for completing development, and (b) to provide a lumenal organ (appropriately connected to feeding and excretory tubes) for storing and digesting food in the adult. There are various morphogenetic strategies by which these goals are achieved in different taxa. In *Drosophila*, for example, cellularization of the syncytial blastoderm generates prospective endodermal cells at the anterior and posterior ends of the embryo, while the bulk of the yolk remains contained in an internal syncytial cell. In later development, the endodermal precursors move inside to cover the yolk, which thus ends up in the lumen of the gut. The situation in oligochaete annelids has not been fully described, but it appears that the non-D macromeres continue dividing more or less equally long after the D quadrant teloblasts have formed, making progressively smaller endodermal cells that eventually form a gut tube. If so, then in this case, the maternally provided nutrients are inside the endodermal cells themselves. Whether or not the oligochaete teloblasts contribute to this or remain separate and transition to form the posterior growth zone for post-embryonic development is one of the great remaining questions in our subfield.

In *Helobdella*, gut formation proceeds by yet another route, involving a series of stepwise cell fusions that lead to a syncytial yolk cell (SYC) with three distinct classes of nuclei (Liu et al. 1998; Desjeux and Price 1999). The endodermal epithelium arises by cellularization of the SYC (Nardelli-Haefliger and Shankland 1993), under the inductive influence of visceral mesoderm (Wedeen and Shankland 1997); this midgut epithelium thus surrounds the yolk within the lumen of the nascent anterior and posterior midgut segments, connected to the foregut and hindgut tubes, neither of which exhibit metamerism.

Cell fusions are monitored and scored by the relatively sudden leakage of injected tracers from one cell to another. In the early embryo, virtually all cells are coupled via gap junctions (Bissen and Weisblat 1989, 1991), but these are impermeable to molecules of great than ~1200 daltons (Simpson et al. 1977), so marker enzymes (e.g., HRP or lacZ) or fluorescently labeled dextrans are initially confined to the cell into which they are injected and its mitotic progeny. The first fusion is between macromeres A''' and B''' to form SYC A/B; tracer injected into either cell remains confined to that cell until near the end of stage 7, at which time the injected tracer is observed to spread bidirectionally between just these two cells. Intriguingly, the A'''–B''' fusion event depends on an inductive signal from the D quadrant blastomeres (Isaksen et al. 1999). Subsequently, macromere C''' fuses with A/B to form SYC A/B/C during late stage 8; and after the teloblasts have formed their entire complements of segmental founder cells, plus a variable number of supernumerary blast cells, the teloblasts and supernumerary blasts cells fuse with the SYC as well. However, since the N and Q teloblasts make roughly twice as many blast cells as do the M and O/P teloblasts, the M and O/P teloblasts have generated their full complements or segmental founder cells and then fuse with the SYC while the N and Q teloblasts are still generating segmental founder cells (Liu et al. 1998; Yoo and Bylsma in preparation).

Thus, the cellular processes by which the segmental repeats in the anterior and posterior midgut (crop and intestine) arise are dramatically different from the stem cell divisions of the PGZ that form segmental ectoderm and mesoderm.

7.16   MESODERM AS A PRIMARY DRIVER OF SEGMENTAL PATTERNING IN LEECHES

In Chapter 4, segmentation was defined as the existence of metameric units in two or more germ layers, with the added requirement that the spatial frequency of the metamers be conserved across the germ layers. This definition raises the question of how spacing of the units in one germ layer is matched to that in the other germ layer(s). One obvious solution to this problem is to have the metameric repeats in one germ layer serve to organize those in the other layer(s).

During cleavage, the four bilateral pairs of ectoteloblasts (N, O/P, O/P, and Q) arise from a bilateral pair of precursor blastomeres, the left and right NOPQ cells. Moreover, as previously described, the segmental mesoderm on each side arises from a single ipsilateral stem cell, the M teloblast. Numerous experiments have documented the fact that leech embryos are essentially lacking in their capacity for regulative replacement of ablated lineages (e.g., Blair and Weisblat 1982, 1984; Stuart et al. 1987, 1989; Torrence et al. 1989). Thus, it was possible to ask if mesoderm is required for segmentation in ectoderm and vice versa by experiments in which individual cells were injected either with fluorescent lineage tracers or cell-restricted killing ablatants, e.g., cytotoxic enzymes such as DNAse I (Blair 1982). When an NOPQ proteloblast was killed on one side of the embryo, thereby removing the ectoderm from that side, two different outcomes were observed. In some embryos, the two m bandlets remained on their respective sides; in this case, the mesodermal bandlet that had no overlying ectodermal bandlets was severely disorganized and failed to generate segments, while the contralateral side developed normally. In other embryos, the m bandlet on the ablated side switched sides, so that both m bandlets were in contact with the remaining ectoderm; in this case, both m bandlets formed somites, but the left and right sides were often out of register, indicating that this process occurs independently on the right and left sides.

When the converse experiments were carried out, killing the mesodermal precursor on one side, the ipsilateral ectodermal blast cells initially exhibited their normal, lineage-specific division patterns (Zackson 1984; Huang and Weisblat 1996) but then became disorganized and failed to generate recognizable segmental structures (Blair 1982). Thus, in *Helobdella*, ectoderm and mesoderm exhibit a mutual interdependence in generating segmental organization of those germ layers.

Other experiments have revealed a requirement for mesodermal signaling in patterning the endodermal midgut as well. Endodermal nuclei arising by cellularization of the SYC nuclei express *lox3*, a parahox gene of the *xlox/pdx* family (Wedeen and Shankland 1997). Prospective endodermal nuclei also express another homeobox gene *lox10* (of the *nk-2* gene family), although the exact timing of the onset of this expression relative to the cellularization of SYC nuclei to form definitive endoderm remains to be determined (Nardelli-Haefliger and Shankland 1993). After labeling the M lineage on one side of the embryo with a photosensitizing lineage tracer (fluorescein dextran), it was possible to photolesion nascent mesodermal somites. In lesioned segments, no *lox3* expression was detected, and the endodermal cells were completely absent from precisely the regions from which the normally overlying mesoderm was absent (Wedeen and Shankland 1997). Thus, it appears that short-range signaling from visceral mesoderm is required for the migration, differentiation, and/or survival of segmentally patterned midgut endoderm.

7.17 SUMMARY AND CONCLUSIONS

Two main conclusions can be drawn from the work presented here, summarizing our current understanding of segmentation in leeches. The first and most obvious is that, as developmental biologists, we are just scratching the surface of understanding how this embryo develops and how it generates segments. The degree of understanding of *Helobodella* development that has been achieved by a handful of research groups over roughly four decades pales in comparison with that achieved for standard model organisms studied by hundreds or thousands or workers, and often for much longer.

Nonetheless, a second and equally significant conclusion is that there is much to be learned from the careful study of non-model organisms such as *Helobdella*, representing comparatively unexplored branches of the phylogenetic tree. The tremendous diversity of animal genomes and animal body plans are linked by a corresponding diversity of developmental mechanisms, and we can only come to understand the full breadth of extant developmental mechanisms by investigating how development has been modified along different evolutionary branches.

Helobdella provides particularly fruitful grounds for such investigations, not only because of its phylogenetic position and its experimental tractability, but also because its highly rearranged genome has effectively accelerated the evolutionary exploration of "developmental hyperspace." Thus, in addition to the retrospective approach of evo-devo, in which developmental comparisons are used primarily to infer ancestral mechanisms of development, studying *Helobdella* and its clitellate annelid allies should provide insights into the possibilities of divergent or novel developmental mechanisms, starting with the mechanisms of genome diversification itself (Simakov et al. 2013; Cho et al. 2014).

In this chapter, we have summarized how studies of leech segmentation have revealed a number of evolutionary developmental novelties: axial growth from a PGZ comprising just five bilateral pairs of lineage-restricted stem cells; the production of segmental founder cells by parental and grandparental stem cell lineages operating in parallel; segmentation by lineage-dependent rather than boundary-dependent mechanisms; non-canonical expression of homologs of the *Drosophila* segmentation genes; dorsoventral patterning by evolutionarily divergent, cell contact-mediated BMP signaling; HOX cluster fragmentation and HOX gene duplication within a paradigmatically segmented animal; production of segmental gut structures by a syncytial yolk cell; and the evolution of a body plan with fixed segment number from an ancestor that presumably underwent indeterminate segmentation. This list of intriguing developmental phenomena is probably far from complete, and the application of emerging techniques in omics, optics, and experimental manipulation to the tractable *Helobdella* embryo should continue to provide insights into the evolution of developmental mechanisms.

REFERENCES

Arai, A., A. Nakamoto, and T. Shimizu. 2001. Specification of ectodermal teloblast lineages in embryos of the oligochaete annelid *Tubifex*: Involvement of novel cell-cell interactions. *Development* 128: 1211–1219.

Astrow, S., B. Holton, and D. A. Weisblat. 1987. Centrifugation redistributes factors determining cleavage patterns in leech embryos. *Dev. Biol.* 120: 270–283.

Bely, A. E., E. E. Zattara, and J. M. Sikes. 2014. Regeneration in spiralians: Evolutionary patterns and developmental processes. *Int. J. Dev. Biol.* 58: 623–634.

Bissen, S. T., and D. A. Weisblat. 1987. Early differences between alternate n blast cells in leech embryo. *J. Neurobiol.* 18: 251–269.

Bissen, S. T., and D. A. Weisblat. 1989. The durations and compositions of cell cycles in embryos of the leech, *Helobdella triserialis*. *Development* 106: 105–118.

Bissen, S. T., and D. A. Weisblat. 1991. Transcription in leech: mRNA synthesis is required for early cleavages in *Helobdella* embryos. *Dev. Biol.* 146: 12–23.

Blair, S. S. 1982. Interactions between mesoderm and ectoderm in segment formation in the embryo of a glossiphoniid leech. *Dev. Biol.* 89: 389–396.

Blair, S. S., and D. A. Weisblat. 1982. Ectodermal interactions during neurogenesis in the glossiphoniid leech *Helobdella triserialis*. *Dev. Biol.* 91: 64–72

Blair, S. S., and D. A. Weisblat. 1984. Cell interactions in the developing epidermis of the leech *Helobdella triserialis*. *Dev. Biol.* 101: 318–325.

Chalfie, M., H. R. Horvitz, and J. E. Sulston. 1981. Mutations that lead to reiterations in the cell lineages of *C. elegans*. *Cell* 24: 59–69.

Cho, S. J., Vallès, Y., and D. A. Weisblat. 2014. Differential expression of conserved germ line markers and delayed segregation of male and female primordial germ cells in a hermaphrodite, the leech *Helobdella*. *Mol. Biol. Evol.* 31: 341–354. Erratum in: *Mol. Biol. Evol.* 32: 833–834.

Cho, S. J., Y. Vallès, V. C. Jr. Giani, E. C. Seaver, and D. A. Weisblat. 2010. Evolutionary dynamics of the *wnt* gene family: A lophotrochozoan perspective. *Mol. Biol. Evol.* 27: 1645–1658.

Desjeux, I., and D. J. Price. 1999. The production and elimination of supernumerary blast cells in the leech embryo. *Dev. Genes Evol.* 209: 284–293.

Dohle, W. 1999. The ancestral cleavage pattern of the clitellates and its phylogenetic deviations. *Hydrobiologia* 402: 267–283.

Erséus, C., and M. Källersjö. 2004. 18S rDNA phylogeny of Clitellata (Annelida). *Zool. Scr.* 33: 187–196.

Evans, A. C. 1946. Distribution of numbers of segments in earthworms and its significance. *Nature* 158: 98–99.

Fernández, J., and G. S. Stent. 1980. Embryonic development of the glossiphoniid leech *Theromyzon rude*: Structure and development of the germinal bands. *Dev. Biol.* 78: 407–434.

Fernández, J., and N. Olea. 1982. Embryonic development of glossiphoniid leeches. In *Developmental Biology of Freshwater Invertebrates*, Eds. F. W. Harrison, and R. R. Cowden, 317–361. New York: A. R. Liss.

Fernández, J., N. Olea, V. Téllez, and C. Matte. 1990. Structure and development of the egg of the glossiphoniid leech *Theromyzon rude*: Reorganization of the fertilized egg during completion of the first meiotic division. *Dev. Biol.* 137: 142–154.

Fernández, J., F. Roegiers, V. Cantillana, and C. Sardet. 1998. Formation and localization of cytoplasmic domains in leech and ascidian zygotes. *Int. J. Dev. Biol.* 42: 1075–1084.

Freeman, G., and J. W. Lundelius. 1992. Evolutionary implications of the mode of D quadrant specification in coelomates with spiral cleavage. *J. Evol. Biol.* 5: 205–247.

Gimlich, R. L., and J. Braun. 1985. Improved fluorescent compounds for tracing cell lineage. *Dev. Biol.* 109: 509–514.

Giribet, G. 2003. Molecules, development and fossils in the study of metazoan evolution; Articulata versus Ecdysozoa revisited. *Zoology* 106: 303–326.

Gleizer, L., and G. S. Stent. 1993. Developmental origin of segmental identity in the leech mesoderm. *Development* 117: 177–189.

Gline, S. E., A. Nakamoto, S. J. Cho, C. Chi, and D. A. Weisblat. 2011. Lineage analysis of micromere 4d, a super-phylotypic cell for Lophotrochozoa, in the leech *Helobdella* and the sludgeworm *Tubifex*. *Dev. Biol.* 353: 120–133.

Goto, A., K. Kitamura, and T. Shimizu. 1999a. Cell lineage analysis of pattern formation in the *Tubifex* embryo. I. Segmentation in the mesoderm. *Int. J. Dev. Biol.* 43: 317–327.

Goto, A., K. Kitamura, A. Arai, and T. Shimizu. 1999b. Cell fate analysis of teloblasts in the *Tubifex* embryo by intracellular injection of HRP. *Dev. Growth Differ.* 41: 703–713.

Ho, R. K., and D. A. Weisblat. 1987. A provisional epithelium in leech embryo: Cellular origins and influence on a developmental equivalence group. *Dev. Biol.* 120: 520–534.

Holton, B., S. H. Astrow, and D. A. Weisblat. 1989. Animal and vegetal teloplasms mix in the early embryo of the leech, *Helobdella triserialis*. *Dev. Biol.* 131: 182–188.

Holton, B., C. J. Wedeen, S. H. Astrow, and D. A. Weisblat. 1994. Localization of polyadenylated RNAs during teloplasm formation and cleavage in leech embryos. *Roux. Arch. Dev. Biol.* 204: 46–53.

Huang, F. Z., and D. A. Weisblat. 1996. Cell fate determination in an annelid equivalence group. *Development* 122: 1839–1847.

Isaksen, D. E., N. J. Liu, and D. A. Weisblat. 1999. Inductive regulation of cell fusion in leech. *Development* 126: 3381–3390.

Janssen, R., M. Le Gouar, M. Pechmann, F. Poulin, R. Bolognesi, E. E. Schwager, C. Hopfen, J. K. Colbourne, G. E. Budd, S. J. Brown, N. M. Prpic, C. Kosiol, M. Vervoort, W. G. Damen, G. Balavoine, and A. P. McGregor 2010. Conservation, loss, and redeployment of Wnt ligands in protostomes: Implications for understanding the evolution of segment formation. *BMC Evol. Biol.* 10: 374.

Kang, D., F. Huang, D. Li, M. Shankland, W. Gaffield, and D. A. Weisblat. 2003. A hedgehog homolog regulates gut formation in leech (*Helobdella*). *Development* 130: 1645–1657.

Kang, D., M. Pilon, and D. A. Weisblat. 2002. Maternal and zygotic expression of a *nanos*-class gene in the leech *Helobdella robusta*: Primordial germ cells arise from segmental mesoderm. *Dev. Biol.* 245: 28–41.

Kourakis, M. J., V. A. Master, D. K. Lokhorst, D. Nardelli-Haefliger, C. J. Wedeen, M. Q. Martindale, and M. Shankland. 1997. Conserved anterior boundaries of *Hox* gene expression in the central nervous system of the leech *Helobdella*. *Dev. Biol.* 190: 284–300.

Kramer, A. P., and D. A. Weisblat. 1985. Developmental neural kinship groups in the leech. *J. Neurosci.* 5: 388–407.

Kuffler, S. W., and D. D. Potter. 1964. Glia in the leech central nervous system: Physiological properties and neuron-glia relationship. *J. Neurophysiol.* 27: 290–320.

Kuo, D. H., and D. A. Weisblat. 2011. A new molecular logic for BMP-mediated dorsoventral patterning in the leech *Helobdella*. *Curr. Biol.* 21: 1282–1288.

Kuo, D. H., and M. Shankland. 2004a. A distinct patterning mechanism of O and P cell fates in the development of the rostral segments of the leech *Helobdella robusta*: Implications for the evolutionary dissociation of developmental pathway and morphological outcome. *Development* 131: 105–115.

Kuo, D. H., and M. Shankland. 2004b. Evolutionary diversification of specification mechanisms within the O/P equivalence group of the leech genus *Helobdella*. *Development* 131: 5859–5869.

Kuo, D. H., M. Shankland, and D. A. Weisblat. 2012. Regional differences in BMP-dependence of dorsoventral patterning in the leech *Helobdella*. *Dev. Biol.* 368: 86–94.

Kutschera, U., H. Langguth, D.-H. Kuo, D. A. Weisblat, and M. Shankland. 2013. Description of a new leech species from North America, *Helobdella austinensis* n. sp. (Hirudinea: Glossiphoniidae), with observations on its feeding behaviour. *Zoosyst. Evol.* 89: 239–246.

Lambert, J. D. 2008. Mesoderm in spiralians: The organizer and the 4d cell. *J. Exp. Zool. B Mol. Dev. Evol.* 310: 15–23.

Lans, D., C. J. Wedeen, and D. A. Weisblat. 1993. Cell lineage analysis of the expression of an *engrailed* homolog in leech embryos. *Development* 117: 857–871.

Liu, N. L., D. E. Isaksen, C. M. Smith, and D. A. Weisblat. 1998. Movements and stepwise fusion of endodermal precursor cells in leech. *Dev. Genes Evol.* 208: 117–127.

Martindale, M. Q., and M. Shankland. 1988. Developmental origin of segmental differences in the leech ectoderm: Survival and differentiation of the distal tubule cell is determined by the host segment. *Dev. Biol.* 125: 290–300.

Martindale, M. Q., and M. Shankland. 1990. Intrinsic segmental identity of segmental founder cells of the leech embryo. *Nature* 347: 672–674.

McHugh, D. 1997. Molecular evidence that echiurans and pogonophorans are derived annelids. *Proc. Natl. Acad. Sci. U.S.A.* 94: 8006–8009.

Meyer, N. P., M. J. Boyle, M. Q. Martindale, and E. C. Seaver. 2010. A comprehensive fate map by intracellular injection of identified blastomeres in the marine polychaete *Capitella teleta*. *Evodevo* 1: 8.

Muller, K. J., J. G. Nicholls, and G. S. Stent. 1981. *Neurobiology of the Leech*. Cold Spring Harbor, NY: Cold Spring Harbor Laboratory Press.

Nardelli-Haefliger, D., A. E. Bruce, and M. Shankland. 1994. An axial domain of HOM/Hox gene expression is formed by morphogenetic alignment of independently specified cell lineages in the leech *Helobdella*. *Development* 120: 1839–1849.

Nardelli-Haefliger, D., and M. Shankland. 1993. *Lox10*, a member of the *NK-2* homeobox gene class, is expressed in a segmental pattern in the endoderm and in the cephalic nervous system of the leech *Helobdella*. *Development* 118: 877–892.

Nelson, B. H., and D. A. Weisblat. 1991. Conversion of ectoderm to mesoderm by cytoplasmic extrusion in leech embryos. *Science* 253: 435–438.

Nelson, B. H., and D. A. Weisblat. 1992. Cytoplasmic and cortical determinants interact to specify ectoderm and mesoderm in the leech embryo. *Development* 115: 103–115.

Nicholls, J. G., and D. A. Baylor. 1968. Specific modalities and receptive fields of sensory neurons in CNS of the leech. *J. Neurophysiol.* 31: 740–756.

Nobes, C. D., and A. Hall. 1995. Rho, rac and cdc42 GTPases: Regulators of actin structures, cell adhesion and motility. *Biochem. Soc. Trans.* 23: 456–459.

Peterson, K. J., J. A. Cotton, J. G. Gehling, and D. Pisani. 2008. The Ediacaran emergence of bilaterians: Congruence between the genetic and the geological fossil records. *Philos. Trans. R. Soc. London B Biol. Sci.* 363: 1435–1443.

Pilon, M., and D. A. Weisblat. 1997. A *nanos* homolog in leech. *Development* 124: 1771–1780.

Purschke, G., W. Westheide, Rohde, D., and R. O. Brinkhurst. 1993. Morphological reinvestigation and phylogenetic relationship of *Acanthobdella peledina* (Annelida, Clitellata). *Zoomorphology* 113: 91–101.

Ramírez, F. A., C. J. Wedeen, D. K. Stuart, D. Lans, and D. A. Weisblat. 1995. Identification of a neurogenic sublineage required for CNS segmentation in an Annelid. *Development* 121: 2091–2097.

Retzius, G. 1891. Zur kenntniss des centralen nervensystem der würmer. *Biol. Unters.* 2: 1–28.

Rivera A. S., and D. A. Weisblat. 2009. And Lophotrochozoa makes three: Notch/Hes signaling in annelid segmentation. *Dev. Genes Evol.* 219: 37–43.

Rivera, A. S., F. C. Gonsalves, M. H. Song, B. J. Norris, and D. A. Weisblat. 2005. Characterization of *Notch*-class gene expression in segmentation stem cells and segment founder cells in *Helobdella robusta* (Lophotrochozoa; Annelida; Clitellata; Hirudinida; Glossiphoniidae). *Evol. Dev.* 7: 588–599.

Sandig, M., and W. Dohle. 1988. The cleavage pattern in the leech *Theromyzon tessulatum* (Hirudinea, Glossiphoniidae). *J. Morphol.* 196: 217–252.

Sawyer, R. T. 1986. *Leech Biology and Behaviour*. Oxford: Clarendon Press.

Seaver, E. C., and M. Shankland. 2001. Establishment of segment polarity in the ectoderm of the leech *Helobdella*. *Development* 128: 1629–1641.

Shain, D. H., F. A. Ramírez-Weber, J. Hsu, and D. A. Weisblat. 1998. Gangliogenesis in leech: Morphogenetic processes leading to segmentation in the central nervous system. *Dev. Genes Evol.* 208: 28–36.

Shain, D. H., D. K.Stuart, F. Z. Huang, and D. A. Weisblat. 2000. Segmentation of the central nervous system in leech. *Development* 127: 735–744.

Shankland, M. 1984. Positional determination of supernumerary blast cell death in the leech embryo. *Nature* 307: 541–543.

Shankland, M. 1987a. Determination of cleavage pattern in embryonic blast cells of the leech. *Dev. Biol.* 120: 494–498.

Shankland, M. 1987b. Differentiation of the O and P cell lines in the embryo of the leech. I. Sequential commitment of blast cell sublineages. *Dev. Biol.* 123: 85–96.

Shankland, M., and D. A. Weisblat. 1984. Stepwise commitment of blast cell fates during the positional specification of the O and P cell lines in the leech embryo. *Dev. Biol.* 106: 326–342.

Shankland, M., M. Q. Martindale, D. Nardelli-Haefliger, E. Baxter, and D. J. Price. 1991. Origin of segmental identity in the development of the leech nervous system. *Dev. Suppl.* 2: 29–38.

Shankland, M., and M. Q. Martindale. 1989. Segmental specificity and lateral asymmetry in the differentiation of developmentally homologous neurons during leech embryogenesis. *Dev. Biol.* 135: 431–448.

Shankland, M., and R. M. Savage. 1997. Annelids, the segmented worms. In *Embryology: Constructing the Organism*, Eds. S. F. Gilbert, and A. M. Raunio, 219–235. Sunderland, MA: Sinauer.

Simakov, O., F. Marletaz, S.-J. Cho, E. Edsinger-Gonzales, P. Havlak, U. Hellsten, …, R. Savage. 2013. Insights into bilaterian evolution from three spiralian genomes. *Nature* 493: 526–531.

Simpson, I., B. Rose, and W. R. Loewenstein. 1977. Size limit of molecules permeating the junctional membrane channels. *Science* 195: 294–296.

Song, M. H., F. Z. Huang, G. Y. Chang, and D. A. Weisblat. 2002. Expression and function of an *even-skipped* homolog in the leech *Helobdella robusta*. *Development* 129: 3681–3692.

Song, M. H., F. Z. Huang, F. C. Gonsalves, and D. A. Weisblat. 2004. Cell cycle-dependent expression of a *hairy* and *Enhancer of split* (*hes*) homolog during cleavage and segmentation in leech embryos. *Dev. Biol.* 269: 183–195.

Storey, K. G. 1989a. Cell lineage and pattern formation in the earthworm embryo. *Development* 107: 519–531.

Storey, K. G. 1989b. The effects of ectoteloblast ablation in the earthworm. *Development* 107: 533–545.

Stuart, D. K., S. S. Blair, and D. A. Weisblat. 1987. Cell lineage, cell death, and the developmental origin of identified serotonin- and dopamine-containing neurons in the leech. *J. Neurosci.* 7: 1107–1122.

Stuart, D. K., S. A. Torrence, and M. I. Law. 1989. Leech neurogenesis. I. Positional commitment of neural precursor cells. *Dev. Biol.* 136: 17–39.

Struck, T. H., C. Paul, N. Hill, S. Hartmann, C. Hösel, M. Kube, B. Lieb, A. Meyer, R. Tiedemann, G. Purschke, and C. Bleidorn. 2011. Phylogenomic analyses unravel annelid evolution. *Nature* 471: 95–98.

Struck, T. H., N. Schult, T. Kusen, E. Hickman, C. Bleidorn, D. McHugh, and K. M. Halanych. 2007. Annelid phylogeny and the status of Sipuncula and Echiura. *BMC Evol. Biol.* 7: 57.

Torrence, S. A., and D. K. Stuart. 1986. Gangliogenesis in leech embryos: Migration of neural precursor cells. *J. Neurosci.* 6: 2736–2746.

Torrence, S. A., M. I. Law, and D. K. Stuart. 1989. Leech neurogenesis. II. Mesodermal control of neuronal patterns. *Dev. Biol.* 136: 40–60.

Verdonschot, P. F. M. 2015. Chapter 20 - Introduction to Annelida and the class Polychaeta. In *Thorp and Covich's Freshwater Invertebrates*, 4th ed., Eds. J. H. Thorp, and D. C. Rogers, 509–528. Academic Press.

Wedeen, C. J., and D. A. Weisblat. 1991. Segmental expression of an *engrailed*-class gene during early development and neurogenesis in an annelid. *Development* 113: 805–814.

Wedeen, C. J., and M. Shankland. 1997. Mesoderm is required for the formation of a segmented endodermal cell layer in the leech *Helobdella*. *Dev. Biol.* 191: 202–214.

Weisblat, D. A., and D.-H., Kuo. 2009. *Helobdella* (leech): A model for developmental studies. In *Emerging Model Organisms*, Vol. 1, 245–274. Cold Spring Harbor, NY: Cold Spring Harbor Laboratory Press.

Weisblat, D. A., and D.-H. Kuo. 2014. Developmental biology of the leech Helobdella. *Int. J. Dev. Biol.* 58: 429–443.

Weisblat, D. A., and M. Shankland. 1985. Cell lineage and segmentation in the leech. *Philos. Trans. R. Soc. London B Biol. Sci.* 312: 39–56.

Weisblat, D. A., S. Y. Kim, and G. S. Stent. 1984. Embryonic origins of cells in the leech *Helobdella triserialis*. *Dev. Biol.* 104: 65–85.

Weisblat, D. A., R. T. Sawyer, and G. S. Stent. 1978. Cell lineage analysis by intracellular injection of a tracer enzyme. *Science* 202: 1295–1298.

Weisblat, D. A., and S. S. Blair. 1984. Developmental interdeterminacy in embryos of the leech *Helobdella triserialis*. *Dev. Biol.* 101: 326–335.

Weisblat, D. A., S. L. Zackson, S. S. Blair, and J. D. Young. 1980. Cell lineage analysis by intracellular injection of fluorescent tracers. *Science* 209: 1538–1541.

Whitman, C. O. 1878. The embryology of *Clepsine*. *Q. J. Microsc. Sci.* 18: 213–315.

Whitman, C. O. 1887. A contribution to the history of the germ-layer in *Clepsine*. *J. Morphol.* 1: 105–182.

Wordeman, L. 1983. Kinetics of primary blast cell production in the embryo of the leech *Helobdella triserialis*. Honors thesis. Department of Molecular Biology. University of California, Berkeley CA.

Zackson, S. L. 1982. Cell clones and segmentation in leech development. *Cell* 31: 761–70.

Zackson, S. L. 1984. Cell lineage, cell-cell interaction, and segment formation in the ectoderm of a glossiphoniid leech embryo. *Dev. Biol.* 104: 143–160.

Zattara, E. E., and A. E. Bely. 2013. Investment choices in post-embryonic development: Quantifying interactions among growth, regeneration, and asexual reproduction in the annelid *Pristina leidyi*. *J. Exp. Zool. B Mol. Dev. Evol.* 320: 471–88.

Zhang, S. O., and D. A. Weisblat. 2005. Applications of mRNA injections for analyzing cell lineage and asymmetric cell divisions during segmentation in the leech *Helobdella robusta*. *Development* 132: 2103–2113.

Zhang, S. O., D. H. Kuo, and D. A. Weisblat. 2009. Grandparental stem cells in leech segmentation: Differences in CDC42 expression are correlated with an alternating pattern of blast cell fates. *Dev. Biol.* 336: 112–121.

8 Segmentation in Motion

Andres F. Sarrazin

CONTENTS

8.1   INTRODUCTION

The construction of the enormous variety of animal body plans that are found in nature is mainly based on the combination of cellular behaviors and tissue rearrangements that must be tightly coordinated, both temporally and spatially, with the corresponding gene expression patterning process that organizes the developing embryo. Thus, the study of these biological events *in vivo* at high resolution emerges as a key aspect that has provided insightful knowledge on developmental morphogenesis.

Over a century ago (1907), the Swiss biologist Julius Ries was one of the first scientists to serially record microscopic images taken at regular time intervals in order to show his students the cellular dynamics of fertilization and early development of the sea urchin egg in a 2-minute film, a process that normally takes 14 hours (Landecker 2009). The original idea to give movement to static images persists until today, given that the incorporation of the temporal dimension to the study of biological processes confers realism and makes often imperceptible phenomena visible when compared to fixed samples.

Since Ries's films we can currently visualize single-cell resolution of many biological processes. Given the increasing variety of microscopy imaging techniques that have been developed, together with a broad set of cell labeling methods, direct visualization of single cells within their context in a wide range of organisms by time-lapse imaging has become an easier and more reliable way to observe cell behavior during development. At present, the phrase "One look (picture) is worth a thousand words" could be easily updated to "One movie is worth a thousand pictures."

8.2   LIVE IMAGING TO STUDY SEQUENTIAL SEGMENTATION IN VERTEBRATES

The extraordinarily dynamic nature of axial elongation and segment formation that take place in all vertebrates analyzed so far has been revealed paradoxically by decades of experiments based mainly on static images. Since the first studies at the end of the 1980s (Wilson et al. 1989; Thorogood and Wood 1987), a modest number of publications have incorporated time-lapse imaging in their analysis. Although the boom of the use of this technique to understand the elongation and segmentation processes is still yet to come, some recent studies in vertebrates (and also arthropods) have demonstrated that live imaging is a powerful tool for investigating the complex relationship between cellular behavior and the molecular machinery underlying germband extension and the ordered formation of body segments.

Despite the small contribution of time-lapse analysis (in relation to the relative number of papers), there are some key points during axial extension and segmentation in vertebrates where the use of live imaging was essential or at least very influential to reach our current understanding of these processes. These include: (1) cellular dynamics during segmental border formation, (2) the segmentation clock including the control of segment size and number, and (3) the migratory behavior of presomitic mesoderm (PSM) cells during axial elongation, especially the comparison between posterior and anterior PSM.

8.2.1   CELL DYNAMICS DURING SOMITE BORDER FORMATION

During somitogenesis, the sequential formation of regularly sized segments implies the passage from an unsegmented tissue composed of undifferentiated cells to an ordered and epithelialized somite. Moreover, at the transition zone, a segment boundary is formed that separates the PSM from the newly formed somite. Although the molecular mechanisms involved in this process are well documented, the molecular dynamics and their cellular context are less understood.

Time-lapse recording using Nomarski differential interference contrast (DIC) microscopy, without any labeling but taking advantage of the transparency of the embryos of the teleost fish *Barbus conchonius*, allowed Peter Thorogood and Andrew Wood to describe for the first time the modest individual cell rearrangements that take place during the formation of a new somite boundary (Thorogood and Wood 1987; Wood and Thorogood 1994). Similar results were obtained years later in zebrafish, when the analysis of time-lapse imaging showed that intersomitic borders are formed both in wild type (WT) as well as in convergent extension mutant embryos, based on local epithelial movements (Henry et al. 2000). The separation between the new somite and the posteriorly located PSM occurred despite the absence of mediolateral convergence in these mutants, maintaining the cellular behaviors observed in the WT. Additionally, the lack of internal mesenchymal cells within mutant somites (who only have border cells) revealed that these cells are dispensable for somite boundary formation (Henry et al. 2000).

When cellular movements were analyzed during segmentation in the chicken embryo, *in ovo* time-lapse imaging showed that far from being an ordered division of a pre-patterned PSM, as had been inferred from fish and *Xenopus* embryos (Thorogood and Wood 1987; Wilson et al. 1989; Henry et al. 2000), avian somite border formation was a dynamic process that involves high cell motility (Kulesa and Fraser 2002). Chick eggs were injected with the vital dye Bodipy ceramide in order to fluorescently label cell membranes and track them at the site of the presumptive somite border formation. The high spatiotemporal resolution that was achieved (10–15 μm thick *z*-stacks captured every 2 minutes) revealed a "ball and socket" separation of the recently formed somite from the anterior part of the PSM (forming a hollow structure of cells that contains the ball-shaped somite). Here, different groups of cells move in opposed directions, exchanging positions in the anterior–posterior axis. In the end, original posterior cells become integrated into the forming somite and cells that begin at a more anterior position end up in the anterior PSM, leaving a gap between the new somite and the presomitic mesoderm, which creates the somite border (Kulesa and Fraser 2002).

At the molecular level, the vertebrate segmental boundary is also defined at the anterior PSM, where posterior-to-anterior traveling waves of gene expression (known as the segmentation clock; see Chapter 5) are arrested by a determination front mainly defined by posterior Wnt/FGF gradients. Since segmentation is coupled to axis elongation, this "wavefront" moves posteriorly as the embryo grows, converting the temporal progression of oscillations into spatially patterned segments (Palmeirim et al. 1997; Dubrulle et al. 2001; Aulehla et al. 2003; Dubrulle and Pourquié 2004). Just anterior to this front, the dynamic expression of a key transcription factor, Mesoderm Posterior 2 (Mesp2 in mice; related to cMeso-1 in chickens, Thylacine 1 in *Xenopus* and Mesp-a/-b in zebrafish), is necessary to arrest oscillations and define the segmental border, as well as to establish the rostrocaudal somite orientation (Saga et al. 1997; Buchberger et al. 1998; Sparrow et al. 1998; Sawada et al. 2000).

In an attempt to reproduce the endogenous expression pattern of Mesp2 and, thus, correlate its localization with the presumptive new somite border, Morimoto and colleagues generated a Mesp2-venus knock-in mouse (Morimoto et al. 2005). They visualized the expression of Mesp2-venus at the anterior border of the future somite *in vivo*, before a morphologically recognizable limit is formed at the PSM. This evidence, combined with Mesp2 and Notch1 activity (based on the presence of the Notch intracellular domain) double immunostaining, allowed them to propose that new somite boundaries are formed at the border between Notch1 activity and Mesp2 expression. However, they only managed to reproduce one cycle of Mesp2-venus expression in their PSM culture conditions, losing part of the dynamic behavior of the Mesp2 endogenous protein (Morimoto et al. 2005).

Years later, by using double destabilized luciferase reporters of Mesp2 and Fgf signaling (Mesp2–UbEluc and Dusp4–UbSLR, respectively) and live imaging, Niwa et al. (2011) showed the periodic onset of Mesp2 expression at the prospective limit between newly formed somites *in vivo* and that this dynamic expression was the consequence of the cyclic cutoff in Fgf–Erk activity oscillation (Niwa et al. 2011). This last part was well supported by time-lapse analysis of the expression of both reporters in mice where oscillations were not operating. For example, in Hes7-null mice, where the oscillatory expression of *Hes7* (a Notch target gene) is abolished, the authors found no sign of phosphorylated Erk (the active form of the Fgf effector Erk) oscillations and a desynchronized onset of Mesp2. A similar effect was found after blocking the Fgf signaling pathway with SU5402 (Fgf receptor inhibitor): no Fgf–Erk activity and a premature and continuous expression of Mesp2 (Niwa et al. 2011). Interestingly, Erk activity was then uncovered in zebrafish as an earlier marker of the future somite boundary than *mesp-b* expression, showing that segmentation was established at least three segments before it was expected (Akiyama et al. 2014). For this they used time-lapse images to track photoconverted cells at the presumptive position of Erk activity, in a transgenic line where the photoconvertible fluorescent protein KikGR is expressed under the *her1* promoter. Several rounds of segmentation later, photoconverted cells were found at the anterior border of the formed somite, confirming p-Erk as an earlier somite boundary marker in zebrafish (Akiyama et al. 2014).

Regarding the segmentation clock, the relationship between oscillations and somite border formation is established at the aforementioned wavefront, where cyclic gene expression is arrested. It was known that oscillations gradually slow down from the posterior-to-anterior PSM, but it was not until live imaging allowed for the measurement of the dynamic expression of an oscillator that the molecular behavior at the determination front was analyzed (Lauschke et al. 2013; Shih et al. 2015). Using the *her1:her1-venus* transgenic zebrafish line, where the regulatory region of the *her1* gene drives the cyclic expression of the yellow fluorescent protein Venus fused to Her1 (Delaune et al. 2012; see later), Shih and colleagues were able to track the oscillation period of the *Her1-Venus* reporter across the PSM, with single-cell resolution (Shih et al. 2015). They confirmed the gradual slowing of oscillations along the PSM, estimating that anterior cells have a period of around 1.5 times longer than posterior cells. They also found that the oscillator at the anterior PSM is expressed in alternate segments, showing a peak at the presumptive new somite position. Since they did not find an abrupt cessation in oscillations at the determination front, as was expected, they proposed a new model of segmentation based on phase (and antiphase) gradients instead of the transition from permissive (clock) to restrictive (wavefront) phases, as the clock and wavefront model proposes (Lauschke et al. 2013; Shih et al. 2015; Pais-de-Azevedo et al. 2018).

8.2.2 REAL-TIME IMAGING OF THE SEGMENTATION CLOCK

The theoretical proposition that the process of somitogenesis (and also segment formation in arthropods) was founded on an interacting "clock" and "wavefront" (the clock and wavefront model; Cooke and Zeeman 1976), occurred more than 20 years before the experimental demonstration by Olivier Pourquié's group in 1997 (Palmeirim et al. 1997). Based on static images of bisected embryos and comparing temporally delayed halves, they showed by *in situ* hybridization that *c-hairy1* gene expression oscillates in the PSM of the chicken embryo for a period of 90 minutes and that each oscillation correlates with the formation of a new somite. Such a segmentation clock operating at the PSM was live-imaged for the first time in the mouse in 2006, almost 10 years after Palmeirim and colleagues' experiments, using a luciferase-based reporter of the cyclic gene *Hes1*, a Notch signaling effector and the mammalian homolog of the *Drosophila hairy* gene (Masamizu et al. 2006).

Since the oscillator period is short, ranging from 25 minutes in zebrafish to approximately 120 minutes in mice (Hubaud and Pourquié 2014), and Hes proteins have short half-lives (around 20 minutes in mice), to visualize the dynamic expression of cyclic genes it is necessary to use reporters with rapid turnover, avoiding the accumulation of the reporter mRNA or protein as somite formation proceeds. For this reason, transgenic animals made for real-time live imaging of cyclic genes had to incorporate sequences that destabilize their mRNA and/or induce the degradation of their protein.

Masamizu and colleagues made use of a mutated Ubiquitin tag resistant to hydrolysis that directed rapid degradation of the Luciferase reporter (Ub-Luc). They replaced the *Hes1* coding region with the *Luciferase* coding region, but maintained the endogenous *Hes1* gene 3'UTR in order to destabilize the reporter mRNA (Masamizu et al. 2006; Soroldoni and Oates 2011). Using this unstable reporter (Hes1-Ub-Luc), they generated transgenic animals and performed time-lapse imaging of *ex utero* cultured explants of the caudal region of mouse embryos. During the 15-hour movie, several (at least 5) successive waves of bioluminescence were propagated from posterior-to-anterior along the PSM. These *in vivo* oscillations showed to be stable in both period and amplitude, in contrast to what was observed *in vitro* with dissociated PSM cells, where period and amplitude varied between cells. Nevertheless, time-lapse analysis of isolated PSM single cells allowed them to demonstrate that oscillations are cell-autonomous and that intercellular communication is important for their stability (Masamizu et al. 2006).

Years later the first fluorescently labeled transgenic mouse was developed by Aulehla and coworkers, who directed the expression of the Venus reporter, an improved version of the yellow fluorescent protein with faster maturation, under the regulation of the Notch cyclic gene *Lunatic fringe* (*Lfng*) promoter region. The *LuVeLu* transgenic mouse, as they named it, also carried a modified PEST domain and the *Lfng* 3'UTR in order to increase protein and mRNA instability, respectively (Aulehla et al. 2008). Using two-photon microscopy and taking advantage of the brightness of the fluorescent Venus protein, the authors obtained higher temporal resolution *in vivo* (8.5 minutes) than what was obtained with the Hes1-Ub-Luc transgenic mouse (20 minutes), while maintaining a tissue-level resolution (Masamizu et al. 2006; Aulehla et al. 2008; Soroldoni and Oates 2011).

Mutant mice carrying a stabilized β-catenin protein that disrupts its posterior-to-anterior PSM expression gradient reported the absence of somite boundaries, as well as the expansion of the posterior PSM. When these mutants were crossed with *LuVeLu* transgenic mice, time-lapse imaging analysis showed that *Lfng-Venus* cyclic expression was not abrogated. On the contrary, multiple oscillations traversed the extended PSM, highlighting that the arrest of oscillations and consequently the formation of the new somite border requires the diminution of β-catenin levels (Aulehla et al. 2008).

The transgenic approach described heretofore includes two kinds of reporters: the bioluminescent activity of the luciferase enzyme and the fluorescent emission of a mutated variant of the jellyfish *Aequorea Victoria* green fluorescent protein (GFP). Both models were developed in mice and their expression was driven by the regulatory region of two genes belonging to the Notch signaling pathway, *Hes1* and *Lfng*.

The first and so far only live reporter of clock oscillations constructed in another model organism was the *her1:her1–venus* transgenic zebrafish line (Delaune et al. 2012; an improved version was then generated by Soroldoni et al. 2014). In order to obtain this cyclic reporter line, Delaune and colleagues tested several constructs and strategies, some of them based on the live cyclic reporters already generated in mice (and commented earlier; Masamizu et al. 2006; Aulehla et al. 2008), but their *Her1-Venus* reporter showed to be the best combination between dynamic instability and detectable expression using confocal microscopy. For real-time imaging of individual PSM cells, they incorporated important features that improved resolution, bringing it to a cellular level (Delaune et al. 2012). They measured and calculated the phase of the oscillations in individual PSM cells for 4–6 hours, starting at the 8-somite stage and acquiring about 30 *z*-stack images every 4 minutes, from crosses between the *her1:her1-venus* line and Notch pathway mutants. They found that cells continued to oscillate but that phase-synchronization between neighbors was lost, indicating that Notch signaling is necessary for neighboring cell synchronization and not for single-cell oscillations. They also discovered by time-lapse analysis that newly divided cells in both WT and mutant backgrounds were more in phase (low phase-shift) with each other than with their neighbor cells. However, only WT sibling cells progressively resynchronized with their neighbors, in contrast to Notch mutant cells that were always desynchronized. In addition, they observed that PSM cells had a marked preference to divide at a certain oscillation phase (off phase in the case of Her1–Venus). Taking into account that 10–15% of the PSM cells are found in M-phase and that each mitosis lasts for a minimum of 15 minutes, as was measured by time-lapse imaging during zebrafish segmentation (Horikawa et al. 2006), this cellular process is considered an important source of noise that can affect cell-to-cell oscillation coupling. Delaune et al. (2012) argued that the tendency of a PSM cell to enter into mitosis at a particular phase would collaborate to minimize the expected developmental noise caused by cell proliferation.

In addition to the crucial role of Notch signaling in cell–cell synchronization during vertebrate segmentation, genes coding for Hes/Her transcription factors that act downstream of the Notch pathway appear to be involved in the establishment of single-cell oscillations by regulating the pace of the clock. Based on mathematical simulations and previous empirical and theoretical work, Julian Lewis (2003) proposed the relationship between the delay in the autoinhibitory feedback loop of Hes/Her proteins and the length of the oscillation period. After that, great progress has been made to uncover the mechanisms underlying the "pacemaker" of the segmentation clock. Time-lapse imaging highly contributed to these advances, particularly by the Kageyama group, who generated transgenic mice carrying the promoter region of the *Hes7* gene fused to the *Luciferase* gene and destabilized by the non-hydrolyzable human ubiquitin variant (G76V) (Masamizu et al. 2006). What they evaluated *in vivo* was the effect of the transcriptional delay on the timing (period) of cyclic gene expression, by removing one, two, or three (all) intron sequences from their Hes7-Ub-Luciferase construct (Takashima et al. 2011; Harima et al. 2013). When the caudal part of transgenic mouse embryos in culture were imaged by time-lapse microscopy, complete elimination of introns (3070 bp transcript) showed the absence of oscillatory expression compared with the transgenic line that contained all three introns (4913 bp transcript) (Takashima et al. 2011), whereas reducing the number of introns to only one accelerates the oscillations and shortens the period of the segmentation clock in 8.8% of cases (Harima et al. 2013).

Using a similar approach, the Kageyama group also generated two mutant versions of a Delta1 (Dll1) knock-in mouse expressing the Dll1-luciferase fusion protein from the endogenous gene locus by the control of a light-inducible expression system; one mutant with only exon sequences (type 1) and the other with all 10 introns plus an extra sequence (type 2) in order to extend the transcript (Shimojo et al. 2016). Time-lapse imaging analysis of the spatiotemporal expression of the Dll1-Luc reporter showed dampened oscillations with smaller amplitudes in both mutants. In addition, type 1 and type 2 mutants exhibited shorter and longer periods, respectively.

As we can see, live imaging of transgenic embryos has clearly shown the role of the Notch signaling pathway in the control of the segmentation clock at two different levels (Oates et al. 2012): single-cell oscillations (Masamizu et al. 2006; Takashima et al. 2011; Harima et al. 2013) and synchronization between neighboring cells (Masamizu et al. 2006; Delaune et al. 2012; Shimojo et al. 2016). Interestingly, the importance of intron delays in the cyclic expression in individual cells seems to be extended to neighboring synchronization, given that the oscillatory expression of the Dll1 protein (also demonstrated by time-lapse imaging by Shimojo and colleagues), a membrane-bound ligand of the Notch signaling pathway, must be related to intercellular communication and the maintenance of the oscillatory rhythm.

Beyond the local control of oscillations, segment formation at the anterior part of the PSM depends on the arrest of the traveling waves of cyclic gene expression that move from posterior-to-anterior regions along the PSM. Real-time imaging analysis has allowed researchers to visualize differences in the oscillating period between posterior and anterior regions within the zebrafish PSM. They found more oscillations near the arrest front (anterior PSM) than close to the posterior PSM and only the anterior cell oscillating period matches the timing of segment formation (Soroldoni et al. 2014). These findings were explained by the existence of a Doppler effect modulating the period of segmentation. To do this, the authors generated a new transgenic line (Her1-Venus or *Looping*), showing oscillatory expression along the entire PSM, improving the transgene construct designed by Delaune and colleagues (2012) that mimicked endogenous oscillations only at anterior and intermediate positions along the PSM. Moreover, Soroldoni and colleagues (2014) used a time-lapse setup developed to perform multidimensional imaging of 20 zebrafish embryos simultaneously with real-time resolution.

Considering the reduction in the number of traveling waves that they also found as the PSM shortens, Soroldoni and coworkers concluded that in zebrafish, both the Doppler effect together with this dynamic wavelength effect (as they call it) cannot account for the scaling of the segment length to the size of the PSM that Lauschke et al. (2013) visualized *ex vivo* in the mouse. Using a monolayer culture of PSM cells (mPSM) from the tail bud mesoderm of *LuVeLu* transgenic mice (Aulehla et al. 2008), the authors analyzed the periodic oscillations of the Lfng-Venus reporter activity by time-lapse imaging. Compared to *in vivo* experiments, *ex vivo* PSM cultured cells showed equivalent oscillation periods, with traveling waves progressing from the center to the periphery (12-15 oscillations), even forming segment boundaries (Lauschke et al. 2013). In order to visualize the scaling process, they measured the oscillation phase of each cell (as Delaune et al. 2012 did *in vivo*) forming part of PSM monolayers of different lengths over time, and thus they could determine the spatial phase changes along the mPSM. First, they found that at the center of the monolayer the period was constant, independent of the mPSM length, and that smaller segments arose from shorter mPSMs in a linear relationship. The quantified slope of the phase gradient (the oscillation phase differences between single mPSM cells) was found to be inversely proportional to the segment size. They concluded that the scaling property of the mPSM relies on the conservation of the phase gradient amplitude independent of the mPSM length (Lauschke et al. 2013).

Taking advantage of the same *ex vivo* culture approaches, two studies performed by the same lab (Alexander Aulehla's group) addressed on the one hand the emergence of the collective synchronization within the PSM (Tsiairis and Aulehla 2016) and on the other the dynamic relationship between Wnt and Notch signaling oscillations (Sonnen et al. 2018), both using time-lapse microscopy.

Tsiairis and Aulehla imaged reaggregated PSM explants where cells were dissociated and mixed in order to lose their original cell–cell interactions and position along the anterior–posterior axis. PSM explants were obtained from *LuVeLu* transgenic embryos (Aulehla et al. 2008) and LuVeLu reporter activity was used to monitor the appearance of *de novo* traveling expression waves. They visualized the formation of regularly spaced multiple foci, or emergent PSM (ePSM) as the authors call them, of synchronized cells oscillating from the center to the periphery. Moreover, the various ePSM formed in a single reaggregated PSM are also synchronized. In addition, Tsiairis and Aulehla performed FACS (fluorescence activated cell sorting) to separate PSM cells from several embryos in two groups (high and low), based on their LuVeLu expression level (intensity). Thus, one cell population was out of phase with respect to the other at their original positions in the intact PSM. When cells were separately mixed, they allowed them to reaggregate and tracked them by time-lapse imaging. The authors found that formed ePSMs retained their oscillatory phases and both were in antiphase with respect to each other (Tsiairis and Aulehla 2016). Together with more interesting and clever experiments combined with the use of time-lapse microscopy, the authors concluded that PSM oscillators have the ability to self-organize forming wave patterns that reflect the integration between the single-cell and the systemic levels.

In the other study, Sonnen and colleagues (2018) generated the first oscillatory Wnt signaling transgenic reporter mouse line, using the promoter of Axin2 (a negative regulator and target of the Wnt/β-catenin pathway). Using this *in vivo* reporter combined with the Notch signaling reporter *LuVeLu* (Aulehla et al. 2008) they performed simultaneous imaging of monolayer PSM cells. Both signaling reporters oscillated out of phase at the center of the mPSM and in phase at the periphery of the monolayer, preceding the region of Mesp2 expression, where Axin2-mediated fluorescence abruptly decreases, which they could visualize by combining the expression of the Axin2 line with a Mesp2-GFP line (Sonnen et al. 2018). In order to study the relative timing of Wnt and Notch signaling oscillations, Sonnen and coworkers set out to control and visualize the rhythm of each oscillator by entraining their cyclic expression on a microfluidic system using pharmacological manipulation combined with time-lapse imaging analysis. Using this approach they found that Notch and Wnt signaling oscillations are coupled, which means that when altering the rhythm of one, the oscillations of the other are also altered, maintaining their relative synchronization. On the other hand, when the authors simultaneously entrained both signaling oscillations, resulting in antiphase oscillations at the "anterior" mPSM, LuVeLu expression arrest was delayed and no sign of a morphological segment appeared, indicating that Wnt/Notch phase shift is crucial for segment formation (Sonnen et al. 2018).

The molecular nature of the segmentation clock forced the use of bioluminescent or fluorescent dynamic reporter systems to visualize and analyze the dynamic expression of cyclic genes, first at the tissue level and later with single-cell resolution (Table 8.1). By using only two animal models, the mouse and zebrafish, some

TABLE 8.1
Oscillatory Reporters Created and Used to Elucidate the Underlying Mechanisms of the Segmentation Clock in Vertebrates

	Signaling Pathway	Reference
In Mice		
Hes1-Ub-Luciferase	Notch-delta	Masamizu et al. 2006*
Hes7-Ub-Luciferase	Notch-delta	Takashima et al. 2011*
(Four variants: one containing all three introns and the		Niwa et al. 2011
others lacking one, two, or all three introns)		González et al. 2013
		Harima et al. 2013
LuVeLu (Lunatic-fringe-Venus/YFP)	Notch-delta	Aulehla et al. 2008*
		Lauschke et al. 2013
		Tsiairis and Aulehla 2016
		Sonnen et al. 2018
Delta1-Ub-Luciferase	Notch-delta	Shimojo et al. 2016*
(And two mutant types: one containing only exon		
sequences and the other having all ten introns plus an		
extra sequence)		
Axin2T2A	Wnt/β-catenin	Sonnen et al. 2018*
(Axin2T2A-mVenus-PEST)		
Axin2T2A-Luci	Wnt/β-catenin	Sonnen et al. 2018*
(Axin2T2A-Luciferase-PEST)		
LfngT2A	Notch-delta	Sonnen et al. 2018*
(Lunatic-fringe-mVenus-PEST)		
In Zebrafish		
Her1-Venus	Notch-delta	Delaune et al. 2012*
		Shih et al. 2015
Looping	Notch-delta	Soroldoni et al. 2014*
(improved version of Her1-Venus)		

*Original article where the transgenic reporter was first described.

research groups have analyzed long-standing questions *in vivo* about the onset of oscillations, their synchronicity, and how cells maintain their rhythm along segmentation. Interestingly, the use of time-lapse imaging in the study of segmentation was not reduced to cell tracking but has been combined with very different approaches, such as intensity kymograph analysis (Aulehla et al. 2008; Takashima et al. 2011; Lauschke et al. 2013; Soroldoni et al. 2014; Tsiairis and Aulehla 2016; Sonnen et al. 2018), FACS (Tsiairis and Aulehla 2016), *ex vivo* and two-dimensional monolayer PSM cultures (Lauschke et al. 2013; Sonnen et al. 2018), mutant backgrounds (Aulehla et al. 2008; Niwa et al. 2011; Takashima et al. 2011; Delaune et al. 2012; Harima et al. 2013; Shimojo et al. 2016) and pharmacological manipulation of signaling pathways in real time and at desired time windows (González et al. 2013; Sonnen et al. 2018), among others.

8.2.3   The Migratory Behavior of the Presomitic Mesoderm Cells

Somites are formed one-by-one from the posterior region of the paraxial mesoderm, the region also known as the presomitic mesoderm. The morphogenetic movements that account for the extension, segmentation, and narrowing of this tissue have been studied using live imaging mainly in the chicken and zebrafish, due to its accessibility, among other technical reasons. Nevertheless, the early studies of Wilson and colleagues performed in *Xenopus* embryos where cell explants of paraxial mesoderm were filmed during full somitogenesis, described similar cell behaviors to what Thorogood and Wood described 2 years before in fish (Thorogood and Wood 1987; Wilson et al. 1989). Convergent extension movements, including first radial intercalation, followed by mediolateral intercalation and finally cell-shape changes that occurred in a sequential fashion that was repeated from posterior-to-anterior PSM over time. Wilson and coauthors found that only the most posterior region of the PSM contributed to tissue elongation and that neither cell-shape changes nor cell division were the driving force of axis extension in *Xenopus* embryos but solely cell rearrangements (Wilson et al. 1989). Staining the membranes of small groups of cells at different positions along the PSM by iontophoretic injection of the fluorescent lipophilic dye DiI in chicken eggs allowed Kulesa and Fraser (2002), using time-lapse imaging, to discover that cells from the posterior PSM move more and exchange neighbors in a more dynamic way, compared to the anterior PSM. This is similar to what Wilson and colleagues found in the frog. When Bénazéraf and colleagues tracked the orientation of these cell movements they found a decreasing posterior-to-anterior gradient of cell motility with an opposed gradient of cell density, where all PSM cells move toward the caudal end of the embryo and only anterior cells show convergent movements directed to the midline (Bénazéraf et al. 2010). Nevertheless, given that embryo elongation by itself causes tissue deformation, the authors cleverly subtracted the movements displayed by the extracellular matrix, stained with a fluorescently labeled anti-fibronectin antibody, from the cell displacements tracked by H2B-GFP. The remaining cell movements showed random instead of directional motility. Moreover, by electroporating an inducible (in order to avoid early developmental events) dominant negative FGF receptor in the PSM and tracking cell movements by time-lapse imaging, Bénazéraf and coworkers found that cell motility and the concomitant elongation depend on FGF signaling. In addition, when they overexpressed FGF8 in the cells of the PSM, cell motility at the anterior PSM increases, disrupting the motility gradient and reducing tissue elongation. Analyzing a long series of 13 time-lapse movies, Bénazéraf and colleagues (2010) proposed a model of PSM cell motility where they concluded that the FGF-dependent gradient of random cell motility along the PSM causes the opposed cell density gradient that directs the elongation toward the posterior end of the embryo. Previous studies from the same group demonstrated years before that the posterior-to-anterior motility gradient in the PSM was regulated by the Mapk/Erk pathway and that this pathway was in turn controlled by FGF signaling (Delfini et al. 2005). For this they performed live imaging of mosaic chicken PSM cells electroporated with the pCIG vector (that contains GFP as a reporter) fused to the MKK1*ca* or MKK1*dn* constructs, that contained a constitutively active or a dominant negative form of MKK1 (the MAPK kinase), respectively. They compared velocity, cell directionality, and cell dispersion using time-lapse imaging and cell tracking, finding that activated MAPK/ERK signaling increases cell velocity and dispersion. On the other hand, MKK1-inhibited cells were slower and formed patches that distinguished them from the WT/non-electroporated cells (Delfini et al. 2005). Taken together, these findings suggest that cell migration during posterior growth in vertebrates is not directed by mechanisms such as cell polarization toward or against chemical stimuli but by an FGF gradient that, via Mapk/Erk, maintains single cell movement without specific directionality but depends on the levels of FGF/Erk activity. This generates more cell displacement at the posterior that triggers a reduction in cell density, which directs collective cell migration from high to low density by "simple" diffusion (Delfini et al. 2005; Bénazéraf et al. 2010; Bénazéraf and Pourquié 2013).

Time-lapse imaging of zebrafish PSM cells has been used mainly to visualize the oscillatory expression of the cyclic gene *her1* (Delaune et al. 2012; Soroldoni et al. 2014; Shih et al. 2015). Some of the cellular processes involved in axial elongation and segmentation were covered using live imaging approaches by at least two studies. Cell division (Horikawa et al. 2006) and cell motility (Lawton et al. 2013) were analyzed in the context of their effect on the synchronization of oscillations (commented before) or cell motility through different regions of the tailbud, respectively. The time-lapses and cell tracking performed by Lawton and colleagues in zebrafish embryos injected with *nlsRFP* mRNA showed a transition in the coherence of cell flow from the dorsal medial zone (DMZ) to the progenitor zone (PZ) and then to the PSM. The velocity during the collective cell migration decreases, revealing changes in tissue fluidity, in addition to the loss of coherence between cells, which means that neighbor cells within the PZ and PSM migrate in different directions compared to the DMZ (Lawton et al. 2013). Moderate reduction of the Wnt and FGF signaling pathways by injecting *notum1a* (Wnt inhibitor) mRNA or incubating embryos with the FGF inhibitor SU5402 allowed authors to visualize the effect of these signaling pathways on cell fluidity within the tailbud using time-lapse microscopy. The reduction in Wnt or FGF signaling mainly showed a loss of cell migration coherence within the DMZ, without affecting this parameter in the PZ or PSM regions.

The question that arises after understanding some of the cell behaviors that take place during vertebrate posterior growth and axial elongation is how they are coordinated with the establishment of the segmental patterning, especially because the same signaling pathways appear to be implied in both processes and cellular dynamics are possibly influencing (and influenced by) the molecular behavior of the cells. Given that the use of live imaging is still in the early stages, the expectations are as high as the growing number of research groups interested in the molecular and cellular mechanisms underlying axial elongation and segmentation in vertebrates (and arthropods, later below).

8.3   LIVE IMAGING DURING ANNELID SEQUENTIAL SEGMENTATION

In addition to vertebrates, both annelids and most arthropods also display sequential segment formation from a posterior region (similar to the PSM), in this case called the posterior growth zone or the segment addition zone (SAZ). Annelids are extremely variable in their morphology, and segmentation in this phylum has been understudied compared to vertebrates and arthropods (Balavoine 2014; see Chapter 4). Few studies have included long-term live imaging in their analysis. For example, Zattara and colleagues (2016; also see Chapter 10) evaluated cell migration during anterior and posterior regeneration of the adult freshwater annelid *Pristina leidyi*, based on cell morphology and migratory characteristics using DIC microscopy on unlabeled tissue. Recently, Özpolat and coworkers (2017) performed time-lapse imaging at single-cell resolution of embryonic and larval stages of the polychaete *Platynereis dumerilii*, by injecting different constructs in order to label cell components such as nuclei (HistoneH2A-mCherry) and the cell membrane (EGFP-caax) or to visualize cell cycle progression. For the latter they developed a fluorescent cell cycle reporter based on FUCCI, a fluorescent ubiquitination-based cell cycle indicator. Pd-FUCCI was made by two constructs, the HistoneH2A-mCherry that directed the expression of mCherry to all cell nuclei and the mVenus-Cdt1(aa1-147), composed by a truncated form of Cdt1 (a cell cycle protein) that lacked sequences for DNA binding domains but contained an endogenous ubiquitination site that conferred the necessary instability for the dynamic expression of the reporter, fused to the fluorescent protein mVenus. The cyclic expression was observed during late-G2/mitosis/G1 phases and allowed the authors to perform germline and mesodermal cell lineage analysis that showed the cellular origin of the early swimming larva mesodermal segments (formed by simultaneous segmentation) as well as the origin of the posterior growth zone, from where the SAZ is formed and sequential segmentation proceeds later during juvenile development (Özpolat et al. 2017).

8.4   LIVE IMAGING DURING ARTHROPOD SEQUENTIAL SEGMENTATION

Among arthropods, most of our small live imaging collection regarding sequential segmentation and axial elongation comes from time-lapse analyses performed on germbands of the isopod crustacean *Porcellio scaber* (Hejnol et al. 2006) and fluorescently labeled embryos of the red flour beetle *Tribolium castaneum* (El-Sherif et al. 2012; Sarrazin et al. 2012; Benton et al. 2013, 2016; Zhu et al. 2017; Benton 2018) and the spider *Parasteatoda tepidariorum* (Hemmi et al. 2018). In addition, some studies in the cricket *Gryllus bimaculatus*, the amphipod crustacean *Parhyale hawaiensis* and also *T. castaneum*, have indirectly covered part of these processes (Nakamura et al. 2010; Strobl and Stelzer 2014; Donoughe and Extavour 2015; Wolff et al. 2018).

A conserved feature of malacostracan, the largest crustacean class that includes the wood louse *P. scaber* and the sand flea *P. hawaiensis*, is that their posterior or post-naupliar segments are formed by cells that divide in a very stereotyped pattern (see Chapter 6). Using 4D-Nomarski microscopy, Hejnol and colleagues (2006) generated a detailed description of the post-naupliar cell division patterns and tracked the lineages of individual cells at the transitional zone between the naupliar and post-naupliar regions of *P. scaber*. Within this zone, they detected irregular cell divisions and unexpected cell movements that allowed them to establish phylogenetic relationships between different orders of malacostracan crustaceans (tanaids and isopods).

In 2012, two groups demonstrated, based on different approaches that included live fluorescent imaging, the presence of a segmentation clock operating in the segment addition zone of *T. castaneum* (El-Sherif et al. 2012; Sarrazin et al. 2012). The use of time-lapse imaging and cell tracking was essential to correlate the expression waves of the pair-rule genes *Tc-even-skipped* (*Tc-eve*; El-Sherif et al. 2012) and *Tc-odd-skipped* (*Tc-odd*; Sarrazin et al. 2012) suggested by the detailed analysis of multiple fixed embryos with changes in the expression levels within cells rather than cell movements. For this, Sarrazin and colleagues (2012) generated the EFA-nGFP transgenic line that expresses a ubiquitous nuclear-localized GFP under the control of the regulatory sequences of the EF1-α gene. Given that during *T. castaneum* elongation, the posterior region of the germband, including the SAZ, is surrounded by opaque yolk cells that interfered with fluorescence visualization, they decided to dissect the embryo and mount it flat prior to performing the time-lapse recording, for which they developed a whole embryo culture. In a one-hour time-lapse fluorescence recording of a dissected *T. castaneum* germband (8 *z*-stacks captured every 3 minutes), Sarrazin and coworkers tracked cell nuclei during the dynamic formation of the forth *Tc-odd* stripe. In the case of El-Sherif and colleagues, they used the same transgenic line as Sarrazin et al. (2012) to track cell movements from blastoderm formation to germband condensation, covering 11 hours (5 *z*-stacks captured every 5 minutes). Since cell displacements within the SAZ and blastoderm could not explain the oscillatory behavior of *Tc-odd* and *Tc-eve* respectively, both groups suggested that a vertebrate-like molecular clock was probably underlying the sequential segmentation in arthropods. It is worth noting that very recently Hemmi and colleagues (2018) demonstrated the oscillatory expression of the segment polarity gene *hedgehog* (*Pt-hh*) in the opisthosomal region (SAZ) of the spider *Parasteatoda tepidariorum* using a similar approach. For that they tracked fluorescently labeled cell clones (injected with mRNAs coding for NLS-tdEosFP or NLS-tdTomato) during the dynamic expression of *Pt-hh* and they found that cells persisted in the area of Pt-hh waves, indicating that cell movements were not responsible for this oscillatory behavior.

Over the following years, three interesting scientific papers were published, which used live imaging and fluorescent labeling approaches in *T. castaneum* (Benton et al. 2013; Strobl and Stelzer 2014, 2016). Strobl and Stelzer performed, for the first time, long-term live imaging of *T. castaneum* EFA-nGFP embryos by light sheet–based fluorescence microscopy (LSFM), covering the complete embryogenesis with a high spatiotemporal resolution and reduced photodamage (Strobl and Stelzer 2014). In addition, they presented a new transgenic line (FNL; Strobl and Stelzer 2016) that together with the combined LAN-GFP line (LifeAct-EGFP plus EFA1-nGFP; Sarrazin et al. 2012; Benton et al. 2013) presented by van Drongelen and coworkers (2018), expanded the transgenic and microscopy resources for the study of axial elongation and segmentation in *T. castaneum* (see Tables 8.2 and 8.3). In the same

TABLE 8.2

Fluorescent Transgenic Lines and Labeling Approaches for Live Imaging Used or Feasible to Use to Study Axial Elongation and Sequential Segmentation in Different Arthropods

	Signaling Pathway	Reference
In *Tribolium* (Hexapoda)		
EFA-nGFP	Nuclei	Sarrazin et al. 2012*
		El-Sherif et al. 2012
		Strobl and Stelzer 2014
H2B-RFP (for transient expression)	Chromatin	Benton et al. 2013*
LifeAct-EGFP (for transient expression)	F-actin	Benton et al. 2013*
GAP43-YFP (for transient expression)	Cell membrane	Benton et al. 2013*
		van der Zee et al. 2015[†]
LAN-GFP (LifeAct-EGFP plus EFA1-nGFP)	F-actin (LifeAct) Nuclei (EFA1)	van Drongelen et al. 2018*[†]
FNL line (Histone2B-EGFP)	Nuclei	Strobl and Stelzer 2016*[†]
H2B-*Venus* (for transient expression)	Nuclei	Benton 2018*
NLS-td*Eos* (nuclear localization signal-tandem *Eos*; for transient expression; photoconvertable)	Nuclei	Benton 2018*
In *Gryllus* (Hexapoda)		
pBGact-*eGFP*	Nuclei and cytoplasm	Nakamura et al. 2010*[†]
pXLBGact-*actin:eGFP*	Cytoplasm	Nakamura et al. 2010*[†]
pXLBGact-*Histone2B:eGFP*	Nuclei	Nakamura et al. 2010*[†]
In *Parasteatoda* (Chelicerata)		
NLS-tdEosFP (for transient expression)	Nuclei	Hemmi et al. 2018*
NLS-td*Tomato* (for transient expression)	Nuclei	Hemmi et al. 2018*
H1-tdEosFP (for transient expression)	Nuclei	Hemmi et al. 2018*
In *Parhyale* (Crustacea)		
PhHsp70-DsRed-NLS	Nuclei	Hannibal et a. 2012
PhHS-H2BmRFPruby (for transient expression)	Nuclei	Wolff et al. 2018*
EGFP-caax (for transient expression)	Cell membrane	Wolff et al. 2018*

*Original article where the transgenic reporter/labeling was first described.
[†]Live imaging feasible to use.

TABLE 8.3
Time-Lapse Movie Links

Akiyama et al. 2014

http://dev.biologists.org/content/develop/suppl/2014/02/12/141.5.1104.DC1/DEV098905.pdf

Aulehla et al. 2008 (mouse)

https://www.nature.com/articles/ncb1679#supplementary-information

Benton 2018 (beetle)

https://journals.plos.org/plosbiology/article?id=10.1371/journal.pbio.2005093#sec013

Benton et al. 2013 (beetle)

http://dev.biologists.org/content/develop/suppl/2013/07/11/140.15.3210.DC1/DEV096271.pdf

Delaune et al. 2012 (zebrafish)

https://www.sciencedirect.com/science/article/pii/S1534580712004200?via%3Dihub#app2

Delfini et al. 2005 (chicken)

http://www.pnas.org/content/102/32/11343/tab-figures-data

El-Sherif et al. 2012 (beetle)

http://dev.biologists.org/content/139/23/4341.supplemental

Harima et al. 2013 (mouse)

https://www.sciencedirect.com/science/article/pii/S2211124712003932?via%3Dihub#mmc1

Hejnol et al. 2006 (crustacean)

https://link.springer.com/article/10.1007/s00427-006-0105-4#SupplementaryMaterial

Horikawa et al. 2006 (zebrafish)

https://www.nature.com/articles/nature04861#supplementary-information

Masamizu et al. 2006 (mouse)

http://www.pnas.org/content/103/5/1313/tab-figures-data#M1

Nakamura et al. 2010 (cricket)

https://www.sciencedirect.com/science/article/pii/S0960982210009462?via%3Dihub#mmc6

Sarrazin et al. 2012 (beetle)

www.sciencemag.org/cgi/content/full/science.1218256/DC1

Shimojo et al. 2016 (mouse)

http://genesdev.cshlp.org/content/30/1/102/suppl/DC1

Sonnen et al. 2018 (mouse)

https://www.sciencedirect.com/science/article/pii/S009286741830103X?via%3Dihub#mmc2

Soroldoni et al. 2014 (zebrafish)

http://science.sciencemag.org/content/suppl/2014/07/09/345.6193.222.DC1

Strobl and Stelzer 2016 (beetle)

https://ars.els-cdn.com/content/image/1-s2.0-S2214574516300955-mmc1.mp4

Takashima et al. 2011 (mouse)

www.pnas.org/lookup/suppl/doi:10.1073/pnas.1014418108/-/DCSupplemental/sm01.avi

www.pnas.org/lookup/suppl/doi:10.1073/pnas.1014418108/-/DCSupplemental/sm02.avi

Wolff et al. 2018 (crustacean)

https://elifesciences.org/articles/34410#fig1video2

path, Benton and colleagues (Benton et al. 2013) presented the combined use of transient fluorescence labeling and RNAi by the coinjection of the different cell marker constructs H2B-RFP, LifeAct-EGFP, and GAP43-YFP to label chromatin, F-actin, and the cell membrane, respectively, and dsRNA. The combination of these labeling strategies with live imaging allowed the authors to track cell movements during blastoderm formation, germband condensation, and elongation both in WT and RNAi (*Tc-caudal* knock-down) embryos. Using the same strategy, Benton et al. (2016) observed how difficult it is for the double knock-down *Tc-Toll7-Toll10* embryos transiently labeled with GAP43-YFP to elongate along the anterior–posterior axis and how cell intercalation was affected during embryo condensation. These results were essential to confirm the involvement of a new subfamily of Toll genes, the Long Toll or Loto clade, in germband elongation in *T. castaneum*, and together with results in six other groups of arthropods, they suggest a conserved role of these genes in axis elongation and that this function was probably present in the last common ancestor of all arthropods (Benton et al. 2016).

Very recently, making use of live cell tracking and a photoconvertible fluorescent construct (the nuclear-localized *nls-tdEos*), Matthew Benton (2018) challenged the preconceived idea that dorsal epithelium during *T. castaneum* germband elongation was formed completely by the amnion (an extraembryonic tissue). By photoconverting small groups of dorsal cells he found that some of these cells became part of the embryonic tissue by moving to the ventral epithelium along the entire germband. This finding changed the current fate map from late blastoderm to the elongating germband stages, increasing the number of ectodermal cells involved in the axial elongation and segmentation. In addition, Benton found mediolateral cell intercalations within the SAZ during elongation, a cellular behavior suggested to be absent during this process by previous work based on static images (Nakamoto et al. 2015).

8.5 CONCLUSIONS

Live imaging has the power to transform our understanding of the cellular and molecular events that are involved in highly dynamic processes, such as morphogenesis in embryonic development. Axial elongation and the sequential formation of segments both in arthropods and vertebrates, as well as in annelids, have proven to be a very complex mix of cell behaviors and molecular mechanisms combined in different ways depending on the developmental phase and the studied organism. In order to capture the dynamic nature of the process, incorporation of real-time *in vivo* analysis has been and will be of central importance.

What the use of live imaging for studying axial elongation and segmentation has achieved can be summarized into these three points:

- *Putting experimental results into a developmental context, bringing the analysis from the cellular and molecular levels to the systemic level.* For example, as was described in Chapter 5 (and briefly in this chapter), the existence of a segmentation clock, both in vertebrates and arthropods, had been demonstrated using series of fixed samples (bisected embryos). Nevertheless, the visualization *in vivo* that each new somite appears precisely after (temporally) and anterior (spatially) to every new wave of gene expression oscillation, provided the necessary evidence that a segmentation clock is operating and showed how the different processes involved are dynamically coordinated. In turn, in arthropods, cell tracking by time-lapse imaging showed beyond doubt that cell movements do not account for the changes in the oscillating expression, demonstrating the existence of a vertebrate-like molecular clock in the beetle *T. castaneum* and in the spider *P. tepidariorum*. The same kind of contextualization occurred with many other experimental results. For example, the demonstration that single-cell oscillations determine their period from the delay on the autoinhibitory feedback loop of Hes/Her proteins (that is, the time that it takes to be transcribed into a mRNA and translated in a ribosome to produce a protein that finally inhibits its own transcription), was elegantly explained using *in vivo* imaging and transgenic animals that showed shorter periods of oscillations with shorter transcripts.

- *Increasing data acquisition details in order to improve temporal and spatial resolution, unveiling cellular and molecular dynamics often hidden behind static sampling.* Technical advances in microscopy and new molecular tools allowed researchers to achieve the single-cell resolution, obtaining (quantifying) among other things, the oscillation phase, signaling activity, migratory behavior, and cell division status, of each cell along the entire PSM during many hours of development, as well as the ectodermal origin of part of the dorsal epithelium and the cell intercalation behavior within the SAZ in Tribolium germbands, leading to the proposition of new models of axis elongation and segmentation both in vertebrates and insects.
- *Matching cell behaviors with the molecular machinery underlying posterior growth and segmentation.* The combination of live imaging and genetic tools also enables the visualization of gene expression dynamics at the same time that segment border formation, PSM cells migration, and cell divisions were occurring, especially in a mutant context or using pharmacologically manipulated embryos. In this way researchers could find the relationship between Notch signaling and the cell-to-cell synchronization of oscillations, how cells avoid the interference of cell division within oscillations, the coupling rhythm between Wnt and Notch signaling pathways and its importance in somite formation, as well as that the nondirectional cell migration that operates at the PSM during vertebrate axis elongation is dependent on FGF signaling.

REFERENCES

Akiyama, R., M. Masuda, S. Tsuge, Y. Bessho, and T. Matsui. 2014. An anterior limit of FGF/Erk signal activity marks the earliest future somite boundary in zebrafish. *Development* 141: 1104–1109.

Aulehla, A., C. Wehrle, B. Brand-Saberi, R. Kemler, A. Gossler, B. Kanzler, …, B. G. Herrmann. 2003. Wnt3a plays a major role in the segmentation clock controlling somitogenesis. *Dev. Cell* 4: 395–406.

Aulehla, A., W. Wiegraebe, V. Baubet, M. B. Wahl, C. Deng, M. Taketo, …, O. Pourquié. 2008. A β-catenin gradient links the clock and wavefront systems in mouse embryo segmentation. *Nat. Cell Biol.* 10(2): 186–193.

Balavoine, G. 2014. Segment formation in annelids: Patterns, processes and evolution. *Int. J. Dev. Biol.* 58(6–8): 469–483.

Bénazéraf, B., P. Francois, R. E. Baker, N. Denans, C. D. Little, and O. Pourquié. 2010. A random cell motility gradient downstream of FGF controls elongation of an amniote embryo. *Nature* 466: 248–252.

Bénazéraf, B., and O. Pourquié. 2013. Formation and segmentation of the vertebrate body axis. *Ann. Rev. Cell Dev. Biol.* 29: 1–26.

Benton, M. A. 2018. A revised understanding of *Tribolium* morphogenesis further reconciles short and long germ development. *PLoS Biol.* 16(7): e2005093.

Benton, M. A., M. Akam, and A. Pavlopoulos. 2013. Cell and tissue dynamics during *Tribolium* embryogenesis revealed by versatile fluorescence labeling approaches. *Development* 140: 3210–3220.

Benton, M. A., M. Pechmann, N. Frey, D. Stappert, K. H. Conrads, Y. T. Chen, …, S. Roth. 2016. *Toll* genes have an ancestral role in axis elongation. *Curr. Biol.* 26: 1–7.

Buchberger, A., K. Seidl, C. Klein, H. Eberhardt, and H. H. Arnold. 1998. cMeso-1, a novel bHLH transcription factor, is involved in somite formation in chicken embryos. *Dev. Biol.* 199: 201–215.

Cooke, J., and E. C. Zeeman. 1976. A clock and wavefront model for control of the number of repeated structures during animal morphogenesis. *J. Theor. Biol.* 58(2): 455–476.

Delaune, E. A., P. François, N. P. Shih, and S. L. Amacher. 2012. Single-cell-resolution imaging of the impact of Notch signaling and mitosis on segmentation clock dynamics. *Dev. Cell* 23: 995–1005.

Delfini, M.-C., J. Dubrulle, P. Malapert, J. Chal, and O. Pourquié. 2005. Control of the segmentation process by graded MAPK/ERK activation in the chick embryo. *Proc. Natl. Acad. Sci. USA* 102(32): 11343–11348.

Donoughe, S., and C. G. Extavour. 2015. Embryonic development of the cricket *Gryllus bimaculatus*. *Dev. Biol.* 411(1): 140–156.

Dubrulle, J., M. J. McGrew, and O. Pourquié. 2001. Fgf signaling controls somite boundary position and regulates segmentation clock control of spatiotemporal *Hox* gene activation. *Cell* 106: 219–232.

Dubrulle, J., and O. Pourquié. 2004. *Fgf8* mRNA decay establishes a gradient that couples axial elongation to patterning in the vertebrate embryo. *Nature* 427: 419–422.

El-Sherif, E., M. Averof, and S. J. Brown. 2012. A segmentation clock operating in blastoderm and germband stages of *Tribolium* development. *Development* 139: 4341–4346.

González, A., I. Manosalva, T. Liu, and R. Kageyama. 2013. Control of *Hes7* expression by Tbx6, the Wnt pathway and the chemical Gsk3 inhibitor LiCl in the mouse segmentation clock. *PLoS One* 8: e53323.

Hannibal, R. L., A. L. Price, and N. H. Patel. 2012. The functional relationship between ectodermal and mesodermal segmentation in the crustacean, *Parhyale hawaiensis*. *Dev. Biol.* 361: 427–438.

Harima, Y., Y. Takashima, Y. Ueda, T. Ohtsuka, and R. Kageyama. 2013. Accelerating the tempo of the segmentation clock by reducing the number of introns in the *Hes7* gene. *Cell Rep.* 3(1): 1–7.

Hejnol, A., R. Schnabel, and G. Scholtz. 2006. A 4D-microscopic analysis of the germ band in the isopod crustacean *Porcellio scaber* (Malacostraca, Peracarida)- developmental and phylogenetic implications. *Dev. Genes Evol.* 216(12): 755–767.

Hemmi, N., Y. Akiyama-Oda, K. Fijimoto, and H. Oda. 2018. A quantitative study of the diversity of stripe-forming processes in an arthropod cell-based field undergoing axis formation and growth. *Dev. Biol.* 437: 84–104.

Henry, C. A., L. A. Hall, M. B. Hille, L. Solnica-Krezel, and M. S. Cooper. 2000. Somites in zebrafish doubly mutant for *knypek* and *trilobite* form without internal mesenchymal cells or compaction. *Curr. Biol.* 10: 1063–1066.

Horikawa, K., K. Ishimatsu, E. Yoshimoto, S. Kondo, and H. Takeda. 2006. Noise-resistant and synchronized oscillation of the segmentation clock. *Nature* 441: 719–723.

Hubaud, A., and O. Pourquié. 2014. Signalling dynamics in vertebrate segmentation. *Nat. Rev. Mol. Cell Biol.* 15: 709–721.

Kulesa, P. M., and S. E. Fraser. 2002. Cell dynamics during somite boundary formation revealed by time-lapse analysis. *Science* 298: 991–995.

Landecker, H. 2009. Seeing things: From microcinematography to live cell imaging. *Nat. Methods* 6(10): 707–709.

Lauschke, V. M., C. D. Tsiairis, P. François, and A. Aulehla. 2013. Scaling of embryonic patterning based on phase-gradient encoding. *Nature* 493: 101–106.

Lawton, A. K., A. Nandi, M. J. Stulberg, N. Dray, M. W. Sneddon, W. Pontius, T. Emonet, and S. A. Holley. 2013. Regulated tissue fluidity steers zebrafish body elongation. *Development* 140: 573–582.

Lewis, J. 2003. Autoinhibition with transcriptional delay: A simple mechanism for the zebrafish somitogenesis oscillator. *Curr. Biol.* 13: 1398–1408.

Masamizu, Y., T. Ohtsuka, Y. Takashima, H. Nagahara, Y. Takenaka, K. Yoshikawa, …, R. Kageyama. 2006. Rel-time imaging of the somite segmentation clock: Revelation of unstable oscillators in the individual presomitic mesoderm cells. *Proc. Natl. Acad. Sci. USA* 103(5): 1313–1318.

Morimoto, M., Y. Takahashi, M. Endo, and Y. Saga. 2005. The Mesp2 transcription factor establishes segmental borders by suppressing Notch activity. *Nature* 435: 354–359.

Nakamoto, A., S. D. Hester, S. J. Constantinou, W. G. Blaine, A. B. Tewksbury, M. T. Matei, …, T. A. Williams. 2015. Changing cell behaviours during beetle embryogenesis correlates with slowing of segmentation. *Nat. Commun.* 6: 6635.

Nakamura, T., M. Yoshizaki, S. Ogawa, H. Okamoto, Y. Shinmyo, T. Bando, …, T. Mito. 2010. Imaging of transgenic cricket embryos reveals cell movements consistent with a syncytial patterning mechanism. *Curr. Biol.* 20: 1641–1647.

Niwa, Y., H. Shimojo, A. Isomura, A. González, H. Miyachi, and M. Kageyama. 2011. Different types of oscillations in Notch and Fgf signaling regulate the spatiotemporal periodicity of somitogenesis. *Genes Dev.* 25: 1115–1120.

Oates, A. C., L. G. Morelli, and S. Ares. 2012. Patterning embryos with oscillations: Structure, function and dynamics of the vertebrate segmentation clock. *Development* 139: 625–639.

Özpolat, B. D., M. Handberg-Thorsager, M. Vervoort, and G. Balavoine. 2017. Cell lineage and cell cycling analyses of the 4d micromere using live imaging in the marine annelid *Platynereis dumerilii*. *eLife* 6: e30463.

Pais-de-Azevedo, T., R. Magno, I. Duarte, and I. Palmeirim. 2018. Recent advances in understanding vertebrate segmentation. *F1000Res* 7: 97.

Palmeirim, I., D. Henrique, D. Ish-Horowicz, and O. Pourquié. 1997. Avian *hairy* gene expression identifies a molecular clock linked to vertebrate segmentation and somitogenesis. *Cell* 91: 639–648.

Saga, Y., N. Hata, H. Koseki, and M. M. Taketo. 1997. *Mesp2*: A novel mouse gene expressed in the presegmented mesoderm and essential for segmentation initiation. *Genes Dev.* 11: 1827–1839.

Sarrazin, A. F., A. D. Peel, and M. Averof. 2012. A segmentation clock with two-segment periodicity in insects. *Science* 336: 338–341.

Sawada, A., A. Fritz, Y.-J. Jiang, A. Yamamoto, K. Yamasu, A. Kuroiwa, Y. Saga, and H. Takeda. 2000. Zebrafish Mesp family genes, *mesp-a* and *mesp-b* are segmentally expressed in the presomitic mesoderm, and Mesp-b confers the anterior identity to the developing somites. *Development* 127: 1691–1702.

Shih, N. P., P. François, E. A. Delaune, and S. L. Amacher. 2015. Dynamics of the slowing segmentation clock reveal alternating two-segment periodicity. *Development* 142: 1785–1793.

Shimojo, H., A. Isomura, T. Ohtsuka, H. Kori, H. Miyachi, and R. Kageyama. 2016. Oscillatory control of *delta-like1* in cell interactions regulates dynamic gene expression and tissue morphogenesis. *Genes Dev.* 30: 102–116.

Sonnen, K. F., Lauschke, V. M., Uraji, J., Falk, H. J., Petersen, Y., Funk, M. C., Beaupeux, M., François, P., Merten, C. A., & Aulehla, A. 2018. Modulation of phase shift between Wnt and Notch signaling oscillations controls mesoderm segmentation. *Cell* 172(5): 1079–1090.e12.

Soroldoni, D., D. J. Jörg, L. G. Morelli, D. L. Richmond, J. Schindelin, F. Jülicher, …, A. C. Oates. 2014. Genetic Oscillations. A Doppler effect in embryonic pattern formation. *Science* 345(6193): 222–225.

Soroldoni, D., and A. C. Oates. 2011. Live transgenic reporters of the vertebrate embryo's segmentation clock. *Curr. Opin. Genet. Dev.* 21: 600–605.

Sparrow, D. B., W.-C. Jen, S. Kotecha, N. Towers, C. Kintner, and T. J. Mohun. 1998. *Thylacine 1* is expressed segmentally within the paraxial mesoderm of the *Xenopus* embryo and interacts with the Notch pathway. *Development* 125: 2041–2051.

Strobl, F., and E. H. K. Stelzer. 2014. Non-invasive long-term fluorescence live imaging of *Tribolium castaneum* embryos. *Development* 141: 1–8.

Strobl, F., and E. H. K. Stelzer. 2016. Long-term fluorescence live imaging of *Tribolium castaneum* embryos: Principles, resources, scientific challenges and the comparative approach. *Curr. Opin. Insect Sci.* 18: 17–26.

Takashima, Y., T. Ohtsuka, A. González, H. Miyachi, and R. Kageyama. 2011. Intronic delay is essential for oscillatory expression in the segmentation clock. *Proc. Natl. Acad. Sci. USA* 108(8): 3300–3305.

Thorogood, P., and A. Wood. 1987. Analysis of in vivo cell movement using transparent tissue systems. *J. Cell Sci.* 8: 395–413.

Tsiairis, C. D., and A. Aulehla. 2016. Self-organization of embryonic genetic oscillators into spatiotemporal wave patterns. *Cell* 164: 656–667.

van der Zee, M., M. A. Benton, T. Vazquez-Faci, G. E. M. Lamers, C. G. C. Jacobs, and C. Rabouille. 2015. Innexin7a forms junctions that stabilize the basal membrane during cellularization of the blastoderm in *Tribolium castaneum*. *Development* 142: 1–11.

van Drongelen, R., T. Vazquez-Faci, T. A. P. M. Huijben, M. van der Zee, and T. Idema. 2018. Mechanics of epithelial tissue formation. *J. Theor. Biol.* 454: 182–189.

Wilson, P. A., G. Oster, and R. Keller. 1989. Cell rearrangement and segmentation in *Xenopus*: Direct observation of cultured explants. *Development* 105: 155–166.

Wolff, C., J.-Y. Tinevez, T. Pietzsch, E. Stamataki, B. Harich, L. Guignard, …, A. Pavlopoulos. 2018. Multi-view light-sheet imaging and tracking with the MaMuT software reveals the cell lineage of a direct developing arthropod limb. *eLife* 7: e34410.

Wood, A., and P. Thorogood. 1994. Patterns of cell behavior underlying somitogenesis and notochord formation in intact vertebrate embryos. *Dev. Dyn.* 201: 151–167.

Zattara, E. E., K. W. Turlington, and A. E. Bely. 2016. Long-term time-lapse imaging reveals extensive cell migration during annelid regeneration. *BMC Dev. Biol.* 16: 6.

Zhu, X., H. Rudolf, L. Healey, P. François, S. J. Brown, M. Klingler, …, E. El-Sherif. 2017. Speed regulation of genetic cascades allows for evolvability in the body plan specification of insects. *Proc. Natl. Acad. Sci. USA* 114(41): e8646–e8655.

Section III

Beyond Segmentation

9 Segmental Traits in Non-Segmented Bilaterians

Bruno C. Vellutini

CONTENTS

9.1   BACKGROUND

Earthworms and millipedes are made of a dazzling series of repeated body parts known as segments. This nearly mathematical body architecture has driven dozens of biologists throughout the last centuries to investigate how segments are built during embryogenesis and how they evolved in the first place—a longstanding and highly debated topic in biology (Sedgwick 1884; Masterman 1899; Hyman 1951a; Clark 1963; Beklemishev 1969c; Willmer 1990). Although arthropods, annelids, and vertebrates are the only groups typically considered to be fully segmented (see caveats in Budd 2001), meticulous studies on comparative morphology have long recognized that repetitive traits are present in many other bilaterians—the group of animals with bilateral symmetry.

The rise of developmental genetics, and with it, the discovery of genes and signaling pathways that regulate segment formation, gave the topic a new breadth. The field revealed that common molecular players can be involved in the segmentation of animals as dissimilar as fruit flies and mice, fueling a debate about the segmental nature of the last common ancestor of bilaterians, and inspiring new perspectives on the homology and co-option of segmentation mechanisms (Kimmel 1996; De Robertis 1997, 2008; Davis and Patel 1999; Scholtz 2002; Balavoine and Adoutte 2003; Seaver 2003; Minelli and Fusco 2004; Tautz 2004; Blair 2008; Couso 2009; Chipman 2010; Arendt 2018).

These discussions, however, have largely revolved around the segmentation mechanisms uncovered in laboratory species such as the fruit fly, zebrafish, and mouse, where tools to unravel genetic and developmental details were available. Although we now comprehend these mechanisms with unprecedented detail, the sole comparison between such distantly related species is not sufficient to understand how segments evolved or how the body of the bilaterian ancestor was organized, because it is hard to distinguish inheritance from convergent evolution over long distance phylogenetic comparisons (Sanger and Rajakumar 2018). Therefore, it is crucial to investigate other related bilaterian groups and their repeated traits to better reconstruct the evolutionary steps that gave rise to the segmented body patterns of today (Budd 2001; Scholtz 2002; Minelli and Fusco 2004).

Fortunately, molecular tools as well as sequencing and imaging technologies are becoming more widely applicable to many groups traditionally outside the range of laboratory species. In addition, improved phylogenies of animal relationships are providing better grounds to map and infer character evolution (Dunn et al. 2014). These advancements lay the foundations to investigate the evolution of segmentation mechanisms in greater depth and in a wider diversity of groups. Nevertheless, a comprehensive overview about the diversity of repeated structures across bilaterians is still lacking.

Even though only a few groups have a segmental organization comparable to that of annelids, arthropods, and vertebrates, the vast majority of bilaterians shows some kind of segmental trait (Figure 9.1).

The most notable examples are tapeworms (i.e., parasitic flatworms) and mud dragons (i.e., kinorhynchs), whose adult bodies are not only subdivided into well-defined external segments, but their internal structures also follow a repetitive

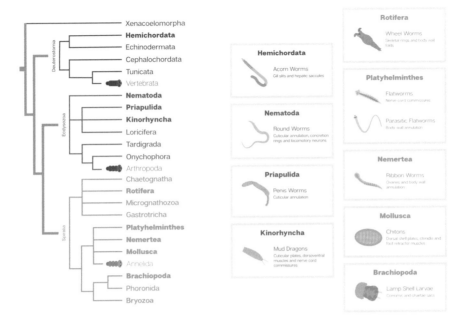

FIGURE 9.1   Examples of segmental traits in bilaterians. Phylogenetic relationships based on recent data sets (Dunn et al. 2014; Kocot et al. 2017). Groups typically considered to be segmented (vertebrates, arthropods, and annelids) are depicted in the tree by a representative drawing. Groups highlighted in bold are illustrated on the right side in featured boxes.

organization. Other groups have a variable set of individual segmental traits—from the dorsal shells of chitons, to the cuticle annulation of nematodes, to the ovaries of nemerteans—which can occur in every organ system, such as the body wall, nervous system, musculature, gonads, or excretory system (Beklemishev 1969c; Willmer 1990; Scholtz 2002; Minelli and Fusco 2004; Schmidt-Rhaesa 2007; Blair 2008; Couso 2009; Hannibal and Patel 2013).

In this chapter, I review the segmental traits of non-segmented bilaterians—namely, every bilaterian group other than arthropods, annelids, and vertebrates—describing their morphology, the developmental processes that give rise to the segmental pattern at the cellular level, and, when available, the genetic interactions patterning these traits. Non-bilaterians are beyond the scope of this chapter even though they also exhibit serially repeated structures (see examples in Beklemishev 1969c).

9.2   WHAT IS SEGMENTAL?

The definition of "segmentation" in biology is not straightforward, but its conceptual grounds have been thoroughly discussed in the literature (Beklemishev 1969c; Newman 1993; Budd 2001; Scholtz 2002, 2010; Minelli and Fusco 2004; Tautz 2004; Fusco 2008; Couso 2009; Hannibal and Patel 2013; also see Chapter 1). Because of the historical load and ambiguity of the word "segmentation," I favor the term "segmental" and adopt a definition that is strictly based on the structure of individual traits rather than whole organisms (Budd 2001; Scholtz 2002, 2010). Thus, throughout this chapter, I use "segmental trait" to refer to any repetitive structures or organs that are arranged in a regular series along the anteroposterior body axis.

To fit this definition, two main conditions must be satisfied: the structures should have a repetitive morphology and be evenly distributed (i.e., in a regular series). For example, a series of transverse commissures between longitudinal nerve cords are segmental when their arrangement and orientation are the same, and they are equidistant to each other. This approach is flexible enough to be applied to individual cells (e.g., neurons), cell bundles (e.g., muscles), tissue boundaries (e.g., epithelial folds), or even to full organs (e.g., nephridia), and each structure or organ system can be accessed individually.

Putting the focus on individual traits bypasses the need to define what a "segment" is and the juggling between what is "true" or "pseudo" segmentation. In this manner, it is more straightforward to identify and compare segmental traits across bilaterians without running into various conceptual dilemmas.

It is also important to note that a trait is "segmental" regardless of how the structures became arranged in a segmental manner during ontogeny. This focus on the structure avoids conflating pattern and process (Scholtz 2010), and helps to explicitly recognize that a segmental organization can be achieved by different cellular and developmental processes even within the same organism.

As a final disclaimer, the presence of a segmental trait is not, by any means, evidence for a hypothetical fully "segmented" ancestor. Evolutionary hypotheses are only discussed trait-by-trait when sufficient comparative and phylogenetic data are available.

9.3   XENACOELOMORPHA

Xenacoelomorphs are small free-living bilaterian worms without a through-gut, coeloms, or excretory system, which might hold a key position in the animal tree of life as the sister group to the remaining bilaterians (Ruiz-Trillo et al. 1999; Jondelius et al. 2002; Hejnol et al. 2009; Rouse et al. 2016; Cannon et al. 2016; Giribet 2016; Laumer et al. 2019; but see Marlétaz et al. (2019) and Philippe et al. (2019) for an alternative view. For this reason, the presence or absence of specific traits in these worms can be highly informative to understand the evolution of bilaterian morphology (Hejnol and Pang 2016) and to what concerns the contents of this chapter, the evolution of segmental traits.

Xenacoelomorpha is comprised of xenoturbellids, nemertodermatids, and acoels. Even though some species are elongated, there are no obvious repetitive structures in the body wall or internal organs in either of these groups. Nevertheless, some aspects about the organization of the nervous system and musculature are worth mentioning.

The nervous system of xenacoelomorphs is greatly variable (Martínez, Hartenstein, and Sprecher 2017), usually consisting of a nerve net, but some lineages independently evolved longitudinal nerve condensations and cords (Hejnol and Pang 2016). In one of these lineages, the acoels, the longitudinal nerve cords are intercepted by transverse commissures along the anteroposterior axis (Semmler et al.

2010; Bery et al. 2010; Hejnol and Martindale 2008). Such commissures, however, are neither symmetric nor regularly spaced, and are restricted to the anterior portion of the body (Bery et al. 2010).

The musculature of xenacoelomorphs is organized in an orthogonal grid with circular (transverse), longitudinal, and often diagonal fibers (Tyler and Hyra 1998; Hooge and Tyler 1999; Ladurner and Rieger 2000; Hooge 2001; Meyer-Wachsmuth, Raikova, and Jondelius 2013; Børve and Hejnol 2014). The circular muscles are usually evenly spaced (Hooge and Tyler 1999), but the density of fibers can change according to the body region (Gschwentner et al. 2003).

Interestingly, the developmental processes that result in a gridlike musculature differ between the Xenacoelomorpha groups. In acoels, muscle progenitors develop progressively from the animal to the vegetal pole, and the emerging fibers are already polarized transversely forming a regular series of circular muscles along the anteroposterior axis (Figure 9.2A) (Ladurner and Rieger 2000; Semmler, Bailly, and Wanninger 2008).

In contrast, muscle progenitors in nemertodermatids appear all around the body, and their projections are irregular, not following a specific orientation in relation to the body axes (Børve and Hejnol 2014). Nevertheless, the nemertodermatid juveniles exhibit a neatly organized series of circular muscles (Børve and Hejnol 2014). Presumably, the musculature grid in nemertodermatids emerges secondarily from the cellular interactions between progenitors, while in acoels a patterning mechanism must impose the polarity of embryonic cells during early development. Which mode reflects the ancestral condition for the group remains to be determined.

9.4 DEUTEROSTOMIA

9.4.1 HEMICHORDATA (ACORN WORMS)

Hemichordates are relatively large vermiform marine invertebrates with a body divided into three unequal regions along the anteroposterior axis—the proboscis (or mouth shield), the collar, and a long unsegmented trunk (Hyman 1959b; Kaul-Strehlow and Röttinger 2015). Nonetheless acorn worms according to Beklemishev (1969d) have "a tendency to multiplication of the organs repeated along its longitudinal axis, and these organs show a tendency to metameric arrangement." The most conspicuous segmental trait of hemichordates are the gills located at the anterior part of the trunk (Bateson 1884).

The gills are arranged in a regular series up to hundreds of pairs with openings to the exterior of the body (Figure 9.2B) (Horst 1930; Hyman 1959b). Gill slits are intercalated by pharyngeal arches that are supported by a series of trident-like acellular collagenous endoskeleton (Gillis, Fritzenwanker, and Lowe 2012). Repeated dorsal outgrowths named tongue bars develop between pharyngeal arches (Hyman 1959b; Gillis, Fritzenwanker, and Lowe 2012). Pharyngeal arches and tongue bars have a single blood vessel each (Pardos and Benito 1988), and thus the arrangement of the circulatory system closely matches that of the gills (Horst 1930). The gonads, which are also serially paired structures in some species, are intercalated with the gills with the genital pores located between gill slits (Horst 1930).

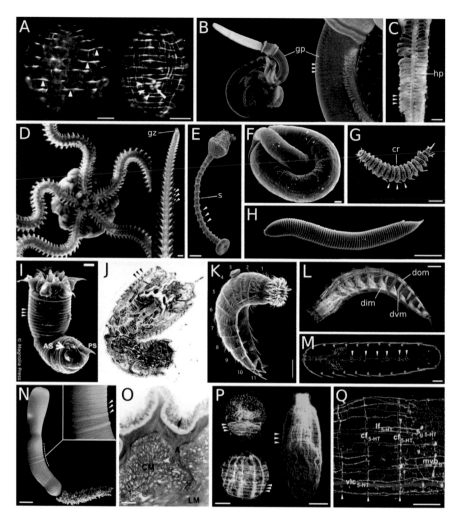

FIGURE 9.2   Selected segmental traits in Xenacoelomorpha, Deuterostomia (Hemichordata and Echinodermata), and Ecdysozoa (Nematoda, Nematomorpha, Kinorhyncha, and Priapulida). Arrowheads indicate serially repeated structures unless otherwise noted. A. Developing musculature in early (left) and late embryo (right) of the acoel *Isodiametra pulchra*. Asterisk and surrounding arrows mark the forming mouth. Scale bars = 20 μm. Images reprinted from Ladurner and Rieger (2000) with permission from Elsevier. B. Gill pores (gp) in the hemichordate *Saccoglossus kowalevskii*. Whole body image (left) reprinted from Dunn (2015) with permission from Springer Nature. Close up image (right) courtesy of Casey Dunn. C. Hepatic saccules (hp) in the hemichordate *Balanoglossus simodensis*. Scale bar = 1 mm. Image reprinted from Miyamoto and Saito (2007) with permission from The Zoological Society of Japan. D. Adult specimen (left) and regenerating arm with skeletal and podial serial elements in the brittle star *Amphiura filiformis*. Scale bar = 100 μm. Images courtesy of Anna Czarkwiani. E. Segmental stalk (s) in the pentacrinoid larva of the feather star *Aporometra wilsoni*. Scale bar = 200 μm. Image reprinted from Haig and Rouse (2008) with permission from John Wiley & Sons. F. Cuticular annulation in the nematode *Caenorhabditis elegans*. Scale bar = 3.8 μm. Image reprinted from Costa, Draper, and Priess

FIGURE 9.2 (CONTINUED)
(1997) with permission from Elsevier. G. Concretion rings (cr) in the nematode *Desmoscolex cosmopolites*. Scale bar = 50 μm. Image reprinted from Lim and Chang (2006) with permission from Taylor & Francis. H. Cuticular annulation in the criconematid nematode *Criconema* sp. Scale bar = 50 μm. Image courtesy of Tom Powers from Powers (2015) and used with permission. I. Cuticular annulation in the larva of the nematomorph *Chordodes janovyi* with terminal anterior (as) and posterior spines (ps). Scale bar = 2 μm. Image reprinted from Bolek et al. (2010) with permission from Magnolia Press. J. Profile of the cuticle annulations in the larva of the nematomorph *Neochordodes occidentalis*. Reprinted from Poinar (2010) with permission from Elsevier. K. Trunk segments in the kinorhynch *Echinoderes hispanicus*. Scale bar = 30 μm. Image by Herranz et al. (2014) licensed under CC-BY. L. Segmental musculature in the kinorhynch *Echinoderes horni* showing dorsal (dom), dorsoventral (dvm), and diagonal (dim) bundles. Scale bar = 20 μm. Image by Herranz et al. (2014) licensed under CC-BY. M. Segmental paired neuronal somata in the nervous system of the kinorhynch *Echinoderes spinifurca*. Scale bar = 20 μm. Image reprinted from Herranz, Pardos, and Boyle (2013) with permission from John Wiley and Sons. N. Cuticular annulation in the priapulid *Priapulus caudatus*. Scale bar = 1 cm. Own work. O. Longitudinal section of the body wall in the priapulid *Halicryptus spinulosus* showing the circular musculature (cm), apodemes (a), and inner longitudinal muscle (lm). Scale bar = 10 μm. Image reprinted from Oeschger and Janssen (1991) with permission from Springer Nature. P. Development of the circular musculature in the priapulid *Priapulus caudatus*. Scale bars = 50 μm. Image by Martín-Durán and Hejnol (2015) licensed under CC-BY. Q. Grid pattern in the abdomen nervous system of the priapulid *Tubiluchus troglodytes* showing serotonin immunoreactive fibers in the ventral longitudinal cord (vlc), circular fibers (cf), longitudinal fibers (lf), and gut bundle (mvb) and somata (s). Scale bar = 75 μm. Image reprinted from Rothe and Schmidt-Rhaesa (2010) with permission from John Wiley & Sons.

More posterior in the trunk, some species exhibit a series of paired dorsal outgrowths named hepatic saccules (Figure 9.2C) (Horst 1930; Hyman 1959b; Beklemishev 1969d). These repeated diverticula of the dorsal gut epithelium deform the adjacent body wall epidermis becoming visible externally (Benito, Fernández, and Pardos 1993; Miyamoto and Saito 2007). Other hemichordate structures such as muscles or neurons have no evident segmental organization—the circular muscles are not evenly spaced, the longitudinal muscles are continuous from end to end, the coelomic cavities are undivided, and the nervous system is a net with no repetitive pattern (Hyman 1959b; Beklemishev 1969d; Kaul-Strehlow et al. 2015).

During development the first pairs of gills can already be formed in the swimming larval stages (Agassiz 1873; Bateson 1884; Morgan 1894; Kaul-Strehlow and Röttinger 2015). They appear progressively from anterior to posterior when endodermal evaginations fuse to the overlying ectoderm creating the gill pores (Hyman 1959b). Each gill follows a stereotypic developmental sequence, first as a small round pore that elongates along the anteroposterior axis and subsequently becomes U-shaped due to the growing dorsal tongue bars (Gillis, Fritzenwanker, and Lowe 2012). The number of gill slits continues to increase during adult life (Hyman 1959b; Kaul-Strehlow and Röttinger 2015).

Gene expression studies suggest the molecules patterning the pharyngeal arches of hemichordates, cephalochordates, and vertebrates are largely conserved, further supporting the homology of deuterostome gills (Ogasawara et al. 1999; Rychel and

Swalla 2007; Gillis, Fritzenwanker, and Lowe 2012; Fritzenwanker et al. 2014; Lowe et al. 2015). Pharyngeal arches are thus a rather ancient deuterostome feature and were probably the first segmental traits to evolve in the vertebrate lineage—before the trunk somites or the hindbrain rhombomeres (Graham et al. 2014).

9.4.2 ECHINODERMATA (SEA URCHINS, SEA STARS, AND OTHERS)

Echinoderms are well-known marine invertebrates that evolved a unique body pattern organized into five symmetrical parts (Hyman 1955). Because of this unorthodox pentamerous body symmetry and the difficulty to identify the position of the ancestral anteroposterior axis (reviewed in Mooi and David 2008), echinoderms are rarely mentioned or often categorized as non-segmented in previous discussions about the evolution of segmentation (Sedgwick 1884; Masterman 1899; Hyman 1955; Minelli and Fusco 2004; Tautz 2004; Couso 2009)—with a few exceptions (Beklemishev 1969a; Balavoine and Adoutte 2003). But echinoderms indeed exhibit a conspicuous set of repetitive serial elements, not along the anteroposterior axis, which is obscured, but through the main axis of each of the five individual rays that make the echinoderm body (Mooi, David, and Wray 2005). Despite not being arranged along the anteroposterior axis, and thus a clear exception to the definition of "segmental" used in this chapter, these traits are worth mentioning given their clear segmental nature.

Adult echinoderms have five growth zones, one at the distal tip of each ray, where new skeletal plates are deposited in a regular series by terminal addition (Mooi, David, and Marchand 1994; Mooi and David 1998). Different than other segmental structures described in this chapter, the skeletal elements are not paired but have an alternating biserial distribution resulting in a zigzag pattern (Mooi, David, and Marchand 1994; Mooi and David 1998; Mooi, David, and Wray 2005). This segmental arrangement is evident by the skeletal plates and is also reflected in the circulatory system, the podia, neurons, and muscles (Beklemishev 1969a). Interestingly, the alternating growth pattern has changed in brittle stars and sea stars, and the elements have lined up side by side forming paired serial structures along the ray (Figure 9.2D) (Mooi and David 2000).

Although genetic interactions have been extensively studied in echinoderm embryos and larvae (Cary and Hinman 2017), the genes involved in axial terminal growth remain less known mainly due to the challenges of obtaining and raising juvenile stages (Byrne et al. 2015). Of the few investigated candidates, the gene Engrailed, known to pattern arthropod segments, is expressed in a segmental manner in the arms of juvenile brittle stars and sea stars (Lowe and Wray 1997; Byrne et al. 2005). However, the gene is likely associated to neurogenesis rather than the generation of repetitive structures (Wray and Lowe 2000; Byrne et al. 2005). Fortunately, a complementary approach to investigate echinoderm terminal growth through arm regeneration experiments is gaining traction, and the molecular and cellular processes patterning these segmental elements are being unraveled in brittle stars and sea stars (Bannister et al. 2005; Czarkwiani, Dylus, and Oliveri 2013; Czarkwiani et al. 2016; Ben Khadra et al. 2017; Ferrario et al. 2018).

Finally, echinoderms have also evolved segmental traits outside the five terminal growth zones. A notable example is the stalk of crinoids, which is built on a repetitive series of internal skeletal elements that can reach 50 cm in extant species (Figure 9.2E) (Hyman 1955). The ossicles that give rise to the stalk appear during larval metamorphosis as five sequential elements (Haig and Rouse 2008; Amemiya et al. 2016), but unlike the pattern of terminal growth in the rays (Mooi, David, and Wray 2005), these elements grow at the proximal region near the main body (Thomson 1865; Lahaye and Jangoux 1987) or by the intercalation of skeletal elements (Amemiya et al. 2016).

9.4.3 CEPHALOCHORDATA (LANCELETS OR AMPHIOXI)

Cephalochordates are swift, fishlike chordates that burrow in sand and filter suspended food particles in the seawater (Ruppert, Fox, and Barnes 2004a). Known as lancelets or amphioxi due to their elongated body with pointy ends, they exhibit several repeated traits along the anteroposterior axis, such as paired coelomic sacs, muscle bundles, gill slits, and gonads (Beklemishev 1969d).

Directly visible through the transparent adult body are the V-shaped muscle bundles arranged in a segmental manner along the anteroposterior axis and the repeated series of gonads located more ventrally (Bertrand and Escriva 2011). Curiously the left and right somites are out of register to each other (Ruppert, Fox, and Barnes 2004a). Underneath lies a pharynx with a series of gill slits separated by cartilaginous bars (Bertrand and Escriva 2011). Associated with the gills there are the paired nephridia also organized in a segmental manner (Goodrich 1902; Holland 2017).

There are no visible segmental traits in the nervous system except for serially repeated dorsal serotonergic neurons (Yasui et al. 1998; Candiani et al. 2001; Wicht and Lacalli 2005) which follow the out-of-register arrangement of the musculature (Wicht and Lacalli 2005). In addition, there are inner cholinergic and GABAergic/glycinergic neurons that are segmentally arranged in the hindbrain (Candiani et al. 2012).

The first segmental structures in the amphioxus embryo are mesodermal constrictions that mark the somite boundaries (Conklin 1932). These evaginations progressively pinch off from anterior to posterior forming 8–12 pairs of somites (Conklin 1932; Holland 2015), while subsequent somites arise instead from the tail bud (Holland 2015). The somites are further regionalized into three populations under distinct regulatory landscapes (Aldea et al. 2019), revealing a hidden complexity of molecular interactions and developmental processes that act in concert to generate a uniform series of repeated structures.

9.4.4 TUNICATA (SEA SQUIRTS AND SALPS)

Tunicates are suspension-feeding, solitary or colonial, marine invertebrates with sessile (e.g., sea squirts) or planktonic habit (e.g., salps). The body is compact like a barrel and covered by a gelatinous tunic (Ruppert, Fox, and Barnes 2004a). Although individuals have no evident segmental traits, a few structures in the adult pharynx, in the larval notochord, and in the circular musculature of some planktonic forms show a segmental arrangement.

In adults, the pharynx is basket-shaped with perforated walls whose function is central for nutrition, gas exchange, and excretion. These openings are often arranged in transverse rows along the main axis of the pharynx (Garstang 1892) and are intercalated by transverse circulatory vessels (Shimazaki, Sakai, and Ogasawara 2006), as well as transverse nerves (Burighel et al. 2001; Zaniolo et al. 2002). Adult zooids of some colonial species can also reproduce asexually by forming segmental epidermal constrictions in the abdomen (Brown and Swalla 2012).

In larval stages we find the notochord, a highly ordered stack of cylindrical cells that support the tail (Miyamoto and Crowther 1985) where each cell shows repeated morphology and anteroposterior polarity (Jiang, Munro, and Smith 2005; Hannibal and Patel 2013). Ascidian larvae also have a series of dorsal and ventral epidermal sensory neurons in the tail, but their arrangement is barely regular (Crowther and Whittaker 1994; Pasini et al. 2006). In general there are no segmental traits in the central or peripheral nervous system of ascidian larvae (Imai and Meinertzhagen 2007a, 2007b).

During development, the notochord is not formed progressively by terminal growth but by the rearrangement of forty progenitor cells that interdigitate and self-organize into an ordered single column (Miyamoto and Crowther 1985; Jeffery and Swalla 1997). This segmental arrangement, however, is transient; it disappears once the intercellular spaces between notochord cells fuse and form a hollow tube (Jeffery and Swalla 1997).

The formation of the first gill slits during development also involves a series of cellular rearrangements at the site of contact between ectodermal and endodermal epithelial sheets (Willey 1893; Manni et al. 2002). These primary gills elongate and subdivide increasing the number of openings per row, and entire rows can also subdivide generating the multiple transverse rows present in adult animals (Willey 1893; Berrill 1947; Shimazaki, Sakai, and Ogasawara 2006).

Additionally, in planktonic tunicates such as doliolids and salps (Thaliacea) the body is marked by a series of evenly spaced bands of circular muscles distributed along the anteroposterior axis. This hooplike musculature pumps water through the pharynx creating a jet thrust used for locomotion and filter-feeding (Ruppert, Fox, and Barnes 2004a). The regular arrangement of these muscle bands is gradually established during asexual budding, a common reproduction mode in thaliaceans. Bud primordia exhibit a dorsolateral mass of muscle tissue that—in concert with bud growth—extend ventrally, separate into initially misaligned individual bands, and finally reach the regular arrangement found in mature zooids (Berrill 1950).

9.5 ECDYSOZOA

9.5.1 Nematoda (Round Worms)

Nematoda is a speciose group of elongated and cylindrical worms with a characteristic unsegmented external appearance (Hyman 1951e). It is not uncommon, however, for nematodes to exhibit regularly spaced circumferential indentations in their cuticle along the anteroposterior axis, which are known as annuli (Lee 1967). The nematode *Caenorhabditis elegans* has a consistent pattern of cuticular annulation throughout its life cycle (Figure 9.2F) (Cox, Staprans, and Edgar 1981; Costa, Draper, and Priess 1997). However, these cuticular rings are more pronounced in two other nematode groups, the Desmoscolecida (Decraemer and Rho 2013) and the Tylenchida (Subbotin 2013).

Desmoscolecida are diminutive and mostly marine nematodes that are characterized by a strongly annulated cuticle ornamented by outer rings of aggregated material (Figure 9.2G) (Decraemer and Rho 2013). These so-called concretion rings are formed by the accumulation of grains, clay, and even bacteria, which are caught up by the differential secretion of mucus between annuli and probably shaped by the body movement (Riemann and Riemann 2010). The cuticle of juveniles is annulated, but it does not exhibit concretion rings; these form during adult life (Lorenzen 1971; Decraemer 1978; Riemann and Riemann 2010). Within Tylenchida, a group of plant-parasitic nematodes of great economic importance (Subbotin 2013), some species show an even more striking pattern of cuticle annulation (Figure 9.2H). These worms often exhibit annuli with crenated margins or scales with evident regularity (Powers et al. 2016). In both Desmoscolecida and Tylenchida, the presence of a strong cuticular annulation might be an adaptation to interstitial sand and soil environments.

The origin of the cuticular annulation is directly associated with the arrangement and differentiation of the hypodermal cells during the early development of the nematode *C. elegans*. Prior to elongation, the embryos secrete an extracellular layer that remains attached to the dorsal and ventral hypodermal cells in regular intervals, forming circumferential ridges along the anteroposterior axis (Priess and Hirsh 1986). There are numerous ridges per hypodermal cell, and each ridge corresponds to the position of a submembranous filamentous actin bundle (Priess and Hirsh 1986). The arrangement is maintained throughout development and the pattern of circumferential actin bundles coincides with the localization of each cuticular annulation in larval and adult stages (Costa, Draper, and Priess 1997; Francis and Waterston 1991; Hardin and Lockwood 2004). The attachment, mediated by catenin and cadherin proteins (Costa et al. 1998), is essential for the changes in cell shape that elongate the embryo and for the mechanical coupling between the hypodermis and the adult cuticle (Priess and Hirsh 1986; Costa, Draper, and Priess 1997; Hardin and Lockwood 2004).

The cuticle annulation of nematodes is a fine example on how an external trait is inherently linked and derived from the segmental organization of intracellular components. In this case, the basis for establishing a repetitive arrangement lies within the realm of protein localization and polarity in the cytoplasm. Many of these cell adhesion proteins have already been identified (Hardin and Lockwood 2004) and may play a role in organizing the circumferential filamentous actin bundles in *C. elegans* (Ding et al. 2003). Nevertheless, the key factors regulating the spacing of these intracellular structures remain unknown.

Another nematode trait exhibiting a certain degree of repetition is the postembryonic locomotory neurons (White et al. 1976; Sulston and Horvitz 1977; Walthall 1995). The ganglia of each class of motoneurons become serially arranged in a repeated manner along the body axis of *C. elegans* during the larval stages (White et al. 1976; Sulston and Horvitz 1977; Walthall 1995). The embryo of *C. elegans* does not show such arrangement (Sulston et al. 1983). The organization of locomotory neurons in segmental pattern also occurs in *Ascaris* and might be a widespread feature of nematodes (Stretton et al. 1978; Johnson and Stretton 1987).

9.5.2 NEMATOMORPHA (HORSEHAIR WORMS)

Horsehair worms are filiform bilaterians with a complex life cycle, where the adult stages are free-living inhabiting water streams, and the larval stages are parasitic on arthropod hosts (Hyman 1951e). The body of adult nematomorphs is covered by a thick cuticle ornamented by areole structures, but there are no annulations (see Bolek et al. 2010).

In contrast, the cuticle of the parasitic larva is markedly annulated (Figure 9.2I) (Montgomery 1904; May 1919; Marchiori, Pereira, and Castro 2009; Bolek et al. 2010; Szmygiel et al. 2014). Similar to nematodes, the annulation encompasses the cuticle and the hypodermal cell layer beneath. However, the rings in nematomorph larvae are somewhat irregular, and the number of folds in each side does not correspond exactly (Figure 9.2J) (Montgomery 1904; Zapotosky 1975). It is also unlikely that these cuticle rings are anchored to actin bundles since no circumferential fibers are visible in the larva (Müller, Jochmann, and Schmidt-Rhaesa 2004).

Overall, the annulation of larval nematomorphs is incipient compared to the cuticular rings of nematodes, but the group is still lacking developmental data (Hejnol 2015a), and earlier embryonic stages need to be investigated further to clarify the morphogenesis of these annulations.

9.5.3 PRIAPULIDA (PENIS WORMS)

Priapulids are marine mud-dwelling scavengers fittingly known as penis worms due to their body shape. Adult priapulids have three well-defined body regions, an anterior proboscis (introvert), an elongated trunk, and a posterior region with caudal appendages (Hyman 1951e). The trunk has a characteristic annulation pattern formed by a series of transverse cuticular rings—between 40 and 180, depending on the species (Figure 9.2N) (McCoy 1845; Hyman 1951e; Sanders and Hessler 1962; Hammond 1970; Morse 1981; Shirley and Storch 1999).

The trunk cuticle in adults consists of a thin outer layer and a thick inner layer with regularly spaced circumferential projections named apodemes (Figure 9.2O). These structures delve deep into the tissue and are tightly intercalated with the rings of circular muscles around the body wall (Oeschger and Janssen 1991). Each cuticular ring corresponds to the position of a circular muscle, which appears to bulge the inner cuticle layer (Oeschger and Janssen 1991). Therefore, the cuticle annulation of priapulids is likely the result of a tight association between the circular musculature and the cuticle of the trunk.

The circular muscles first appear in the embryos of *Priapulus caudatus* during gastrulation with a few transversely oriented fibers located at the introvert–trunk boundary (Figure 9.2P) (Martín-Durán and Hejnol 2015). Additional circular muscle cells get distributed along the whole trunk region, and form a grid pattern with the longitudinal musculature (Figure 9.2P). The circular musculature is well developed in the hatching larvae (Figure 9.2P) (Martín-Durán and Hejnol 2015), but no obvious annulation is visible in this first larval stage and, unfortunately, the trunk is hidden beneath a lorica in later stages (Higgins, Storch, and Shirley 1993; Wennberg, Janssen, and Budd 2009). Nevertheless, juvenile stages can already display cuticular rings (Sørensen et al. 2012), and further developmental studies might corroborate the idea that priapulid trunk annulation emerges from the tight association between the circular musculature and the cuticle.

In addition, the peripheral nervous system of adult priapulids displays circular neurite bundles arranged regularly along the anteroposterior axis (Figure 9.2Q) (Rothe and Schmidt-Rhaesa 2010). However, this orthogonal pattern must develop during postembryonic stages and juvenile growth, since it is not present in the larval stages (Martín-Durán et al. 2016).

9.5.4 Kinorhyncha (Mud Dragons)

Mud dragons are minute and elongated marine invertebrates with a spiny appearance (Hyman 1951e). Different than other groups covered in this chapter, kinorhynchs have an evident segmental organization with the trunk divided into eleven segments, known as zonites (Figure 9.2K) (Zelinka 1908; Hyman 1951e). These are demarcated by a series of overlapping cuticular plates connected by thin transverse joints repeated along the anteroposterior axis (Hyman 1951e; Neuhaus and Higgins 2002). This segmental arrangement is not only external but also occurs internally in the musculature and the nervous system.

For instance, the dorsoventral and diagonal muscles are arranged in a series of pairs along the body that match the subdivisions of the body wall (Figure 9.2L) (Müller and Schmidt-Rhaesa 2003; Schmidt-Rhaesa and Rothe 2006; Herranz et al. 2014; Altenburger 2016). Similarly, the nervous system exhibits two transverse commissures per segment between the longitudinal nerve cords (Nebelsick 1993; Neuhaus and Higgins 2002; Herranz et al. 2019) and a series of paired ventral neuronal somata is present in some species (Figure 9.2M) (Herranz, Pardos, and Boyle 2013; Herranz et al. 2019). Thus, in adult kinorhynchs the segmental traits of distinct organ systems are well correlated to a degree comparable with annelids or arthropods, at least in the species studied so far. Nevertheless, additional data is needed to corroborate if this integrated segmental organization is a feature shared by the entire group (Herranz et al. 2019).

The ontogeny of kinorhynch segmental structures remains elusive due to limited developmental data (Hejnol 2015a)—most observations are restricted to the heroic efforts of E.N. Kozloff (1972, 2007) in a species of the Pacific Northwest: *Echinoderes kozloffi*. What we know about kinorhynch embryogenesis is that the trunk segments of *E. kozloffi* first become visible midway through development after the embryo elongates in the anteroposterior axis (Kozloff 2007). The developmental processes involved in the elongation (e.g., cellular rearrangements or growth zone) are not known. Development is direct and the juvenile hatches with at least eight of the future eleven adult segments (Kozloff 1972; Neuhaus and Higgins 2002; Schmidt-Rhaesa and Rothe 2006). Additional segments are added at the posterior end during postembryonic development, suggesting the presence of a subcaudal growth zone in kinorhynchs (Neuhaus 1995; Lemburg 2002; Neuhaus and Higgins 2002; Sørensen, Accogli, and Hansen 2010).

An interesting aspect revealed by kinorhynch developmental studies is that the correspondence between the number of dorsoventral muscles and trunk segments present in adults is not observed in juveniles. At their earliest stage, juveniles have only eight trunk segments but already display their final number of ten pairs of dorsoventral muscles (Schmidt-Rhaesa and Rothe 2006). This suggests that kinorhynch segmental traits might develop with a certain degree of independence during embryogenesis before becoming well integrated in the adult stage.

9.5.5 LORICIFERA (GIRDLE WEARERS)

Loriciferans are microscopic interstitial bilaterians related to priapulids and kinorhynchs that were only discovered in the 1980s (Kristensen 1983). Their body is divided into an introvert, thorax, and abdomen, and partially enclosed by a sclerotized cuticle—the lorica (Kristensen 1983).

The adult thorax is divided into two segments, one anterior portion with appendages and a posterior portion without which are covered by accordion-like rows of plates in the larval stages (Kristensen 1983; Higgins and Kristensen 1986). There are no internal structures that match the arrangement of these external thorax traits.

In the abdomen, however, a variable number of transverse circular or paired lateral muscle bundles are serially arranged in different species (Gad 2005; Kristensen, Neves, and Gad 2013; Neves et al. 2013), even though the morphology of these transverse muscle bands is not perfectly repeated (Neves et al. 2013). There is no evidence so far of segmental traits in the nervous system of loriciferans (e.g., serial ganglia), but their neural structures have not yet been studied by immunohistochemistry (Herranz et al. 2019).

The development of loriciferans is barely known (Hejnol 2015a). Embryos have only been observed in one aberrant species with a viviparous paedogenetic life cycle and develop inside a cystlike mega-larva (Heiner and Kristensen 2009). Nevertheless, it is interesting to note that the transverse musculature present in adult loriciferans is not present in the larval stages (Neves et al. 2013).

9.5.6 TARDIGRADA (WATER BEARS)

Tardigrades are common dwellers in aquatic and moist terrestrial habitats, widely recognized for their bearlike appearance and for the ability to survive harsh conditions (Ruppert, Fox, and Barnes 2004c; Møbjerg et al. 2011). The body has a segmented appearance with a defined head and four trunk segments, each bearing a pair of jointless legs.

A cuticle secreted by the epidermis covers the whole body and can exhibit transverse folds, which correspond to the flexion zones of the body where the cuticle is thinner (Greven 1984). The cuticle folding does not correspond to the trunk segments (Ruppert, Fox, and Barnes 2004c). Due to the reduced body size, tardigrades lack coelomic cavities, respiratory or excretory organs, and thus show no obvious internal segmental organs (Beklemishev 1969b).

For instance, the musculature of tardigrades is rather complex (e.g., Marchioro et al. 2013) but only a few muscle sets have a segmental pattern (Schmidt-Rhaesa and Kulessa 2007; Halberg et al. 2009; Smith and Jockusch 2014). Leg muscles in segments two and three can be nearly identical (Smith and Jockusch 2014; Smith and Goldstein 2017) but the morphology is usually distinct for each pair of legs (Schmidt-Rhaesa and Kulessa 2007; Halberg et al. 2009). In addition, even though the muscle attachment points are conserved among tardigrades (Marchioro et al. 2013) they do not correspond to segment borders (Schmidt-Rhaesa and Kulessa 2007).

On the other hand, the ventral nervous system of tardigrades has a rope ladder organization with a series of four trunk ganglia linked by a longitudinal pair of connectives (Zantke, Wolff, and Scholtz 2008; Persson et al. 2012; Mayer, Kauschke, et al. 2013a; Mayer, Martin, et al. 2013b; Schulze and Schmidt-Rhaesa 2013; Schulze, Neves, and Schmidt-Rhaesa 2014; Smith and Jockusch 2014; Smith and Goldstein 2017). Upon a closer inspection, however, it is evident that each ventral ganglion has a unique (non-repeated) morphology (Mayer, Kauschke, et al. 2013; Smith and Jockusch 2014; Smith and Goldstein 2017).

During embryogenesis, the trunk ganglia differentiate already at their final adult positions relative to the anteroposterior axis from individual clusters of four neural progenitors (Hejnol and Schnabel 2005). Depending on the species, these four ganglia develop either simultaneously (Hejnol and Schnabel 2005) or following an anteroposterior progression (Gross and Mayer 2015). The longitudinal connectives are only established later in development (Gross and Mayer 2015).

The mesoderm of developing tardigrade embryos is subdivided into repeated somites that correspond to the trunk segments (Gross, Treffkorn, and Mayer 2015). Interestingly, the developmental processes generating the segmental organization in the mesoderm differ between species. In one species, after mesodermal progenitors enter the blastocoel, they migrate and proliferate with no apparent anteroposterior polarity or growth zone, and the resulting mesodermal bands later subdivide into four paired groups of cells composing each somite (Hejnol and Schnabel 2005). In contrast, somites can form by sequentially pinching off the presumptive endomesodermal tube (Gabriel et al. 2007).

At the molecular level, Engrailed protein is expressed in rows of dorsolateral ectodermal cells before any morphological boundary is visible on the ectoderm, but after somite boundaries have been established (Gabriel and Goldstein 2007). The expression matches the posterior border of each somite suggesting that *engrailed* could be involved in setting up these ectodermal boundaries (Gabriel and Goldstein 2007). In addition, anterior Hox genes have been shown to be expressed in a segment-specific manner and could have a role in establishing the segment identities (Smith et al. 2016). Nevertheless, gene expression data on tardigrade development remains limited (Gross, Treffkorn, and Mayer 2015).

9.5.7 ONYCHOPHORA (VELVET WORMS)

Onychophorans are multilegged predators with velvety skin and nocturnal habit that live in humid forest soil (Ruppert, Fox, and Barnes 2004c). Because they exhibit segmental traits in several organ systems—with noteworthy exceptions—and are the sister group to arthropods (Giribet and Edgecombe 2017), velvet worms are a key group for understanding the evolution of the canonical arthropod segmental traits (Mayer and Whitington 2009; Whitington and Mayer 2011; Martin, Gross, Hering, et al. 2017a; Martin, Gross, Pflüger, et al. 2017b; Smith and Goldstein 2017).

There are no folds demarcating external trunk boundaries and, even though minute segmental papillae can be present in the cuticle (Franke and Mayer 2014), the trunk segments are only evident from the series of paired appendages used for locomotion (Ruppert, Fox, and Barnes 2004c). Internally, onychophorans exhibit a mix of segmental and non-segmental traits in the nervous, circulatory, and muscle systems (Franke and Mayer 2014; Martin, Gross, Pflüger, et al. 2017).

In the peripheral nervous system, for example, the leg and nephridial nerves branching from the ventrolateral nerve cords are distributed in a segmental manner in register with the pair of appendages per segment (Mayer 2015). However, the nerve cords themselves lack segmental ganglia (i.e., the neuronal bodies are distributed along the cord) and the numerous ring and median commissures are neither evenly spaced nor match the leg innervations (Mayer and Harzsch 2007, 2008; Whitington and Mayer 2011; Mayer, Martin, et al. 2013; Mayer 2015; Martin, Gross, Hering, et al. 2017; Martin, Gross, Pflüger, et al. 2017). In the same manner, the musculature does not show any obvious segmental arrangement, except for the muscles associated with the legs (Hoyle and Williams 1980; Mayer, Franke, et al. 2015) and gonads are paired non-segmental structures (Storch and Ruhberg 1990; Brockmann et al. 1999). Additional segmental traits include one pair of nephridia per trunk segment (Mayer 2006) and serially repeated valves (i.e., ostia) in the circulatory system through which the blood enters the heart (Ruppert, Fox, and Barnes 2004c).

During embryonic development onychophorans also produce somites in the mesoderm (Mayer, Franke, et al. 2015). They are formed after gastrulation from bilateral bands of mesoderm that elongate and subdivide into segmental blocks of solid tissue, which progressively hollow out (Mayer, Franke, et al. 2015). The somitogenesis process occurs without a posterior growth zone (Mayer et al. 2010). The nervous system also develops from head to tail, but without any sign of segmental organization. First, the neuronal precursors giving rise to the ventrolateral nerve cords delaminate from the ectoderm, and only later in development the leg nerves and commissures are established (Mayer and Whitington 2009; Mayer, Franke, et al. 2015; Martin, Gross, Hering, et al. 2017). Onychophoran embryos have in addition a conspicuous but transient segmental trait made from thickenings of the ventral ectoderm, known as the ventral organs (Whitington and Mayer 2011), which serve as attachment points for the developing leg muscles (Oliveira et al. 2013).

The molecular patterning of onychophoran embryos has been extensively investigated, particularly regarding the expression of known arthropod segmentation genes (Mayer, Franke, et al. 2015). The segment polarity genes *engrailed*, *wingless*, and *hedgehog* are expressed in a segmental manner with a pattern similar to arthropods, but are only detected after the morphological furrows have formed (Eriksson et al. 2009; Janssen and Budd 2013; Franke and Mayer 2014). The upstream factors regulating onychophoran somitogenesis remain unclear (Janssen 2017).

9.6 SPIRALIA

9.6.1 Chaetognatha (Arrow Worms)

Chaetognaths are small torpedo-like marine predators, known as arrow worms, with a well-defined head region and a slender trunk showing no marked subdivisions (Hyman 1959a). Even so, some traits of the chaetognath nervous system are organized in a segmental manner. For example, the central neuropil—a dense synaptic area of the ventral ganglion of chaetognaths—comprises an ordered series of transverse fibers that form around 80 microcompartments along the anteroposterior axis (Figure 9.3A) (Bone and Pulsford 1984; Harzsch and Müller 2007). At the flanks of the neuropil, there are neuronal cell bodies, which are also organized in a serial manner, forming a highly organized grid pattern (Perez et al. 2013). Experimental data from cell proliferation assays suggest the organized pattern emerges from the iterated asymmetric divisions of neuronal progenitors (Perez et al. 2013).

Alongside the neuropil lies another distinct segmental trait of chaetognaths: a repetitive series of paired neurons with RFamide-like immunoreactive somata (Figure 9.3B) (Bone et al. 1987; Goto et al. 1992; Harzsch and Müller 2007; Harzsch et al. 2009; Rieger et al. 2011). The arrangement is widely conserved to the point that these neurons can be homologized between the different species, indicating that such a segmental trait is likely part of the chaetognath ground pattern (Harzsch et al. 2009). It is unclear, however, how they become arranged in a segmental manner during embryonic development. Hatchlings already exhibit the four anteriormost pairs of neurons (Rieger et al. 2011), suggesting that the differentiation progresses from anterior to posterior. But the embryonic origin of these neurons and the cellular processes responsible for their segmental arrangement remain unknown.

9.6.2 Rotifera (Wheel Worms)

Wheel worms are aquatic bilaterians characterized by a prominent ciliary organ, hence their common name, and a highly specialized masticatory apparatus known as the mastax (Hyman 1951e). The body is divided into head, trunk, and foot, all of which can be segmented along the anteroposterior axis by skeletal plates and folds in the tegument body wall (Segers 2004; Fontaneto and Smet 2014). The plates are often articulated forming conspicuous telescopic rings that can be retracted–extended at will by the animals (Figure 9.3C) (Fontaneto and Smet 2014). Unlike kinorhynchs, the skeletal plates of rotifers are intracytoplasmic and formed by a dense lamina within the syncytial integument layer (Fontaneto and Smet 2014).

Rotifers also exhibit a transverse series of circular muscles along the anteroposterior axis (Figure 9.3D) (Hochberg and Litvaitis 2000; Sørensen 2005; Leasi and Ricci 2010; Fontaneto and Smet 2014). The number, width, and configuration of these muscles—which are often incomplete, forming semicircular paired lateral bands—varies between species, although the circular arrangement is considered the ancestral condition (Leasi and Ricci 2010). Regardless, the number and position of these circular muscles do not seem to correspond with the organization of the skeletal plates, when present (e.g., Sørensen 2005; Hochberg and Lilley 2010; Leasi and Ricci 2010). The ontogeny of the circular musculature has not yet been described

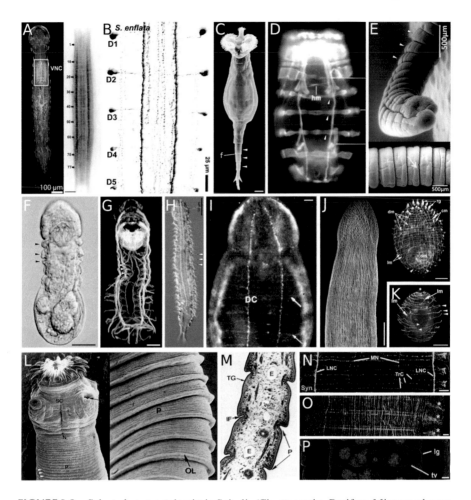

FIGURE 9.3 Selected segmental traits in Spiralia (Chaetognatha, Rotifera, Micrognathozoa, Gastrotricha, and Platyhelminthes). Arrows indicate serially repeated structures unless otherwise noted. A. The ventral nervous center (vnc) in the chaetognath *Spadella cephaloptera* (left) and the segmental microcompartments in the central neuropil of *Sagitta setosa* revealed by synapsin (right). Scale bars = 100 μm (left) and 50 μm (right). Left image reprinted from Rieger et al. (2011) with permission from John Wiley & Sons. Right image by Harzsch and Müller (2007) licensed under CC-BY. B. Repetitive series of paired neurons with RFamide-positive somata (D1-5) in the chaetognath *Sagitta enflata*. Scale bar = 25 μm. Image reprinted from Harzsch et al. (2009) with permission from Springer Nature. C. Telescopic rings in the foot (f) of the bdelloid rotifer *Rotaria macrura*. Scale bar = 20 μm. Image by Diego Fontaneto from Gross (2007) licensed under CC-BY. D. Circular musculature in the rotifer *Philodina* sp. Magnified ×630. Image reprinted from Hochberg and Litvaitis (2000) with permission from Springer Nature. E. Body wall annulations in the acanthocephalan *Mediorhynchus africanus*. Scale bars = 500 μm. Image reprinted from Amin et al. (2013) with permission from Springer Nature. F. The accordion-like thorax in the micrognathozoan *Limnognathia maerski*. Scale bar = 20 μm. Image reprinted from Giribet et al. (2004) with permission from John Wiley & Sons. G. Serially repeated dorsoventral musculature in *L. maerski*. Scale bar = 10 μm. Image by Bekkouche et al. (2014) licensed under CC-BY. H. Series of lateral spines in the

FIGURE 9.3 (CONTINUED)

gastrotrich *Xenodasys riedli*. Scale bar = 50 µm. Image reprinted from Schuster et al. (2018) with permission from Springer Nature. I. Transverse commissures (arrows) intercepting the dorsal nerve cords in the catenulid flatworm *Promonotus schultzei*. Scale bar = 10 µm. Image reprinted from Reuter et al. (1995) with permission from Springer Nature. J. Lattice-like musculature in the proseriate flatworm *Monocelis* sp. (left) and in the polyclad flatworm *Melloplana ferruginea* (right) showing the circular muscles (cm), diagonal muscles (dm), longitudinal muscles (lm), oral muscles (om), and rhabdite glands (rg). Scale bars = 100 µm (left) and 25 µm (right). Left image by Girstmair et al. (2014) licensed under CC-BY. Right image reprinted from Bolaños and Litvaitis (2009) with permission from John Wiley & Sons. K. Early circular myoblast processes in the polyclad flatworm embryos *Maritigrella crozieri* with few developing longitudinal fibers (lm). Scale bar = 50 µm. Image reprinted from Bolaños and Litvaitis (2009) with permission from John Wiley & Sons. L. Anterior region (left) and body wall infoldings (right) in the tapeworm *Taenia taeniaeformis* and *Hymenolepis nana* showing the proglottids (p) and their overlapping margins (ol), neck region (n), scolex (sc), suckers (s), and the rostellum (r). Magnified ×120 and ×80, respectively. Images reprinted from Mehlhorn et al. (1981) with permission from Springer Nature. M. Longitudinal section of the body wall in the tapeworm *H. nana* showing the infoldings (if) in the syncytial tegument layer (tg) of the proglottids (p) with transverse canals of the excretory system (e). Magnified ×200. Image reprinted from Mehlhorn et al. (1981) with permission from Springer Nature. N. Transverse commissures (trc) intercepting the lateral nerve cords (lnc) and median nerves (mn) in the nervous system of the tapeworm *Hymenolepis diminuta*. Scale bar = 100 µm. Image reprinted from Rozario and Newmark (2015) with permission from Elsevier. O. Transverse muscle fibers (arrowheads) and terminal genitalia (asterisks) in the proglottids of tapeworm *H. diminuta*. Scale bar = 100 µm. Image reprinted from Rozario and Newmark (2015) with permission from Elsevier. P. Transverse (tv) and longitudinal (lg) excretory canals in the tapeworm *H. diminuta*. Scale bar = 100 µm. Image reprinted from Rozario and Newmark (2015) with permission from Elsevier.

in detail. Muscle bands are visible in the embryo before hatching (Boschetti et al. 2005), but the mechanisms and processes that lead to the serial arrangement of these circular muscle cells in rotifers still need to be elucidated.

Acanthocephalans are parasitic worms affiliated to rotifers (Hejnol 2015c). They have a hooked proboscis and no gut (Hyman 1951d; Nicholas 1967). Several species have a trunk segmented by regular annular constrictions (Figure 9.3E) (Southwell and Macfie 1925; Amin et al. 2013). The epidermis at each constriction is attached to the longitudinal muscles isolating the outer circular muscles into rings, which correspond to the annuli (Hyman 1951d). The trunk can also exhibit circular girdles of spines along the body. Internally, a fluid-filled lacunar system can be organized in an orthogonal pattern with a series of circular channels along the anteroposterior axis of the body (Hyman 1951d).

9.6.3 MICROGNATHOZOA (JAW ANIMALS)

Limnognathia maerski—the sole representative species of the Micrognathozoa—is a small bilaterian inhabiting mossy environments in Greenland with morphological features that resemble rotifers and other gnathiferans (Kristensen and Funch 2000). The adult body is regionalized into a head, an accordion-like thorax, and an ovoid abdomen

(Figure 9.3F). The thorax is subdivided into five flexible annulations supported by dorsal and lateral plates, which are composed of an intracellular matrix layer (Kristensen and Funch 2000). *L. maerski* exhibits 13 pairs of serially repeated dorsoventral muscles along the anteroposterior axis (Figure 9.3G) (Bekkouche et al. 2014). The muscle bundles occur in the thorax and abdomen, but their position does not correspond to the arrangement of the thorax annulations (see Bekkouche et al. 2014). There are two pairs of lateral protonephridia in the trunk (Kristensen and Funch 2000) and no segmental traits in the nervous system (Bekkouche and Worsaae 2016b).

9.6.4 GASTROTRICHA (HAIRYBACKS)

The minute hairybacks are elongated free-living bilaterians covered with scales, spines, and bristles (Hyman 1951e). The arrangement of the gastrotrich scales on the body surface is highly ordered, but a segmental organization is more evident in the distribution of spines, bristles, and adhesive tube pairs along the lateral of the body (Figure 9.3H) (Hyman 1951e). Internally, there are no obvious segmental structures in the gastrotrich nervous (Rothe and Schmidt-Rhaesa 2009; Rothe, Schmidt-Rhaesa, and Kieneke 2011; Bekkouche and Worsaae 2016a) or muscular systems (Kieneke, Martínez Arbizu, and Riemann 2008b; Bekkouche and Worsaae 2016a). The only exception is the excretory system, where some species can have between two and eleven pairs of protonephridia along the body (Teuchert 1967; Bekkouche and Worsaae 2016a). Nonetheless, because many gastrotrich species have only a single pair of protonephridia, it is not yet resolved if the segmental organization of the excretory system is an ancestral feature (Kieneke, Arbizu, and Ahlrichs 2007; Kieneke et al. 2008a). One other segmental trait that is likely part of the ground pattern of Gastrotricha is the lateral pairs of adhesive tubes (Kieneke, Riemann, and Ahlrichs 2008c).

Developmental studies on gastrotrichs are scarce (Hejnol 2015b). However, Teuchert (1968) provides important insights about the ontogeny of the aforementioned segmental traits. Adults of *Turbanella cornuta* have several pairs of lateral adhesive tubes and tactile bristles, and four pairs of protonephridia that are serially arranged and evenly spaced along the anteroposterior axis of the animal (Teuchert 1967, 1968). The tactile bristles and adhesive tubes develop at the same locations and almost in concert during embryogenesis, and the juveniles hatch with four lateral pairs of each (Teuchert 1968). Additional pairs of bristles and adhesive tubes are added during postembryonic development by intercalation between the initial four pairs, and not at the posterior end as one might expect (Teuchert 1968). In contrast, the formation of new protonephridia pairs follows an anterior-to-posterior progression. *T. cornuta* juveniles only have a single protonephridia pair located anteriorly, and additional pairs are sequentially added until the fourth pair is formed near the posterior end (Teuchert 1968). This suggests once more that the segmental organization of different traits is achieved by distinct, partly independent developmental mechanisms.

9.6.5 PLATYHELMINTHES (FLATWORMS)

Platyhelminthes is a diverse group of dorsoventrally flat bilaterians with complicated life history strategies and unusual morphologies in parasitic species (Hyman 1951b).

Flatworms also have segmental traits of different kinds and degrees, such as an orthogonal nervous system or the proglottids in tapeworms.

In free-living flatworms, the longitudinal nerve cords—which vary from a single pair to multiple pairs—are intercepted by ladder-like transverse commissures forming an orthogonal grid pattern historically referred to as the *orthogon* (Figure 9.3I) (Reisinger 1925; Reuter and Gustafsson 1995; Halton and Gustafsson 1996; Reuter, Mäntylä, and Gustafsson 1998). The arrangement was initially identified in classical platyhelminth works (e.g., Lang 1881; Wheeler 1894; Wilhelmi 1909), but as more species were analyzed, a whole range of variation within the *orthogon* basic pattern was revealed, from dense and regular to sparse and uneven grids (Reuter and Gustafsson 1995; Reuter, Mäntylä, and Gustafsson 1998). Notably, the arrangement is also absent in planktonic larval stages (Rawlinson 2010).

The *orthogon* is formed during embryonic development with the longitudinal nerve cords usually developing first, and the transverse commissures appearing later (Younossi-Hartenstein, Jones, and Hartenstein 2001; Cardona, Hartenstein, and Romero 2005; Monjo and Romero 2015). However, it remains unclear if these commissures are formed progressively from anterior to posterior, and which cellular processes involved. One hypothesis is that neurons migrate from the brain while the longitudinal connectives are extending, and periodically settle to form the transverse projections. A second possibility is that totipotent stem cells present in the mesenchyme (i.e., neoblasts) differentiate into neurons at regularly spaced loci, perhaps guided by the longitudinal nerve cords, to produce the commissures (Younossi-Hartenstein, Jones, and Hartenstein 2001). The latter—also known as "interstitial neural progenitor mode" (Hartenstein and Stollewerk 2015)—has been shown to occur during the regeneration of the planarian nervous system (e.g., Nishimura et al. 2011), but both mechanisms likely play a role in the patterning of these commissures.

Another featured trait of flatworms is the lattice-like musculature with serially arranged circular muscles (Figure 9.3J) (Rieger et al. 1991, 1994; Reiter et al. 1996; Cardona, Hartenstein, and Romero 2005; Bolaños and Litvaitis 2009; Rawlinson 2010; Semmler and Wanninger 2010; Krupenko and Dobrovolskij 2015). This organization originates during early embryogenesis, from an irregular network of cells projecting myofilaments in random orientation, until the first transverse circular muscles are formed (Reiter et al. 1996; Cardona, Hartenstein, and Romero 2005; Bolaños and Litvaitis 2009; Rawlinson 2010; Semmler and Wanninger 2010). The position of the first circular muscles varies between species. In some cases, a primary fiber is present (e.g., Rawlinson 2010), but in general, they do not form progressively from the anterior to the posterior end (Figure 9.3K). During ontogeny, circular muscles can be added via two different cellular processes—by duplication of the whole muscle resulting in double stranded fibers or by the branching off of one end of the fiber (Bolaños and Litvaitis 2009; Semmler and Wanninger 2010). The putative mechanisms in place to control the even distribution of these dynamic populations of circular muscles remain unknown, but it is hypothesized that interactions with the nervous system might play an important role (Rawlinson 2010).

Free-living flatworms also display some less common segmental traits, such as paired series of lateral gut diverticula alternating with gonads (Lang 1881). But perhaps one of the most striking examples of segmental organization in flatworms occurs in some of the parasitic forms—the tapeworms. Their adult body is notably divided into serially repeated morphological units delimited by external constrictions of the body wall (Figure 9.3L), which contain serially arranged internal structures, such as neurons, muscles, and gonads (Hyman 1951b; Mehlhorn et al. 1981).

These segments, or proglottids, are demarcated by annular infoldings of the syncytial tegument layer that form the tapeworm body wall (Figure 9.3M) (Mehlhorn et al. 1981). The folds are superficial and there are no internal membranous structures separating the proglottids (Mehlhorn et al. 1981; Koziol 2017). Unlike the typical segmented animals, the tapeworm segments form at a germinative zone behind the "head" (Rozario and Newmark 2015), so that the oldest proglottids are located toward the posterior end of the animal (Hyman 1951b; Mehlhorn et al. 1981). This growth zone likely undergoes a periodic accumulation of proliferative cells during segmentation (Koziol et al. 2010). Pioneering work on a tapeworm provided initial evidence that the annular infoldings of the tegument are under tight genetic control (Holy et al. 1991). The processes that establish the proglottid boundaries and how they compare to the segmentation mechanisms in other animals remain elusive, but are currently under active investigation (Koziol et al. 2016; Koziol 2017; Olson et al. 2018).

In addition to the body wall segmentation, tapeworms also display internal segmental traits that in most cases are in register with external segmental features. For example, the nervous system arrangement is stereotypic with three transverse commissures per proglottid intercepting the longitudinal nerve cords (Figure 9.3N); the circular musculature is evenly spaced along the anteroposterior axis and inner transverse cortical fibers can be present at each proglottid boundary (Figure 9.3O); the excretory canals of the osmoregulatory system have a ladder-like organization (Figure 9.3P); and finally, each proglottid contains a stereotypical set of reproductive organs and a single genital pore (Rozario and Newmark 2015).

In fact, some tapeworm lineages have no proglottid boundaries and only exhibit internal segmental traits. Strikingly, the phylogeny of tapeworms suggests the segmental arrangement of internal traits evolved before the external annular infoldings in the clade (Olson et al. 2001). This indicates that the segmental organization of tapeworms was established stepwise, possibly involving independent developmental mechanisms. If and how the patterning of internal traits might have influenced the evolution of the external morphology remains an open question, and more detailed data about the development of these internal structures is needed (Koziol 2017). Given that tapeworms evolved *de novo* a complex segmental organization (Olson et al. 2001), the group can provide interesting insights about molecular evolution, for example, how known molecular pathways might have been coopted to pattern tapeworm segmental traits, or potentially reveal novel developmental mechanisms involved in generating repeated structures.

9.6.6 NEMERTEA (RIBBON WORMS)

Nemerteans are slender hunters with a typical unsegmented external appearance, but that can exhibit some internal segmental traits (Hyman 1951c). Examples of nemertean segmental traits are serially arranged ovary pairs, lateral gut diverticula, transverse nerve commissures, and one single case of annular constrictions in the trunk.

A few interstitial nemertean species have a conspicuous segmented appearance where the external body wall is subdivided along the anteroposterior axis into trunk segments (Berg 1985; Norenburg 1988; Chernyshev and Minichev 2004). Each segment is demarcated by epidermal constrictions that coincide with a rather unusual trait for this group—a segmented digestive tract (Berg 1985). In these species, the gut is divided into barrel-shaped segments delimited by an internal membrane with a funnel aperture connecting the compartments (Berg 1985). Both the body wall and gut segmental traits are apomorphies of this particular group of nemerteans, and possibly associated to their interstitial habit (Chernyshev and Minichev 2004; Sundberg and Strand 2007). Given the predictable nature of these traits, a tight developmental control must be involved in patterning these segmental tissue boundaries.

The presence of transverse commissures arranged in regular intervals between the longitudinal nerve cords of nemerteans has been described in classical works (e.g., Hubrecht 1887; Bürger 1895). However, this ladder-like appearance varies across species (Beckers, Faller, and Loesel 2011; Beckers, Loesel, and Bartolomaeus 2013; Chernyshev and Magarlamov 2013; Beckers, Krämer, and Bartolomaeus 2018). Serially arranged circular nerves can also be present (Figure 9.4A) (Beckers, Loesel, and Bartolomaeus 2013; Chernyshev and Magarlamov 2013). These neural structures probably develop either in adults or late juveniles, since they are not detected in larval or early juvenile stages of nemerteans (Hay-Schmidt 1990; Chernyshev and Magarlamov 2010; Maslakova 2010; Hindinger, Schwaha, and Wanninger 2013; Döhren 2016; Martín-Durán et al. 2018). Furthermore, the regular arrangement of the nervous system is likely not part of the nemertean ground pattern, but a derived feature associated with a pelagic habitat (Beckers, Krämer, and Bartolomaeus 2018).

Another and more widespread nemertean segmental trait is the serial arrangement of ovary pairs along the body axis (Figure 9.4B) (Coe 1905; Stricker et al. 2002). The number of pairs ranges from fewer than ten to several thousands, and each ovary can be connected to a correspondent gonoduct and genital opening through the body wall (Stricker et al. 2002). In addition, the ovaries are tightly intercalated by lateral gut diverticula (Figure 9.4C), indicating a certain morphological integration between these traits. Developmental mechanisms responsible for the segmental organization of nemertean ovaries are still unknown.

Finally, nemerteans usually have a single anterior pair of protonephridia, but some groups can have serially arranged protonephridia reaching up to 300 pairs—a condition that probably evolved secondarily (Bartolomaeus and Döhren 2010). Along the body wall and proboscis there are series of densely packed, but not so regularly distributed, circular muscles (Chernyshev 2015, 2010; Chernyshev and Kajihara 2019).

9.6.7 MOLLUSCA (SNAILS AND SQUIDS)

Mollusca is the group of bilaterians comprised of snails, oyster, squids, and other animals with a great diversity of larval and adult body patterns (Ruppert, Fox, and Barnes 2004b). Two particular mollusk groups, the Monoplacophora and the Polyplacophora (e.g., chitons), show a rather notable set of segmental traits, such as

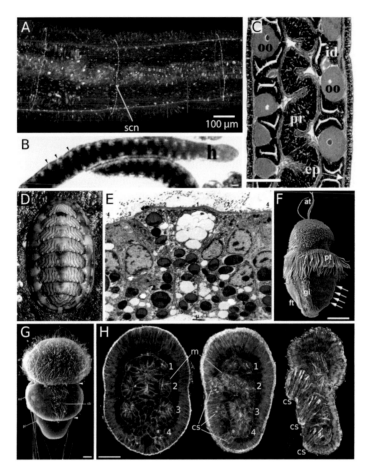

FIGURE 9.4 Selected segmental traits in Spiralia (Nemertea, Mollusca, and Brachiopoda). Arrows indicate serially repeated structures unless otherwise noted. A. Ring nerves (scn) in the nemertean *Procephalothrix filiformis*. Scale bar = 100 μm. Image by Beckers, Loesel, and Bartolomaeus (2013) licensed under CC-BY. B. Serially repeated ovaries in the nemertean *Carcinonemertes epialti*. Scale bar = 100 μm. Image reprinted from Stricker et al. (2002) with permission from John Wiley & Sons. C. Oocytes (oo) intercalated by gut diverticula (id) in the nemertean *Tetrastemma phyllospadicola* showing the proboscis (pr) and body wall epidermis (ep). Scale bar = 400 μm. Image reprinted from Stricker et al. (2002) with permission from John Wiley & Sons. D. Series of dorsal shell plates in the chiton *Tonicella lineata*. Image by Jerry Kirkhart from Kirkhart (2008) licensed under CC-BY. E. Cell types in the shell field epithelium of the chiton *Ischnochiton rissoa*. The surface of each cell is color coded. Scale bar = 1 μm. Image reprinted from Kniprath (1980) with permission from Springer Nature. F. Larva of the chiton *Mopalia muscosa* with developing shell field (arrows), mantle fold (gi), apical tuft (at), and foot (ft). Scale bar = 50 μm. Image reprinted from Wanninger and Wollesen (2015) with permission from Springer Nature. G. Transverse body wall furrows in the trilobed larva of the brachiopod *Macandrevia cranium* showing the apical lobe (al), the mantle lobe (ml) with ciliary band (cb), and the pedicle lobe (pl). Scale bar = 10 μm. Image reprinted from Zakrzewski, Suh, and Lüter (2012) with permission from Elsevier. H. Segmental mesoderm (m) with partitions numbered from 1 to 4 and dorsal chaetae sacs (cs) in the larva of the brachiopod *Novocrania anomala*. Scale bar = 20 μm. Image by Vellutini and Hejnol (2016) licensed under CC-BY.

serially arranged dorsal shells and repeated pairs of dorsoventral muscles (Lemche and Wingstrand 1959; Wingstrand 1985; Willmer 1990). Compared to other groups covered in this chapter, mollusks have been relatively well-studied in terms of cellular, developmental, and genetic mechanisms during the ontogeny of these segmental traits.

Several mollusk groups exhibit a repeated sequence of paired dorsoventral muscles along the anteroposterior axis (Haszprunar and Wanninger 2000; Wanninger and Wollesen 2015). In general, the numbers vary between seven pairs in larval aplacophorans; seven to eight pairs in adult polyplacophorans; eight pairs in monoplacophorans; three to eight pairs in bivalves; and one to two pairs in gastropods, scaphopods, and cephalopods (Wingstrand 1985; Haszprunar and Wanninger 2000; Scherholz et al. 2013). In chitons, the first muscles appear in the embryo as a series of transverse fibers that develop in an anteroposterior progression along the trunk, but the dorsoventral muscles only become arranged in bundles after metamorphosis (Wanninger and Haszprunar 2002).

Concerning other internal traits of mollusks, both mono- and polyplacophorans have gills and nephridiopores arranged in series along the lateral sides of the body. Not much is known about the development of these structures, but there is evidence that their growth and differentiation is not paired, because the number of structures between left and right sides is not symmetric (Hunter and Brown 1965; Wingstrand 1985; Russel-Hunter 1988). In the nervous system, while transverse commissures can be present (Wingstrand 1985), they are not regularly spaced (Voronezhskaya, Tyurin, and Nezlin 2002; Friedrich et al. 2002).

Chitons have a set of eight dorsal shell plates separated by transverse intersegmental ridges (Figure 9.4D). These structures originate from embryonic blastomeres in the second and third quartet micromeres (Heath 1899; Henry, Okusu, and Martindale 2004), and are formed through the differential specialization of epithelial cells (Kowalevsky 1883a; Heath 1899). For instance, the intersegmental ridges are created by a transverse row of mucus-producing cells (known as goblet cells or type-1), while a nearby row of type-4 cells secrete the shell plates (Figure 9.4E) (Kniprath 1980). The goblet cells are surrounded by cells with dense cytoplasm and long microvilli (type-2), while type-4 shell-secreting cells are surrounded by another cell kind (type-3) (Kniprath 1980). Thus, the shell field of chitons is the product of a highly organized epithelium patterned during embryogenesis, and progressively established during larval development (Figure 9.4F), or after metamorphosis in some species (Leise 1984). Albeit gradual, the shell formation does not follow an anteroposterior progression—the central plates appear first and are shortly followed by the first, sixth, and seventh plates, while the eighth plate is formed several days later (Kniprath 1980).

Although we do not yet comprehend the developmental mechanisms regulating the segmental organization of the chiton shell field, gene expression studies began to uncover the molecular identities of these epithelial cells. For instance, the intersegmental cells type-1 and type-2 are known to express the gene *engrailed* (Jacobs et al. 2000), while the shell-secreting type-4 likely express the genes *pax2/5/8* (Wollesen et al. 2015) and *gbx* (Wollesen et al. 2017) (Figure 9.5A). These genes are usually involved in establishing developmental boundaries (e.g., Raible and Brand 2004), and thus might play a role in the patterning of the shell boundaries in mollusks

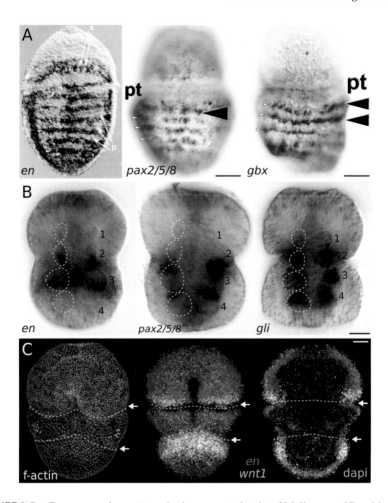

FIGURE 9.5   Gene expression patterns in the segmental traits of Mollusca and Brachiopoda. A. Expression of *engrailed*, *pax2/5/8*, and *gbx* in the developing dorsal shell field of the chiton *Lepidochitona caverna* (left) and *Acanthochitona crinita* (center and right). Scale bars = 20 μm. Left image reprinted from Jacobs et al. (2000) with permission from John Wiley & Sons. Center image by Wollesen et al. (2015) licensed under CC-BY. Right image by Wollesen et al. (2017) licensed under CC-BY. B. Expression of *engrailed*, *pax2/5/8*, and *gli* in the segmental larval mesoderm of the brachiopod *Novocrania anomala*. Scale bar = 20 μm. Images by Vellutini and Hejnol (2016) licensed under CC-BY. C. Cell outlines (left) and abutting domains of *wnt1* (green) and *engrailed* (magenta) at the non-segmental boundaries (arrows) in the larva of the brachiopod *Terebratalia transversa*. Scale bar = 20 μm. Images by Vellutini and Hejnol (2016) licensed under CC-BY.

(Nederbragt, Loon, and Dictus 2002). This is supported by the observations that the expression of *engrailed*, *pax2/5/8*, and *gbx* is also associated to shell development in a wide range of mollusks (Moshel, Levine, and Collier 1998; Jacobs et al. 2000; Wanninger and Haszprunar 2001; Nederbragt, Loon, and Dictus 2002; O'Brien and Degnan 2003; Iijima et al. 2008; Kin, Kakoi, and Wada 2009; Hashimoto,

Kurita, and Wada 2012; Wanninger and Wollesen 2015; Wollesen et al. 2015, 2017; Scherholz et al. 2017). It remains to be determined if these genes could be involved in the generation of the dorsal shell series in chitons, or if instead they have roles in downstream processes such as boundary formation and cell differentiation.

9.6.8 Brachiopoda (Lamp Shells)

Brachiopods are bivalved marine invertebrates of sessile habit that use an intricate ciliated tentacular-crown known as the lophophore to capture food (Hyman 1959c). Despite the enclosed adult body and reduced morphology, larval brachiopods exhibit a combination of putative segmental traits in different germ layers, such as transverse body wall folds, repeated coelomic pouches, and serially arranged chaetae sacs.

The body of brachiopod larvae can be subdivided into two, three, or four lobes along the anteroposterior axis, each delimited by transverse epithelial furrows (Figures 9.4G and 9.5C) (Morse 1873; Kowalevsky 1883b; Lacaze-Duthiers 1861). Even though this segmented appearance has stirred much debate in the late 19th century (Balfour 1880; Shipley 1883; Masterman 1899; Conklin 1902), these ectodermal boundaries are not in fact segmental and each furrow has a distinct morphology (Vellutini and Hejnol 2016). Interestingly, the furrow at the head–trunk boundary is delimited by abutting stripes of *engrailed* and *wnt1* (Figure 9.5C) (Vellutini and Hejnol 2016), a hallmark pattern of the segment polarity cascade in arthropods (Ingham 1991).

On the other hand, the most prominent segmental trait of brachiopods occurs in the mesoderm—a regular series of four paired coelomic pouches distributed along the anteroposterior axis of a larval stage (Nielsen 1991). The three posterior pairs are also associated with correspondent pairs of dorsal chaetae sacs (Figure 9.4H) (Nielsen 1991).

These coelomic pouches are sequentially formed during gastrulation from evaginations of the archenteron (Nielsen 1991) with the exception of the anteriormost pair, which might be formed by cell delamination (Freeman 2000). The partitioning, however, is incomplete, and the tissue remains unsegmented on its ventral side (Figure 9.4H) (Vellutini and Hejnol 2016). In this region, a series of transverse muscle fibers aggregate and form three distinct transverse bundles connecting the pair of mediolateral longitudinal muscles (Altenburger and Wanninger 2010). A segmental mesoderm, however, is not a widespread feature among larval brachiopods. The mesoderm morphology varies considerably between species and the tissues can also be unsegmented or simply split into anterior and posterior regions (Hyman 1959c), suggesting the paired coelomic pouches is a derived trait.

Before the morphological boundaries are visible in the embryonic mesoderm, *engrailed* transcripts are detected in two pairs of stripes at the posterior region of the second and third coelomic pouches (Figure 9.5B) (Vellutini and Hejnol 2016). After the mesodermal boundaries are established, the gene *pax2/5/8* is detected in paired stripes between the same pouches, while components of the Hedgehog pathway such as the transcription factor *gli* become expressed during pouch formation (Figure 9.5B) (Vellutini and Hejnol 2016). However, neither gene exhibited a reiterated pattern throughout the four pairs of coeloms as one would expect of a typical segmental patterning, suggesting that genes relevant for segmenting the brachiopod mesoderm have not yet been identified.

In adult brachiopods, the presence of two pairs of lateral mesenteries and two nephridial pairs have been interpreted as evidence for segmentation in brachiopods (Malakhov and Kuzmina 2006; Temereva and Malakhov 2011), however the segmental nature of these traits remains controversial.

9.6.9 PHORONIDA (HORSESHOE WORMS)

Phoronids are sessile marine invertebrates closely related to brachiopods (Kocot et al. 2017). The adult body does not show any external segmental trait and there are only a few internal traits with presumptive segmental organization. The ventral nerve cord is intercepted by a series of transverse commissures (Temereva 2012; Temereva and Wanninger 2012) and circular muscles are distributed along the body (Santagata and Zimmer 2002; Santagata 2004; Temereva and Tsitrin 2013). However, the spatial regularity is not completely evident and further work is required to verify the segmental nature of these neurons and muscles. In addition, despite having a trimeric coelomic organization in adults (Masterman 1899; Siewing 1973) and larvae (Temereva and Malakhov 2006, 2011), these compartments are not arranged in a segmental manner along the anteroposterior axis.

9.6.10 BRYOZOA (MOSS ANIMALS)

Bryozoans are discrete but ubiquitous bilaterians in marine and freshwater environments that form colonies with complex and intricate arrangements (Hyman 1959d). A colony is formed by functional units named zooids. Their body is reduced to a lophophore, gut, and spacious mesoderm accommodating the gonads. There is no trace of segmental traits in individual zooids, but the colony of one abyssal Mediterranean species shows a unique segmental trait. The colony has a root, a peduncle with modified zooids, and tip with regular zooids (d'Hondt 1976). The peduncle is composed by a series of annular zooids forming a peculiar segmented stalk. This stacked arrangement is not common in other bryozoan colonies and is an unusual case of supraorganismal segmental organization in bilaterians.

9.7 DISCUSSION

The segmental traits covered in this chapter have diverse embryonic origins, molecular fingerprints, developmental mechanisms, and end morphologies. This diversity indicates that a segmental organization can evolve in any germ layer or organ system of bilaterians and do so via various ontogenetic pathways. In most cases, however, these traits are often confined to a particular organ or developmental stage, and are not in register with other repetitive structures present in the same organism (Table 9.1).

For example, chaetognaths only have segmental traits in the nervous system, nematomorphs only exhibit cuticular annulations in the larval stages, and the musculature of rotifers and chitons does not match the position of the skeletal plates

TABLE 9.1

Segmental Traits in Bilaterians (Except Arthropods, Annelids, and Vertebrates)

Group	Body Wall	Neurons	Muscles	Gonads	Excretion	Gut	Features
Xenacoelomorpha	—	—	—*	—	—	—	Circular muscles can be regularly distributed.
Hemichordata	—	—	—	✓	—	✓*	Series of gill pairs intercalated by gonads. Paired series of gut diverticula (hepatic saccules).
Echinodermata	—	✓*	✓*	—	—	—	Repeated serial elements (ossicles, neurons, muscles, and vascular channels) along body rays.
Cephalochordata	—	✓*	✓	✓	✓	—	Somites and repeated muscle bundles. Series of gonads and nephridia. Some dorsal neurons.
Tunicata	—	—*	✓*	—	—	—	Regularly spaced pharynx structures (gill slits, neurons, and vessels). Larval notochord. Circular muscles in planktonic forms. Asexual buds formed by strobilation.
Nematoda	✓	✓	✓*	—	—	—	Cuticular annulation and corresponding series of circular F-actin bundles. Series of concretion rings. Serially arranged locomotory neurons.
Nematomorpha	✓	—	—	—	—	—	Annulated cuticle in the larva.
Priapulida	✓	—*	—*	—	—	—	Cuticular annulation and corresponding circular muscles. Series of evenly spaced circular neurite bundles.
Kinorhyncha	✓	✓	✓	—	—	—	Cuticular plates (zonites) in the trunk. Paired series of repeated muscle bundles. Transverse commissures and paired ventral neuronal somata.
Loricifera	—*	—	—*	—	—	—	Thorax cuticle divided in two segments. Not so evenly spaced circular muscle bundles in the abdomen.
Tardigrada	✓*	✓	—*	—	—	—	Trunk segments and mesodermal pouches in embryos. Paired appendages and few corresponding repeated muscle sets. Series of trunk ganglia.
Onychophora	—	—*	—*	—	✓	—	Trunk segments and paired ventral organs in embryos. Paired appendages and corresponding nerves and muscles. Series of paired nephridia. Series of blood valves.

(Continued)

TABLE 9.1 (CONTINUED)
Segmental Traits in Bilaterians (Except Arthropods, Annelids, and Vertebrates)

Group	Body Wall	Neurons	Muscles	Gonads	Excretion	Gut	Features
Chaetognatha	—	✓	—	—	—	—	Series of paired neurons and neuropil microcompartments.
Rotifera	—*	—	✓	—	—	—	Telescopic skeletal rings. Series of circular or paired lateral muscles.
Acanthocephala	✓	—	✓	—	—	—	Annular constrictions in the trunk with corresponding series of circular muscles. Series of circular channels.
Micrognathozoa	—	—	✓	—	✓	—	Paired series of dorsoventral muscles. Two pairs of lateral protonephridia.
Gastrotricha	—*	—	—	—	✓*	—	Series of paired cuticular spines, bristles, and adhesive tubes. Paired series of protonephridia.
Platyhelminthes	✓*	✓	✓	✓	—	—*	Repeated trunk annular infoldings in tapeworms with corresponding series of transverse neuronal commissures, circular musculature, transverse excretory channels, and set of gonads. Series of transverse neuronal commissures and circular musculature, and at times repeated gut diverticula with alternating gonads in free-living flatworms.
Nemertea	✓*	✓*	—*	✓	✓*	—*	Series of circular nerves or transverse commissures and densely packed circular muscles. Paired ovary series intercalated by lateral gut diverticula. Series of paired protonephridia can occur. Annular epidermal constrictions with corresponding gut compartments in one group.
Mollusca	✓*	—	✓*	✓*	✓*	—	Series of dorsal shell plates. Paired series of dorsoventral retractor muscles. Series of gills, gonads, and nephridiopores can occur.
Brachiopoda	—*	—	—	—	—	—	Serial pairs of coelomic and chaetae sacs in larval stage.
Phoronida	—	—*	—	—	—	—	Not so regular series of neural commissures and circular musculature in larval stage.
Bryozoa	—	—	—	—	—	—	Stalk formed by a series of annular zooids in abyssal species.

Labels: ✓ Presence of segmental trait(s); ✓* Presence of segmental trait(s) but limited to subgroup or life stage; —* Presence of trait(s) with incipient segmental organization; — Absence of segmental trait(s).

and shells. In addition, some segmental traits are apomorphies of specific clades. These likely evolved as adaptations to particular environments, such as the association between strong body wall annulation and the occupation of interstitial habitats in nematodes and nemerteans. Given these observations, it seems reasonable to suggest that bilaterian structures with a segmental organization have diverse origins and behave as independent evolutionary units (Scholtz 2010).

In spite of that, segmental traits can be integrated to various degrees in different bilaterians. The cases range from a direct link between circular musculature and cuticular annulation, as seen in priapulids, to the most-notable coordination between external folds, musculature, nervous system, and other segmental structures present in kinorhynchs or tapeworms. This level of integration, generally considered a landmark of the so-called true segmentation, is on par at least in qualitative terms with that of arthropods and annelids. Even though kinorhynchs and tapeworms display some mismatches between segmental traits, this also occurs in arthropods and annelids (Budd 2001; Fusco 2008), suggesting that these arrangements are rather plastic in evolutionary terms.

The fact that one organism can have multiple segmental traits with different evolutionary histories (e.g., Graham et al. 2014), and that these traits can be integrated or disassembled with relative evolutionary ease, dissolves the artificial division between the so-called segmented and non-segmented animals. The evolution of tapeworm segmentation is an example of the gradual, organ-based attainment of an integrated segmental body organization (Olson et al. 2001), which further supports the hypothesis that the overt segmentation in the typical fully segmented animals evolved "system by system" (Budd 2001; Chipman 2019).

Considering all the traits covered in this chapter, the most common organ systems of bilaterians to exhibit a segmental organization are the body wall (e.g., annulations), the nervous system (e.g., transverse commissures), and the musculature (e.g., circular fibers).

The body wall is usually segmented by infoldings of the epidermis, such as the ones found in tapeworms, or by the presence of extracellular or intracellular plates, as seen in kinorhynchs and rotifers, respectively. For each case, the cellular processes that establish these boundaries are rather different. For instance, on the body wall of kinorhynchs and chitons, segmentation is achieved by the differential secretion of cellular materials, which produce the skeletal plates/shells and the intersegmental cuticle that joins each segment. In contrast, to create a fold in a sheet of epithelial cells, coordinated changes in cell shape (i.e., cytoskeleton remodeling) and the modulation of cell–cell and cell–matrix adhesion are required (Schock and Perrimon 2002).

The segmental traits in the nervous system consist mainly of serially repeated transverse commissures, paired neurons, and nerve cord ganglia. The transverse commissures are present in flatworms and nemerteans, but the regularity of the pattern varies considerably, and the architecture of the nervous system is likely correlated with the life habits of the species. In general, the processes that neuronal cells undergo to form a grid pattern have not been investigated in depth.

The circular musculature is a seemingly ubiquitous and plastic trait of bilaterians, but its arrangement can be strikingly regular, and in some cases, directly responsible for the segmental nature of adjacent traits. This is the case for the cuticular annulations of priapulids, which match the position of the circular musculature. An analogous situation occurs in nematodes, which despite not having actual circular muscles, the subcellular transverse f-actin fibers of the hypodermal cells are directly associated with the cuticle annulations (Priess and Hirsh 1986).

The segmental organization of circular muscles can arise through different cellular processes. Often, the initial set of muscle cells is positioned irregularly during embryonic development, and progressively self-organizes into an evenly spaced set of circular bundles. In a few cases, however, these initial fibers are already oriented, and the muscle cells differentiate from anterior to posterior (e.g., acoels). Because the frequency of circular muscles is usually group- or species-specific and can vary between different body regions (e.g., Tyler and Hyra 1998), there must be identifiable mechanisms, such as signaling molecules and cell-to-cell interactions, by which these specific arrangements emerge. But what controls the spatial and temporal information to orient and organize these muscle cells remains an open question.

Indeed the cascades of signal transduction-triggering effector molecules in these tissues remain obscure, but some earlier steps during embryonic development, which define the identity and polarity of the cells of each segmental domain, are beginning to be uncovered. The groups that should bring further insights about these molecular mechanisms are chitons and tapeworms (e.g., Wollesen et al. 2015; Koziol 2017) where a combination of molecular techniques and live imaging of developing chiton shells or tapeworm proglottids will provide crucial insights about the cellular processes involved.

The traits examined in this chapter provide only a glance into the diversity and the multitude of ways that a segmental organization can be achieved in the different bilaterian organ systems. Uncovering this rich repertoire of cellular and developmental processes will not only reveal novel segmental patterning mechanisms, but perhaps enable unforeseen insights about the typical segmented groups and bring a much needed comparative light into how these diverse segmental traits of bilaterians have evolved.

ACKNOWLEDGMENTS

I thank Ariel D. Chipman for the invitation to write this chapter; Netta Kasher for the illustrations in Figure 9.1; Eduardo E. Zattara for comments and constructive feedback; Juliana G. Roscito for text suggestions; Thomas Schwaha and Rich Mooi for sharing references; María Herranz, Chema Martín-Durán, Anna Czarkwiani, Nicolas Bekkouche, Cheon Young Chang, Tom Powers, and Casey Dunn for sharing original image files; and the Biodiversity Heritage Library for access to classical literature. This work was supported by an EMBO Long-Term Fellowship (ALTF 74-2018).

REFERENCES

Agassiz, A. 1873. The history of *Balanoglossus* and Tornaria. *Mem. Am. Acad. Arts Sci.* 9: 421–436.
Aldea, D., L. Subirana, C. Keime, L. Meister, I. Maeso, S. Marcellini, …, H. Escriva. 2019. Genetic regulation of amphioxus somitogenesis informs the evolution of the vertebrate head mesoderm. *Nat. Ecol. Evol.* 3: 1233–1240.

Altenburger, A. 2016. The neuromuscular system of *Pycnophyes kielensis* (Kinorhyncha: Allomalorhagida) investigated by confocal laser scanning microscopy. *EvoDevo* 7: 25.

Altenburger, A., and A. Wanninger. 2010. Neuromuscular development in *Novocrania anomala*: Evidence for the presence of serotonin and a spiralian-like apical organ in lecithotrophic brachiopod larvae. *Evol. Dev.* 12: 16–24.

Amemiya, S., A. Omori, T. Tsurugaya, T. Hibino, M. Yamaguchi, R. Kuraishi, …, T. Minokawa. 2016. Early stalked stages in ontogeny of the living isocrinid sea lily *Metacrinus rotundus*. *Acta Zool.* 97: 102–116.

Amin, O. M., P. Evans, R. A. Heckmann, and A. M. El-Naggar. 2013. The description of *Mediorhynchus africanus* n. sp. (Acanthocephala: Gigantorhynchidae) from galliform birds in Africa. *Parasitol. Res.* 112: 2897–2906.

Arendt, D. 2018. Hox genes and body segmentation. *Science* 361: 1310–1311.

Balavoine, G., and A. Adoutte. 2003. The segmented *Urbilateria*: A testable scenario. *Integr. Comp. Biol.* 43: 137–147.

Balfour, F. M. 1880. *A Treatise on Comparative Embryology*. Macmillan and Company.

Bannister, R., I. M. McGonnell, A. Graham, M. C. Thorndyke, and P. W. Beesley. 2005. *Afuni*, a novel transforming growth factor-β gene is involved in arm regeneration by the brittle star *Amphiura filiformis*. *Dev. Genes Evol.* 215: 393–401.

Bartolomaeus, T., and J. von Döhren. 2010. Comparative morphology and evolution of the nephridia in Nemertea. *J. Nat. Hist.* 44: 2255–2286.

Bateson, W. 1884. Note on the later stages in the development of *Balanoglossus kowalevskii* (Agassiz), and on the affinities of the Enteropneusta. *Proc. R. Soc. London* 38: 23–30.

Beckers, P., S. Faller, and R. Loesel. 2011. Lophotrochozoan neuroanatomy: An analysis of the brain and nervous system of *Lineus viridis* (Nemertea) using different staining techniques. *Front. Zool.* 8: 17.

Beckers, P., D. Krämer, and T. Bartolomaeus. 2018. The nervous systems of Hoplonemertea (Nemertea). *Zoomorphology* 137: 473–500.

Beckers, P., R. Loesel, and T. Bartolomaeus. 2013. The nervous systems of basally branching nemertea (Palaeonemertea). *PLoS One* 8: e66137.

Bekkouche, N., R. M. Kristensen, A. Hejnol, M. V. Sørensen, and K. Worsaae. 2014. Detailed reconstruction of the musculature in *Limnognathia maerski* (Micrognathozoa) and comparison with other Gnathifera. *Front. Zool.* 11: 71.

Bekkouche, N., and K. Worsaae. 2016a. Nervous system and ciliary structures of Micrognathozoa (Gnathifera): Evolutionary insight from an early branch in Spiralia. *R. Soc. Open Sci.* 3: 160289.

Bekkouche, N., and K. Worsaae. 2016b. Neuromuscular study of early branching *Diuronotus aspetos* (Paucitubulatina) yields insights into the evolution of organs systems in Gastrotricha. *Zool. Lett.* 2: 21.

Beklemishev, V. N. 1969a. Architectonics of echinodermata. In *Principles of Comparative Anatomy of Invertebrates, Vol. 1: Promorphology*, Kabata, Z. (ed.) MacLennan, J. M. (translator), 415–448. Oliver & Boyd.

Beklemishev, V. N. 1969b. Heteronomous metamerism in Articulata. In *Principles of Comparative Anatomy of Invertebrates, Vol. 1: Promorphology*, 243–349. Oliver & Boyd.

Beklemishev, V. N. 1969c. Metamerism as a special type of symmetry. In *Principles of Comparative Anatomy of Invertebrates, Vol. 1: Promorphology*, 188–242. Oliver & Boyd.

Beklemishev, V. N. 1969d. Origin of bilateral symmetry in Deuterostomia. Architectonics of lower Chordata. In *Principles of Comparative Anatomy of Invertebrates, Vol. 1: Promorphology*, 374–414. Oliver & Boyd.

Benito, J., I. Fernández, and F. Pardos. 1993. Fine structure of the hepatic sacculations of *Glossobalanus minutus* (Enteropneusta, Hemichordata). *Acta Zool.* 74: 77–86.

Ben Khadra, Y., M. Sugni, C. Ferrario, F. Bonasoro, A. V. Coelho, P. Martinez, and M. D. Candia Carnevali. 2017. An integrated view of asteroid regeneration: Tissues, cells and molecules. *Cell Tissue Res.* 370: 13–28.

Berg, G. 1985. *Annulonemertes* gen. nov., a new segmented hoplonemertean. In *The Origins and Relationships of Lower Invertebrates*, Ed. S. C. Morris, 200–209. Oxford University Press.

Berrill, N. J. 1947. The development and growth of *Ciona. J. Mar. Biol. Assoc. UK* 26: 616–625.

Berrill, N. J. 1950. Budding and development in Salpa. *J. Morphol.* 87: 553–606.

Bertrand, S., and H. Escriva. 2011. Evolutionary crossroads in developmental biology: Amphioxus. *Development* 138: 4819–4830.

Bery, A., A. Cardona, P. Martínez, and V. Hartenstein. 2010. Structure of the central nervous system of a juvenile acoel, *Symsagittifera roscoffensis. Dev. Genes Evol.* 220: 61–76.

Blair, S. S. 2008. Segmentation in animals. *Curr. Biol.* 18: R991–R995.

Bolaños, D. M., and M. K. Litvaitis. 2009. Embryonic muscle development in direct and indirect developing marine flatworms (Platyhelminthes, Polycladida). *Evol. Dev.* 11: 290–301.

Bolek, M. G., A. Schmidt-Rhaesa, B. Hanelt, and D. J. Richardson. 2010. Redescription of the african *Chordodes albibarbatus* Montgomery 1898, and description of *Chordodes janovyi* n. sp. (Gordiida, Nematomorpha) and its non-adult stages from Cameroon, Africa. *Zootaxa* 2631: 36–50.

Bone, Q., C. J. P. Grimmelikhuijzen, A. Pulsford, and K. P. Ryan. 1987. Possible transmitter functions of acetylcholine and an RFamide-like substance in *Sagitta* (Chaetognatha). *Proc. R. Soc. London B Biol. Sci.* 230: 1–14.

Bone, Q., and A. Pulsford. 1984. The sense organs and ventral ganglion of *Sagitta* (Chaetognatha). *Acta Zool.* 65: 209–220.

Børve, A., and A. Hejnol. 2014. Development and juvenile anatomy of the nemertodermatid *Meara stichopi* (Bock) Westblad 1949 (Acoelomorpha). *Front. Zool.* 11: 50.

Boschetti, C., C. Ricci, C. Sotgia, and U. Fascio. 2005. The development of a bdelloid egg: A contribution after 100 years. *Hydrobiologia* 546: 323–331.

Brockmann, C., R. Mummert, H. Ruhberg, and V. Storch. 1999. Ultrastructural investigations of the female genital system of *Epiperipatus biolleyi* (Bouvier 1902) (Onychophora, Peripatidae). *Acta Zool.* 80: 339–349.

Brown, F. D., and B. J. Swalla. 2012. Evolution and development of budding by stem cells: Ascidian coloniality as a case study. *Dev. Biol.* 369: 151–162.

Budd, G. E. 2001. Why are arthropods segmented? *Evol. Dev.* 3: 332–342.

Bürger, O. 1895. *Fauna und flora des golfes von Neapel, Vol. 22: Die Nemertinen des Golfes von Neapel und der angrenzenden Meeres-Abschnitte.* Berlin: Verlag von R. Friedländer & Sohn.

Burighel, P., M. Sorrentino, G. Zaniolo, M. C. Thorndyke, and L. Manni. 2001. The peripheral nervous system of an ascidian, *Botryllus schlosseri*, as revealed by cholinesterase activity. *Invertebr. Biol.* 120: 185–198.

Byrne, M., P. A. Cisternas, L. Elia, and B. Relf. 2005. *Engrailed* is expressed in larval development and in the radial nervous system of *Patiriella* sea stars. *Dev. Genes Evol.* 215: 608–617.

Byrne, M., D. Koop, P. Cisternas, D. Strbenac, J. Y. Yang, and G. A. Wray. 2015. Transcriptomic analysis of Nodal- and BMP-associated genes during juvenile development of the sea urchin *Heliocidaris erythrogramma. Mar. Genomics* 24(Pt 1): 41–45.

Candiani, S., A. Augello, D. Oliveri, M. Passalacqua, R. Pennati, F. De Bernardi, and M. Pestarino. 2001. Immunocytochemical localization of serotonin in embryos, larvae and adults of the lancelet, *Branchiostoma floridae. Histochem. J.* 33: 413–420.

Candiani, S., L. Moronti, P. Ramoino, M. Schubert, and M. Pestarino. 2012. A neurochemical map of the developing amphioxus nervous system. *BMC Neurosci.* 13: 59.

Cannon, J. T., B. C. Vellutini, J. Smith 3rd, F. Ronquist, U. Jondelius, and A. Hejnol. 2016. Xenacoelomorpha is the sister group to Nephrozoa. *Nature* 530: 89–93.

Cardona, A., V. Hartenstein, and R. Romero. 2005. The embryonic development of the triclad *Schmidtea polychroa*. *Dev. Genes Evol.* 215: 109–131.

Cary, G. A., and V. F. Hinman. 2017. Echinoderm development and evolution in the post-genomic era. *Dev. Biol.* 427: 203–211.

Chernyshev, A. V. 2010. Confocal laser scanning microscopy analysis of the phalloidin-labelled musculature in nemerteans. *J. Nat. Hist.* 44: 2287–2302.

Chernyshev, A. V. 2015. CLSM analysis of the Phalloidin-Stained muscle system of the nemertean proboscis and rhynchocoel. *Zool. Sci.* 32: 547–560.

Chernyshev, A. V., and H. Kajihara. 2019. Comparative muscular morphology in Archinemertea (Nemertea: Palaeonemertea). *Zoomorphology* 138: 193–207.

Chernyshev, A. V., and T. Y. Magarlamov. 2010. The first data on the nervous system of hoplonemertean larvae (Nemertea, Hoplonemertea). *Dokl. Biol. Sci.* 430: 48–50.

Chernyshev, A. V., and T. Y. Magarlamov. 2013. Metameric structures in the subepidermal nervous system of the nemerteans with review of the metamerism in Nemertea. *Invert. Zool.* 10: 245–254.

Chernyshev, A. V., and Y. S. Minichev. 2004. First finding of segmented nemerteans of the genus *Annulonemertis* (Nemertea, Enopla) in the Arctic. *Russ. J. Mar. Biol.* 30: 135–137.

Chipman, A. D. 2010. Parallel evolution of segmentation by co-option of ancestral gene regulatory networks. *BioEssays* 32: 60–70.

Chipman, A. D. 2019. Becoming segmented. In *Perspectives on Evolutionary and Developmental Biology*, Ed. G. Fusco, 235–244. Padova University Press.

Clark, R. B. 1963. The evolution of the celom and metameric segmentation. In *The Lower Metazoa: Comparative Biology and Phylogeny*, Eds. E. C. Dougherty, Z. N. Brown, E. D. Hanson, and W. D. Hartman, 91–107. Berkeley: University of California Press.

Coe, W. R. 1905. Nemerteans of the west and northwest coasts of America. *Bull. Mus. Comp. Zool. Harvard College* 47: 1–318.

Conklin, E. G. 1902. The embryology of a brachiopod, *Terebratulina septentrionalis* Couthouy. *P. Am. Philos. Soc.* 41: 41–76.

Conklin, E. G. 1932. The embryology of Amphioxus. *J. Morphol.* 54: 69–151.

Costa, M., B. W. Draper, and J. R. Priess. 1997. The role of actin filaments in patterning the *Caenorhabditis elegans* cuticle. *Dev. Biol.* 184: 373–384.

Costa, M., W. Raich, C. Agbunag, B. Leung, J. Hardin, and J. R. Priess. 1998. A putative Catenin–Cadherin system mediates morphogenesis of the *Caenorhabditis elegans* embryo. *J. Cell Biol.* 141: 297–308.

Couso, J. P. 2009. Segmentation, metamerism and the Cambrian explosion. *Int. J. Dev. Biol.* 53: 1305–1316.

Cox, G. N., S. Staprans, and R. S. Edgar. 1981. The cuticle of *Caenorhabditis elegans*. II. Stage-specific changes in ultrastructure and protein composition during postembryonic development. *Dev. Biol.* 86: 456–470.

Crowther, R. J., and J. R. Whittaker. 1994. Serial repetition of cilia pairs along the tail surface of an ascidian larva. *J. Exp. Zool.* 268: 9–16.

Czarkwiani, A., D. V. Dylus, and P. Oliveri. 2013. Expression of skeletogenic genes during arm regeneration in the brittle star *Amphiura filiformis*. *Gene Exp. Patterns* 13: 464–472.

Czarkwiani, A., C. Ferrario, D. V. Dylus, M. Sugni, and P. Oliveri. 2016. Skeletal regeneration in the brittle star *Amphiura filiformis*. *Front. Zool.* 13: 18.

Davis, G. K., and N. H. Patel. 1999. The origin and evolution of segmentation. *Trends Cell. Biol.* 9: M68–M72.

Decraemer, W. 1978. Morphological and taxonomic study of the genus *Tricoma* Cobb (Nematoda: Desmoscolecida), with the description of new species from the Great Barrier Reef of Australia. *Aust. J. Zool. Supps.* 26: 1–121.

Decraemer, W. I., and H. S. Rho. 2013. 7.11 Order Desmoscolecida. In *Nematoda*, Ed. A. Schmidt-Rhaesa, 351–372. Berlin, Boston: De Gruyter.

De Robertis, E. M. 1997. Evolutionary biology. The ancestry of segmentation. *Nature* 387: 25–26.

De Robertis, E. M. 2008. The molecular ancestry of segmentation mechanisms. *Proc. Natl. Acad. Sci. U S A* 105: 16411–16412.

d'Hondt, J. L. 1976. Bryozoaires cténostomes bathyaux et abyssaux de l'Atlantique nord. *Doc. Lab. Geol. Fac. Sci. Lyon* 3: 311–333.

Ding, M., A. Goncharov, Y. Jin, and A. D. Chisholm. 2003. *C. elegans* ankyrin repeat protein VAB-19 is a component of epidermal attachment structures and is essential for epidermal morphogenesis. *Development* 130: 5791–5801.

Dunn, C. W. 2015. Genomics: Acorn worms in a nutshell. *Nature* 527: 448–449.

Dunn, C. W., G. Giribet, G. D. Edgecombe, and A. Hejnol. 2014. Animal phylogeny and its evolutionary implications. *Annu. Rev. Ecol. Evol. Syst.* 45: 371–395.

Eriksson, B. J., N. N. Tait, G. E. Budd, and M. Akam. 2009. The involvement of *engrailed* and *wingless* during segmentation in the onychophoran *Euperipatoides kanangrensis* (Peripatopsidae: Onychophora) (Reid 1996). *Dev. Genes Evol.* 219: 249–264.

Ferrario, C., Y. Ben Khadra, A. Czarkwiani, A. Zakrzewski, P. Martinez, G. Colombo, …, M. Sugni. 2018. Fundamental aspects of arm repair phase in two echinoderm models. *Dev. Biol.* 433: 297–309.

Fontaneto, D., and W. H. D. Smet. 2014. 4. Rotifera. In *Gastrotricha and Gnathifera*, Ed. A. Schmidt-Rhaesa, 217–300. Berlin, München, Boston: De Gruyter.

Francis, R., and R. H. Waterston. 1991. Muscle cell attachment in *Caenorhabditis elegans*. *J. Cell Biol.* 114: 465–479.

Franke, F. A., and G. Mayer. 2014. Controversies surrounding segments and parasegments in Onychophora: Insights from the expression patterns of four "segment polarity genes" in the peripatopsid *Euperipatoides rowelli*. *PLoS One* 9: e114383.

Freeman, G. 2000. Regional specification during embryogenesis in the craniiform brachiopod *Crania anomala*. *Dev. Biol.* 227: 219–238.

Friedrich, S., A. Wanninger, M. Brückner, and G. Haszprunar. 2002. Neurogenesis in the mossy chiton, *Mopalia muscosa* (Gould) (Polyplacophora): Evidence against molluscan metamerism. *J. Morphol.* 253: 109–117.

Fritzenwanker, J., J. Gerhart, R. Freeman, and C. Lowe. 2014. The *Fox / Forkhead* transcription factor family of the hemichordate *Saccoglossus kowalevskii*. *EvoDevo* 5: 17.

Fusco, G. 2008. Morphological nomenclature, between patterns and processes: Segments and segmentation as a paradigmatic case. *Zootaxa* 102: 96–102.

Gabriel, W. N., and B. Goldstein. 2007. Segmental expression of *pax3/7* and *engrailed* homologs in tardigrade development. *Dev. Genes Evol.* 217: 421–433.

Gabriel, W. N., R. McNuff, S. K. Patel, T. R. Gregory, W. R. Jeck, C. D. Jones, …, B. Goldstein. 2007. The tardigrade *Hypsibius dujardini*, a new model for studying the evolution of development. *Dev. Biol.* 312: 545–559.

Gad, G. 2005. A parthenogenetic, simplified adult in the life cycle of *Pliciloricus pedicularis* sp. n. (Loricifera) from the deep sea of the Angola Basin (Atlantic). *Org. Divers. Evol.* 5: 77–103.

Garstang, W. 1892. IV. On the development of the stigmata in ascidians. *Proc. R. Soc. London* 51: 505–513.

Gillis, J. A., J. H. Fritzenwanker, and C. J. Lowe. 2012. A stem-deuterostome origin of the vertebrate pharyngeal transcriptional network. *Proc. R. Soc. London B Biol. Sci.* 279: 237–246.

Giribet, G. 2016. Genomics and the animal tree of life: Conflicts and future prospects. *Zool. Scr.* 45: 14–21.

Giribet, G., and G. D. Edgecombe. 2017. Current understanding of Ecdysozoa and its internal phylogenetic relationships. *Integr. Comp. Biol.* 57: 455–466.

Giribet, G., M. V. Sorensen, P. Funch, R. M. Kristensen, and W. Sterrer. 2004. Investigations into the phylogenetic position of Micrognathozoa using four molecular loci. *Cladistics* 20: 1–13.

Girstmair, J., R. Schnegg, M. J. Telford, and B. Egger. 2014. Cellular dynamics during regeneration of the flatworm *Monocelis* sp. (Proseriata, Platyhelminthes). *EvoDevo* 5: 37.

Goodrich, E. S. 1902. On the structure of the excretory organs of Amphioxus. Part I. *J. Cell Sci.* s2-45: 493–501.

Goto, T., Y. Katayama-Kumoi, M. Tohyama, and M. Yoshida. 1992. Distribution and development of the serotonin-and RFamide-like immunoreactive neurons in the arrowworm, *Paraspadella gotoi* (Chaetognatha). *Cell Tissue Res.* 267: 215–222.

Graham, A., T. Butts, A. Lumsden, and C. Kiecker. 2014. What can vertebrates tell us about segmentation? *EvoDevo* 5: 24.

Greven, H. 1984. Tardigrada. In *Biology of the Integument: Invertebrates*, Eds. J. Bereiter-Hahn, A. G. Matoltsy, and K. S. Richards, 714–727. Berlin, Heidelberg: Springer.

Gross, L. 2007. Who needs sex (or males) anyway? *PLoS Biol.* 5: e99.

Gross, V., and G. Mayer. 2015. Neural development in the tardigrade *Hypsibius dujardini* based on anti-acetylated α-tubulin immunolabeling. *EvoDevo* 6: 12.

Gross, V., S. Treffkorn, and G. Mayer. 2015. Tardigrada. In *Evolutionary Developmental Biology of Invertebrates 3*, Ed. A. Wanninger, 35–52. Vienna: Springer.

Gschwentner, R., J. Mueller, P. Ladurner, R. Rieger, and S. Tyler. 2003. Unique patterns of longitudinal body-wall musculature in the Acoela (Plathelminthes): The ventral musculature of *Convolutriloba longifissura. Zoomorphology* 122: 87–94.

Haig, J. A., and G. W. Rouse. 2008. Larval development of the featherstar *Aporometra wilsoni* (Echinodermata: Crinoidea). *Invertebr. Biol.* 127: 460–469.

Halberg, K. A., D. Persson, N. Møbjerg, A. Wanninger, and R. M. Kristensen. 2009. Myoanatomy of the marine tardigrade *Halobiotus crispae* (Eutardigrada: Hypsibiidae). *J. Morphol.* 270: 996–1013.

Halton, D. W., and M. K. S. Gustafsson. 1996. Functional morphology of the platyhelminth nervous system. *Parasitology* 113: S47–S72.

Hammond, R. A. 1970. The surface of *Priapulus caudatus* (Lamarck, 1816). *Z. Morphol. Tiere* 68: 255–268.

Hannibal, R. L., and N. H. Patel. 2013. What is a segment? *EvoDevo* 4: 35.

Hardin, J., and C. Lockwood. 2004. Skin tight: Cell adhesion in the epidermis of *Caenorhabditis elegans. Curr. Opin. Cell Biol.* 16: 486–492.

Hartenstein, V., and A. Stollewerk. 2015. The evolution of early neurogenesis. *Dev. Cell* 32: 390–407.

Harzsch, S., and C. H. G. Müller. 2007. A new look at the ventral nerve centre of *Sagitta*: Implications for the phylogenetic position of Chaetognatha (arrow worms) and the evolution of the bilaterian nervous system. *Front. Zool.* 4: 14.

Harzsch, S., C. H. G. Müller, V. Rieger, Y. Perez, S. Sintoni, C. Sardet, and B. Hansson. 2009. Fine structure of the ventral nerve centre and interspecific identification of individual neurons in the enigmatic Chaetognatha. *Zoomorphology* 128: 53–73.

Hashimoto, N., Y. Kurita, and H. Wada. 2012. Developmental role of *dpp* in the gastropod shell plate and co-option of the *dpp* signaling pathway in the evolution of the operculum. *Dev. Biol.* 366: 367–373.

Haszprunar, G., and A. Wanninger. 2000. Molluscan muscle systems in development and evolution. *J. Zool. Syst. Evol. Res.* 38: 157–163.

Hay-Schmidt, A. 1990. Catecholamine-containing, serotonin-like and neuropeptide FMRFamide-like immunoreactive cells and processes in the nervous system of the pilidium larva (Nemertini). *Zoomorphology* 109: 231–244.

Heath, H. 1899. The development of *Ischnochiton*. *Zool. Jahrb. Abt. Anat. Ontog. Tiere* 12: 567–656.

Heiner, I., and R. M. Kristensen. 2009. *Urnaloricus gadi* nov. gen. et nov. sp. (Loricifera, Urnaloricidae nov. fam.), an aberrant Loricifera with a viviparous pedogenetic life cycle. *J. Morphol.* 270: 129–153.

Hejnol, A. 2015a. Cycloneuralia. In *Evolutionary Developmental Biology of Invertebrates 3*, Ed. A. Wanninger, 1–13. Vienna: Springer.

Hejnol, A. 2015b. Gastrotricha. In *Evolutionary Developmental Biology of Invertebrates 2*, Ed. A. Wanninger, 13–19. Vienna: Springer.

Hejnol, A. 2015c. Gnathifera. In *Evolutionary Developmental Biology of Invertebrates 2*, Ed. A. Wanninger, 1–12. Vienna: Springer.

Hejnol, A., and M. Q. Martindale. 2008. Acoel development supports a simple planula-like urbilaterian. *Philos. Trans. R. Soc. London B Biol. Sci.* 363: 1493–1501.

Hejnol, A., M. Obst, A. Stamatakis, M. Ott, G. W. Rouse, G. D. Edgecombe, …, C. W. Dunn. 2009. Assessing the root of bilaterian animals with scalable phylogenomic methods. *Proc. R. Soc. London B Biol. Sci.* 276: 4261–4270.

Hejnol, A., and K. Pang. 2016. Xenacoelomorpha's significance for understanding bilaterian evolution. *Curr. Opin. Genet. Dev.* 39: 48–54.

Hejnol, A., and R. Schnabel. 2005. The eutardigrade *Thulinia stephaniae* has an indeterminate development and the potential to regulate early blastomere ablations. *Development* 132: 1349–1361.

Henry, J. Q., A. Okusu, and M. Q. Martindale. 2004. The cell lineage of the polyplacophoran, *Chaetopleura apiculata*: Variation in the spiralian program and implications for molluscan evolution. *Dev. Biol.* 272: 145–160.

Herranz, M., M. J. Boyle, F. Pardos, and R. C. Neves. 2014. Comparative myoanatomy of *Echinoderes* (Kinorhyncha): A comprehensive investigation by CLSM and 3D reconstruction. *Front. Zool.* 11: 31.

Herranz, M., B. S. Leander, F. Pardos, and M. J. Boyle. 2019. Neuroanatomy of mud dragons: A comprehensive view of the nervous system in *Echinoderes* (Kinorhyncha) by confocal laser scanning microscopy. *BMC Evol. Biol.* 19: 86.

Herranz, M., F. Pardos, and M. J. Boyle. 2013. Comparative morphology of serotonergic-like immunoreactive elements in the central nervous system of kinorhynchs (Kinorhyncha, Cyclorhagida). *J. Morphol.* 274: 258–274.

Higgins, R. P., and R. M. Kristensen. 1986. New Loricifera from southeastern United States coastal waters. *Smithson Contrib. Zool.* 1–70.

Higgins, R. P., V. Storch, and T. C. Shirley. 1993. Scanning and transmission electron microscopical observations on the larvae of *Priapulus caudatus* (Priapulida). *Acta Zool.* 74: 301–319.

Hindinger, S., T. Schwaha, and A. Wanninger. 2013. Immunocytochemical studies reveal novel neural structures in nemertean pilidium larvae and provide evidence for incorporation of larval components into the juvenile nervous system. *Front. Zool.* 10: 31.

Hochberg, R., and G. Lilley. 2010. Neuromuscular organization of the freshwater colonial rotifer, *Sinantherina socialis*, and its implications for understanding the evolution of coloniality in Rotifera. *Zoomorphology* 129: 153–162.

Hochberg, R., and M. K. Litvaitis. 2000. Functional morphology of the muscles in *Philodina* sp. (Rotifera: Bdelloidea). *Hydrobiologia* 432: 57–64.

Holland, L. Z. 2015. Cephalochordata. In *Evolutionary Developmental Biology of Invertebrates 6*, Ed. A. Wanninger, 91–133. Vienna: Springer.

Holland, N. D. 2017. The long and winding path to understanding kidney structure in amphioxus - a review. *Int. J. Dev. Biol.* 61: 683–688.

Holy, J. M., J. A. Oaks, M. Mika-Grieve, and R. Grieve. 1991. Development and dynamics of regional specialization within the syncytial epidermis of the rat tapeworm, *Hymenolepis diminuta. Parasitol. Res.* 77: 161–172.

Hooge, M. D. 2001. Evolution of body-wall musculature in the Platyhelminthes (Acoelomorpha, Catenulida, Rhabditophora). *J. Morphol.* 249: 171–194.

Hooge, M. D., and S. Tyler. 1999. Body-wall musculature of *Praeconvoluta tornuva* n. sp. (Acoela, Platyhelminthes) and the use of muscle patterns in taxonomy. *Invertebr. Biol.* 118: 8–17.

Horst, C. J. van der 1930. Metamerism in Enteropneusta. *J. Cell Sci.* s2-73: 393–402.

Hoyle, G., and M. Williams. 1980. The musculature of *Peripatus* and its innervation. *Philos. Trans. R. Soc. London B Biol. Sci.* 288: 481–510.

Hubrecht, A. S. W. 1887. Report on the Nemertea collected by H. M. S. Challenger during the years 1873–76. Edinburgh: Neil and Company.

Hunter, W. R., and S. C. Brown. 1965. Ctenidial number in relation to size in certain chitons, with a discussion of its phyletic significance. *Biol. Bull.* 128: 508–521.

Hyman, L. H. 1951a. Introduction to the bilateria. In *The Invertebrates, Vol. II: Platyhelminthes and Rhynchocoela*, Ed. E. J. Boell, 1–51. New York: McGraw-Hill Book Company, Inc.

Hyman, L. H. 1951b. The acoelomate bilateria—phylum Platyhelminthes. In *The Invertebrates, Vol. II: Platyhelminthes and Rhynchocoela*, Ed. E. J. Boell, 52–458. New York: McGraw-Hill Book Company, Inc.

Hyman, L. H. 1951c. The acoelomate bilateria—phylum Rhynchocoela. In *The Invertebrates, Vol. II: Platyhelminthes and Rhynchocoela*, Ed. E. J. Boell, 459–531. New York: McGraw-Hill Book Company, Inc.

Hyman, L. H. 1951d. The pseudocoelomate bilateria—phylum Acanthocephala. In *The Invertebrates, Vol. III: Acanthocephala, Aschelminthes, and Entoprocta*, Ed. E. J. Boell, 1–52. New York: McGraw-Hill Book Company, Inc.

Hyman, L. H. 1951e. The pseudocoelomate bilateria—phylum Aschelminthes. In *The Invertebrates, Vol. III: Acanthocephala, Aschelminthes, and Entoprocta*, Ed. E. J. Boell, 53–520. New York: McGraw-Hill Book Company, Inc.

Hyman, L. H. 1955. *The Invertebrates, Vol. IV: Echinodermata*. New York: McGraw-Hill Book Company, Inc.

Hyman, L. H. 1959a. The enterocoelous coelomates—phylum Chaetognatha. In *The Invertebrates, Vol. V: Smaller Coelomate Groups*, Ed. E. J. Boell, 1–71. New York: McGraw-Hill Book Company, Inc.

Hyman, L. H. 1959b. The enterocoelous coelomates—phylum Hemichordata. In *The Invertebrates, Vol. V: Smaller Coelomate Groups*, Ed. E. J. Boell, 72–207. New York: McGraw-Hill Book Company, Inc.

Hyman, L. H. 1959c. The lophophorate coelomates—phylum Brachiopoda. In *The Invertebrates, Vol. V: Smaller Coelomate Groups*, Ed. E. J. Boell, 516–609. New York: McGraw-Hill Book Company, Inc.

Hyman, L. H. 1959d. The lophophorate coelomates—phylum Ectoprocta. In *The Invertebrates, Vol. V: Smaller Coelomate Groups*, Ed. E. J. Boell, 275–501. New York: McGraw-Hill Book Company, Inc.

Iijima, M., T. Takeuchi, I. Sarashina, and K. Endo. 2008. Expression patterns of *engrailed* and *dpp* in the gastropod *Lymnaea stagnalis. Dev. Genes. Evol.* 218: 237–251.

Imai, J. H., and I. A. Meinertzhagen. 2007a. Neurons of the ascidian larval nervous system in *Ciona intestinalis*: I. Central nervous system. *J. Comp. Neurol.* 501: 316–334.

Imai, J. H., and I. A. Meinertzhagen. 2007b. Neurons of the ascidian larval nervous system in *Ciona intestinalis*: II. Peripheral nervous system. *J. Comp. Neurol.* 501: 335–352.

Ingham, P. W. 1991. Segment polarity genes and cell patterning within the *Drosophila* body segment. *Curr. Opin. Genet. Dev.* 1: 261–267.

Jacobs, D. K., C. G. Wray, C. J. Wedeen, R. Kostriken, R. DeSalle, J. L. Staton, …, D. R. Lindberg. 2000. Molluscan *engrailed* expression, serial organization, and shell evolution. *Evol. Dev.* 2: 340–347.

Janssen, R. 2017. A molecular view of onychophoran segmentation. *Arthropod Struct. Dev.* 46: 341–353.

Janssen, R., and G. E. Budd. 2013. Deciphering the onychophoran 'segmentation gene cascade': Gene expression reveals limited involvement of pair rule gene orthologs in segmentation, but a highly conserved segment polarity gene network. *Dev. Biol.* 382: 224–234.

Jeffery, W. R., and B. J. Swalla. 1997. Tunicates. In *Embryology: Constructing the Organism*, Ed. S. F. Gilbert, and A. M. Raunio, 331–364. Sinauer Associates, Inc.

Jiang, D., E. M. Munro, and W. C. Smith. 2005. Ascidian *prickle* regulates both mediolateral and anterior-posterior cell polarity of notochord cells. *Curr. Biol.* 15: 79–85.

Johnson, C. D., and A. O. Stretton. 1987. GABA-immunoreactivity in inhibitory motor neurons of the nematode *Ascaris*. *J. Neurosci.* 7: 223–235.

Jondelius, U., I. Ruiz-Trillo, J. Baguñà, and M. Riutort. 2002. The Nemertodermatida are basal bilaterians and not members of the Platyhelminthes. *Zool. Scr.* 31: 201–215.

Kaul-Strehlow, S., and E. Röttinger. 2015. Hemichordata. In *Evolutionary Developmental Biology of Invertebrates 6*, Ed. A. Wanninger, 59–89. Vienna: Springer.

Kaul-Strehlow, S., M. Urata, T. Minokawa, T. Stach, and A. Wanninger. 2015. Neurogenesis in directly and indirectly developing enteropneusts: Of nets and cords. *Org. Divers. Evol.* 15: 405–422.

Kieneke, A., W. H. Ahlrichs, P. M. Arbizu, and T. Bartolomaeus. 2008a. Ultrastructure of protonephridia in *Xenotrichula carolinensis syltensis* and *Chaetonotus maximus* (Gastrotricha: Chaetonotida): Comparative evaluation of the gastrotrich excretory organs. *Zoomorphology* 127: 1–20.

Kieneke, A., P. M. Arbizu, and W. H. Ahlrichs. 2007. Ultrastructure of the protonephridial system in *Neodasys chaetonotoideus* (Gastrotricha: Chaetonotida) and in the ground pattern of Gastrotricha. *J. Morphol.* 268: 602–613.

Kieneke, A., P. M. Arbizu, and O. Riemann. 2008b. Body musculature of *Stylochaeta scirtetica* Brunson, 1950 and *Dasydytes* (*Setodytes*) *tongiorgii* (Balsamo, 1982) (Gastrotricha: Dasydytidae): A functional approach. *Zool. Anz.* 247: 147–158.

Kieneke, A., O. Riemann, and W. H. Ahlrichs. 2008c. Novel implications for the basal internal relationships of Gastrotricha revealed by an analysis of morphological characters. *Zool. Scr.* 37: 429–460.

Kimmel, C. B. 1996. Was *Urbilateria* segmented? *Trends Genet.* 12: 329–331.

Kin, K., S. Kakoi, and H. Wada. 2009. A novel role for *dpp* in the shaping of bivalve shells revealed in a conserved molluscan developmental program. *Dev. Biol.* 329: 152–166.

Kirkhart, J. 2008. Lined chiton (*Tonicella lineata*). https://www.flickr.com/photos/jkirkhart35/2388553245 (accessed August 31, 2019).

Kniprath, E. 1980. Ontogenetic plate and plate field development in two chitons, *Middendorffia* and *Ischnochiton*. *Roux Arch Dev. Biol.* 189: 97–106.

Kocot, K. M., T. H. Struck, J. Merkel, D. S. Waits, C. Todt, P. M. Brannock, …, K. M. Halanych. 2017. Phylogenomics of Lophotrochozoa with consideration of systematic error. *Syst. Biol.* 66: 256–282.

Kowalevsky, M. A. 1883a. Embryogénie du Chiton Polii (Philippi) avec quelques remarques sur le développement des autres chitons. *Ann. Mus. Hist. Nat. Marseille* 5: 1–46.

Kowalevsky, M. A. 1883b. Le Développement des Brachiopodes. *Arch Zool. Exp. Gen.* ser.2 T.1: 57–76.

Koziol, U. 2017. Evolutionary developmental biology (evo-devo) of cestodes. *Exp. Parasitol.* 180: 84–100.

Koziol, U., M. F. Domínguez, M. Marín, A. Kun, and E. Castillo. 2010. Stem cell proliferation during in vitro development of the model cestode *Mesocestoides corti* from larva to adult worm. *Front. Zool.* 7: 22.

Koziol, U., F. Jarero, P. D. Olson, and K. Brehm. 2016. Comparative analysis of Wnt expression identifies a highly conserved developmental transition in flatworms. *BMC Biol.* 14: 10.

Kozloff, E. N. 1972. Some aspects of development in *Echinoderes* (Kinorhyncha). *Trans. Am. Microsc. Soc.* 91: 119–130.

Kozloff, E. N. 2007. Stages of development, from first cleavage to hatching, of an *Echinoderes* (Phylum Kinorhyncha: Class Cyclorhagida). *Cah. Biol. Mar.* 48: 199–206.

Kristensen, R. M. 1983. Loricifera, a new phylum with Aschelminthes characters from the meiobenthos. *J. Zool. Syst. Evol. Res.* 21: 163–180.

Kristensen, R. M., and P. Funch. 2000. Micrognathozoa: A new class with complicated jaws like those of Rotifera and Gnathostomulida. *J. Morphol.* 246: 1–49.

Kristensen, R. M., R. C. Neves, and G. Gad. 2013. First report of Loricifera from the Indian Ocean: A new *Rugiloricus*-species represented by a hermaphrodite. *Cah. Biol. Mar.* 54: 161–171.

Krupenko, D. Y., and A. A. Dobrovolskij. 2015. Somatic musculature in trematode hermaphroditic generation. *BMC Evol. Biol.* 15: 189.

Lacaze-Duthiers, H. 1861. Histoire de la Thécidie. *Ann. Sci. Nat. Zool.* 4: 262–330.

Ladurner, P., and R. Rieger. 2000. Embryonic muscle development of *Convoluta pulchra* (Turbellaria–Acoelomorpha, Platyhelminthes). *Dev. Biol.* 222: 359–375.

Lahaye, M.-C., and M. Jangoux. 1987. The skeleton of the stalked stages of the comatulid crinoid *Antedon bifida* (Echinodermata). *Zoomorphology* 107: 58–65.

Lang, A. 1881. Der Bau von *Gunda segmentata* und die Verwandtschaft der Plathelminthen mit Coelenteraten und Hirudineen. *Mitt. Zool. Stn. Neapel* 3: 187–251.

Laumer, C. E., R. Fernández, S. Lemer, D. Combosch, K. M. Kocot, A. Riesgo, …, G. Giribet. 2019. Revisiting metazoan phylogeny with genomic sampling of all phyla. *Proc. R. Soc. London B Biol. Sci.* 286: 20190831.

Leasi, F., and C. Ricci. 2010. Musculature of two bdelloid rotifers, *Adineta ricciae* and *Macrotrachela quadricornifera*: Organization in a functional and evolutionary perspective. *J. Zool. Syst. Evol. Res.* 48: 33–39.

Lee, D. L. 1967. The structure and composition of the helminth cuticle. In *Advances in Parasitology, Vol. 4*, Ed. B. Dawes, 187–254. Academic Press.

Leise, E. M. 1984. Chiton integument: Metamorphic changes in *Mopalia muscosa* (Mollusca, Polyplacophora). *Zoomorphology* 104: 337–343.

Lemburg, C. 2002. A new kinorhynch *Pycnophyes australensis* sp. n. (Kinorhyncha: Homalorhagida: Pycnophyidae) from Magnetic Island, Australia. *Zool. Anz.* 241: 173–189.

Lemche, H., and K. G. Wingstrand. 1959. The anatomy of *Neopilina galatheae* Lemche, 1957. *Galathea. Rep.* 3: 9–71.

Lim, H. W., and C. Y. Chang. 2006. First record of *Desmoscolex* Nematoda (Desmoscolecida: Desmoscolecidae) from Korea. *Integr. Biosci.* 10: 219–225.

Lorenzen, S. 1971. Jugendstadien von *Desmoscolex*-Arten (Nematoda, Desmoscolecidae) und deren Bedeutung für die Taxonomie. *Mar. Biol.* 10: 343–345.

Lowe, C. J., D. N. Clarke, D. M. Medeiros, D. S. Rokhsar, and J. Gerhart. 2015. The deuterostome context of chordate origins. *Nature* 520: 456–465.

Lowe, C. J., and G. A. Wray. 1997. Radical alterations in the roles of homeobox genes during echinoderm evolution. *Nature* 389: 718–721.

Malakhov, V. V., and T. V. Kuzmina. 2006. Metameric origin of lateral mesenteries in Brachiopoda. *Dokl. Biol. Sci.* 409: 340–342.

Manni, L., N. J. Lane, G. Zaniolo, and P. Burighel. 2002. Cell reorganisation during epithelial fusion and perforation: The case of ascidian branchial fissures. *Dev. Dyn.* 224: 303–313.

Marchiori, N. C., J. Pereira Jr, and L. A. S. Castro. 2009. Morphology of larval *Gordius dimorphus* (Nematomorpha: Gordiida). *J. Parasitol.* 95: 1218–1220.

Marchioro, T., L. Rebecchi, M. Cesari, J. G. Hansen, G. Viotti, and R. Guidetti. 2013. Somatic musculature of Tardigrada: Phylogenetic signal and metameric patterns. *Zool. J. Linn. Soc.* 169: 580–603.

Marlétaz, F., K. T. C. A. Peijnenburg, T. Goto, N. Satoh, and D. S. Rokhsar. 2019. A new spiralian phylogeny places the enigmatic arrow worms among gnathiferans. *Curr. Biol.*

Martin, C., V. Gross, L. Hering, B. Tepper, H. Jahn, I. de Sena Oliveira, …, G. Mayer. 2017a. The nervous and visual systems of onychophorans and tardigrades: Learning about arthropod evolution from their closest relatives. *J. Comp. Physiol. A Neuroethol. Sens. Neural Behav. Physiol.* 203: 565–590.

Martin, C., V. Gross, H.-J. Pflüger, P. A. Stevenson, and G. Mayer. 2017b. Assessing segmental versus non-segmental features in the ventral nervous system of onychophorans (velvet worms). *BMC Evol. Biol.* 17: 3.

Martín-Durán, J. M., and A. Hejnol. 2015. The study of *Priapulus caudatus* reveals conserved molecular patterning underlying different gut morphogenesis in the Ecdysozoa. *BMC Biol.* 13: 29.

Martín-Durán, J. M., K. Pang, A. Børve, H. S. Lê, A. Furu, J. T. Cannon, …, A. Hejnol. 2018. Convergent evolution of bilaterian nerve cords. *Nature* 553: 45–50.

Martín-Durán, J. M., G. H. Wolff, N. J. Strausfeld, and A. Hejnol. 2016. The larval nervous system of the penis worm *Priapulus caudatus* (Ecdysozoa). *Philos. Trans. R. Soc. London B Biol. Sci.* 371: 20150050.

Martínez, P., V. Hartenstein, and S. G. Sprecher. 2017. Xenacoelomorpha nervous systems. In *Oxford Research Encyclopedia of Neuroscience*. Oxford University Press. Accessed 18 Feb. 2020. https://oxfordre.com/neuroscience/view/10.1093/acrefore/9780190264086.001.0001/acrefore-9780190264086-e-203.

Maslakova, S. A. 2010. Development to metamorphosis of the nemertean pilidium larva. *Front. Zool.* 7: 30.

Masterman, A. T. 1899. On the theory of archimeric segmentation and its bearing upon the phyletic classification of the *Cœlomata. Proc. R. Soc. Edinb.* 22: 270–310.

May, H. G. 1919. *Contributions to the Life Histories of* Gordius Robustus *(Leidy) and* Paragordius Varius *(Leidy).* University of Illinois.

Mayer, G. 2006. Origin and differentiation of nephridia in the Onychophora provide no support for the Articulata. *Zoomorphology* 125: 1–12.

Mayer, G. 2015. Onychophora. In *Structure and Evolution of Invertebrate Nervous Systems*, Ed. A. Schmidt-Rhaesa, S. Harzsch, and G. Purschke, 390–401. Oxford University Press.

Mayer, G., F. A. Franke, S. Treffkorn, V. Gross, and Sena Oliveira, I. de. 2015. Onychophora. In *Evolutionary Developmental Biology of Invertebrates 3*, Ed. A. Wanninger, 53–98. Vienna: Springer.

Mayer, G., and S. Harzsch. 2007. Immunolocalization of serotonin in Onychophora argues against segmental ganglia being an ancestral feature of arthropods. *BMC Evol. Biol.* 7: 118.

Mayer, G., and S. Harzsch. 2008. Distribution of serotonin in the trunk of *Metaperipatus blainvillei* (Onychophora, Peripatopsidae): Implications for the evolution of the nervous system in Arthropoda. *J. Comp. Neurol.* 507: 1196–1208.

Mayer, G., C. Kato, B. Quast, R. H. Chisholm, K. A. Landman, and L. M. Quinn. 2010. Growth patterns in Onychophora (velvet worms): Lack of a localised posterior proliferation zone. *BMC Evol. Biol.* 10: 339–339.

Mayer, G., S. Kauschke, J. Rüdiger, and P. A. Stevenson. 2013a. Neural markers reveal a one-segmented head in tardigrades (water bears). *PloS One* 8: e59090.

Mayer, G., C. Martin, J. Rüdiger, S. Kauschke, P. A. Stevenson, I. Poprawa, …, M. Schlegel. 2013b. Selective neuronal staining in tardigrades and onychophorans provides insights into the evolution of segmental ganglia in panarthropods. *BMC Evol. Biol.* 13: 230.

Mayer, G., and P. M. Whitington. 2009. Neural development in Onychophora (velvet worms) suggests a step-wise evolution of segmentation in the nervous system of Panarthropoda. *Dev. Biol.* 335: 263–275.

McCoy, F. 1845. XL.—Contributions to the fauna of Ireland. *Ann. Mag. Nat. Hist.* 15: 270–274.

Mehlhorn, H., B. Becker, P. Andrews, and H. Thomas. 1981. On the nature of the proglottids of cestodes: A light and electron microscopic study on *Taenia*, *Hymenolepis*, and *Echinococcus*. *Z. Parasitenkd.* 65: 243–259.

Meyer-Wachsmuth, I., O. I. Raikova, and U. Jondelius. 2013. The muscular system of *Nemertoderma westbladi* and *Meara stichopi* (Nemertodermatida, Acoelomorpha). *Zoomorphology* 132: 239–252.

Minelli, A., and G. Fusco. 2004. Evo-devo perspectives on segmentation: Model organisms, and beyond. *Trends Ecol. Evol.* 19: 423–429.

Miyamoto, D. M., and R. J. Crowther. 1985. Formation of the notochord in living ascidian embryos. *J. Embryol. Exp. Morphol.* 86: 1–17.

Miyamoto, N., and Y. Saito. 2007. Morphology and development of a new species of *Balanoglossus* (Hemichordata: Enteropneusta: Ptychoderidae) from Shimoda, Japan. *Zool. Sci.* 24: 1278–1285.

Møbjerg, N., K. A. Halberg, A. Jørgensen, D. Persson, M. Bjørn, H. Ramløv, …, R. M. Kristensen. 2011. Survival in extreme environments - on the current knowledge of adaptations in tardigrades. *Acta physiol.* 202: 409–420.

Monjo, F., and R. Romero. 2015. Embryonic development of the nervous system in the planarian *Schmidtea polychroa*. *Dev. Biol.* 397: 305–319.

Montgomery, T. H. 1904. The development and structure of the larva of *Paragordius*. *Proc. Acad. Nat. Sci. Phila.* 738–755.

Mooi, R., and B. David. 1998. Evolution within a bizarre phylum: Homologies of the first echinoderms. *Integr. Comp. Biol.* 38: 965–974.

Mooi, R., and B. David. 2000. What a new model of skeletal homologies tells us about asteroid evolution. *Integr. Comp. Biol.* 40: 326–339.

Mooi, R., and B. David. 2008. Radial symmetry, the anterior/posterior axis, and echinoderm Hox genes. *Annu. Rev. Ecol. Evol. Syst.* 39: 43–62.

Mooi, R., B. David, and D. Marchand. 1994. Echinoderm skeletal homologies: Classical morphology meets modem phylogenetics. In *Echinoderms through Time*, Ed. B. David, A. Guille, and J.-P. Feral, 87–95. A. A. Balkema.

Mooi, R., B. David, and G. A. Wray. 2005. Arrays in rays: Terminal addition in echinoderms and its correlation with gene expression. *Evol. Dev.* 7: 542–555.

Morgan, T. H. 1894. The development of *Balanoglossus*. *J. Morphol.* 9: 1–86.

Morse, E. S. 1873. Embryology of *Terebratulina*. *Mem. Read Boston Soc. Nat. Hist.* 2: 249–264.

Morse, M. P. 1981. *Meiopriapulus fijiensis* n. gen., n. sp.: An interstitial priapulid from coarse sand in Fiji. *Trans. Am. Microsc. Soc.* 100: 239–252.

Moshel, S. M., M. Levine, and J. R. Collier. 1998. Shell differentiation and *engrailed* expression in the *Ilyanassa* embryo. *Dev. Genes. Evol.* 208: 135–141.

Müller, M. C. M., R. Jochmann, and A. Schmidt-Rhaesa. 2004. The musculature of horsehair worm larvae (*Gordius aquaticus, Paragordius varius,* Nematomorpha): F-actin staining and reconstruction by cLSM and TEM. *Zoomorphology* 123: 45–54.

Müller, M. C. M., and A. Schmidt-Rhaesa. 2003. Reconstruction of the muscle system in *Antygomonas* sp. (Kinorhyncha, Cyclorhagida) by means of phalloidin labeling and cLSM. *J. Morphol.* 256: 103–110.

Nebelsick, M. 1993. Introvert, mouth cone, and nervous system of *Echinoderes capitatus* (Kinorhyncha, Cyclorhagida) and implications for the phylogenetic relationships of Kinorhyncha. *Zoomorphology* 113: 211–232.

Nederbragt, A. J., A. E. Loon van, and W. J. A. G. Dictus, 2002. Expression of *patella vulgata* orthologs of *engrailed* and *dpp-BMP2/4* in adjacent domains during molluscan shell development suggests a conserved compartment boundary mechanism. *Dev. Biol.* 246: 341–355.

Neuhaus, B. 1995. Postembryonic development of *Paracentrophyes praedictus* (Homalorhagida): Neoteny questionable among the Kinorhyncha. *Zool. Scr.* 24: 179–192.

Neuhaus, B., and R. P. Higgins. 2002. Ultrastructure, biology, and phylogenetic relationships of Kinorhyncha. *Integr. Comp. Biol.* 42: 619–632.

Neves, R. C., X. Bailly, F. Leasi, H. Reichert, M. V. Sørensen, and R. M. Kristensen. 2013. A complete three-dimensional reconstruction of the myoanatomy of Loricifera: Comparative morphology of an adult and a Higgins larva stage. *Front. Zool.* 10: 19.

Newman, S. A. 1993. Is segmentation generic? *BioEssays* 15: 277–283.

Nicholas, W. L. 1967. The biology of the Acanthocephala. In *Advances in Parasitology,* Ed. B. Dawes, 205–246. Academic Press.

Nielsen, C. 1991. The development of the brachiopod *Crania* (*Neocrania*) *anomala* (O. F. Müller) and its phylogenetic significance. *Acta Zool.* 72: 7–28.

Nishimura, K., T. Inoue, K. Yoshimoto, Y. Taniguchi, Y. Kitamura, and K. Agata. 2011. Regeneration of dopaminergic neurons after 6-hydroxydopamine-induced lesion in planarian brain. *J. Neurochem.* 119: 1217–1231.

Norenburg, J. L. 1988. Nemertina. In *Introduction to the Study of Meiofauna,* Eds. R. P. Higgins, and H. Thiel, 287–292. Washington: Smithsonian Institution Press.

O'Brien, E. K., and B. M. Degnan. 2003. Expression of *Pax258* in the gastropod statocyst: Insights into the antiquity of metazoan geosensory organs. *Evol. Dev.* 5: 572–578.

Oeschger, R., and H. H. Janssen. 1991. Histological studies on *Halicryptus spinulosus* (Priapulida) with regard to environmental hydrogen sulfide resistance. *Hydrobiologia* 222: 1–12.

Ogasawara, M., H. Wada, H. Peters, and N. Satoh. 1999. Developmental expression of *Pax1/9* genes in urochordate and hemichordate gills: Insight into function and evolution of the pharyngeal epithelium. *Development* 126: 2539–2550.

Oliveira, I. de S., N. N. Tait, I. Strübing, and G. Mayer. 2013. The role of ventral and preventral organs as attachment sites for segmental limb muscles in Onychophora. *Front. Zool.* 10: 73.

Olson, P. D., D. T. Littlewood, R. A. Bray, and J. Mariaux. 2001. Interrelationships and evolution of the tapeworms (Platyhelminthes: Cestoda). *Mol. Phylogenet. Evol.* 19: 443–467.

Olson, P. D., M. Zarowiecki, K. James, A. Baillie, G. Bartl, P. Burchell, ..., M. Berriman. 2018. Genome-wide transcriptome profiling and spatial expression analyses identify signals and switches of development in tapeworms. *EvoDevo* 9: 21.

Pardos, F., and J. Benito. 1988. Blood vessels and related structures in the gill bars of *Glossobalanus minutus* (Enteropneusta). *Acta Zool.* 69: 87–94.

Pasini, A., A. Amiel, U. Rothbächer, A. Roure, P. Lemaire, and S. Darras. 2006. Formation of the ascidian epidermal sensory neurons: Insights into the origin of the chordate peripheral nervous system. *PLoS Biol.* 4: e225.

Perez, Y., V. Rieger, E. Martin, C. H. G. Müller, and S. Harzsch. 2013. Neurogenesis in an early protostome relative: Progenitor cells in the ventral nerve center of chaetognath hatchlings are arranged in a highly organized geometrical pattern. *J. Exp. Zool. B Mol. Dev. Evol.* 320: 179–193.

Persson, D. K., K. A. Halberg, A. Jørgensen, N. Møbjerg, and R. M. Kristensen. 2012. Neuroanatomy of *Halobiotus crispae* (Eutardigrada: Hypsibiidae): Tardigrade brain structure supports the clade Panarthropoda. *J. Morphol.* 273: 1227–1245.

Philippe, H., A. J. Poustka, M. Chiodin, K. J. Hoff, C. Dessimoz, B. Tomiczek, …, M. J. Telford. 2019. Mitigating anticipated effects of systematic errors supports sister-group relationship between xenacoelomorpha and ambulacraria. *Curr. Biol.* 29: 1818–1826.e6.

Poinar, G. O. 2010. Nematoda and nematomorpha. In *Ecology and Classification of North American Freshwater Invertebrates (Third Edition)*, Ed. J. H. Thorp, and A. P. Covich, 237–276. San Diego: Academic Press.

Powers, T. O. 2015. Criconematid project. https://nematode.unl.edu/CriconematidProject.htm (accessed September 4, 2019).

Powers, T. O., P. Mullin, R. Higgins, T. Harris, and K. S. Powers. 2016. Description of *Mesocriconema ericaceum* n. sp. (Nematoda: Criconematidae) and notes on other nematode species discovered in an ericaceous heath bald community in Great Smoky Mountains National Park, USA. *Nematology* 18: 879–903.

Priess, J. R., and D. I. Hirsh. 1986. *Caenorhabditis elegans* morphogenesis: The role of the cytoskeleton in elongation of the embryo. *Dev. Biol.* 117: 156–173.

Raible, F., and M. Brand. 2004. *Divide et Impera*—the midbrain-hindbrain boundary and its organizer. *Trends Neurosci.* 27: 727–734.

Rawlinson, K. A. 2010. Embryonic and post-embryonic development of the polyclad flatworm *Maritigrella crozieri*; implications for the evolution of spiralian life history traits. *Front. Zool.* 7: 12.

Reisinger, E. 1925. Untersuchungen am Nervensystem der *Bothrioplana semperi* Braun. (Zugleich ein Beitrag zur Technik der vitalen Nervenfärbung und zur vergleichenden Anatomie des Plathelminthennervensystems.). *Z. Morph. u. Okol. Tiere* 5: 119–149.

Reiter, D., P. Ladurner, G. Mair, W. Salvenmoser, R. Rieger, and B. Boyer. 1996. Differentiation of the body wall musculature in *Macrostomum hystricinum marinum* and *Hoploplana inquilina* (Plathelminthes), as models for muscle development in lower Spiralia. *Rouxs Arch Dev. Biol.* 205: 410–423.

Reuter, M., and M. K. S. Gustafsson. 1995. The flatworm nervous system: Pattern and phylogeny. In *The Nervous Systems of Invertebrates: An Evolutionary and Comparative Approach*, Eds. O. Breidbach, and W. Kutsch, 25–59. Basel: Birkhäuser Basel.

Reuter, M., K. Mäntylä, and M. K. S. Gustafsson. 1998. Organization of the orthogon—main and minor nerve cords. *Hydrobiologia* 383: 175–182.

Reuter, M., A. G. Maule, D. W. Halton, M. K. S. Gustafsson, and C. Shaw. 1995. The organization of the nervous system in Plathelminthes. The neuropeptide F-immunoreactive pattern in Catenulida, Macrostomida, Proseriata. *Zoomorphology* 115: 83–97.

Rieger, R. M., W. Salvenmoser, A. Legniti, and S. Tyler. 1994. Phalloidin-rhodamine preparations of *Macrostomum hystricinum marinum* (Plathelminthes): Morphology and post-embryonic development of the musculature. *Zoomorphology* 114: 133–147.

Rieger, R., W. Salvenmoser, A. Legniti, S. Reindl, H. Adam, P. Simonsberger, and S. Tyler. 1991. Organization and differentiation of the body-wall musculature in *Macrostomum* (Turbellaria, Macrostomidae). *Hydrobiologia* 227: 119–129.

Rieger, V., Y. Perez, C. H. G. Müller, T. Lacalli, B. S. Hansson, and S. Harzsch. 2011. Development of the nervous system in hatchlings of *Spadella cephaloptera* (Chaetognatha), and implications for nervous system evolution in Bilateria. *Dev. Growth Differ.* 53: 740–759.

Riemann, F., and O. Riemann. 2010. The enigmatic mineral particle accumulations on the cuticular rings of marine desmoscolecoid nematodes structure and significance explained with clues from live observations. *Meiofauna Marina* 18: 1–10.

Rothe, B. H., and A. Schmidt-Rhaesa. 2009. Architecture of the nervous system in two *Dactylopodola* species (Gastrotricha, Macrodasyida). *Zoomorphology* 128: 227–246.

Rothe, B. H., and A. Schmidt-Rhaesa. 2010. Structure of the nervous system in *Tubiluchus troglodytes* (Priapulida). *Invertebr. Biol.* 129: 39–58.

Rothe, B. H., A. Schmidt-Rhaesa, and A. Kieneke. 2011. The nervous system of *Neodasys chaetonotoideus* (Gastrotricha: *Neodasys*) revealed by combining confocal laserscanning and transmission electron microscopy: Evolutionary comparison of neuroanatomy within the Gastrotricha and basal Protostomia. *Zoomorphology* 130: 51–84.

Rouse, G. W., N. G. Wilson, J. I. Carvajal, and R. C. Vrijenhoek. 2016. New deep-sea species of *Xenoturbella* and the position of Xenacoelomorpha. *Nature* 530: 94–97.

Rozario, T., and P. A. Newmark. 2015. A confocal microscopy-based atlas of tissue architecture in the tapeworm *Hymenolepis diminuta*. *Exp. Parasitol.* 158: 31–41.

Ruiz-Trillo, I., M. Riutort, D. T. Littlewood, E. A. Herniou, and J. Baguñà. 1999. Acoel flatworms: Earliest extant bilaterian metazoans, not members of Platyhelminthes. *Science* 283: 1919–1923.

Ruppert, E. E., R. S. Fox, and R. D. Barnes. 2004a. Chordata. In *Invertebrate Zoology: A Functional Evolutionary Approach*, 930–963. Brooks/Cole.

Ruppert, E. E., R. S. Fox, and R. D. Barnes. 2004b. Mollusca. In *Invertebrate Zoology: A Functional Evolutionary Approach*, 283–412. Brooks/Cole.

Ruppert, E. E., R. S. Fox, and R. D. Barnes. 2004c. Onychophora and Tardigrada. In *Invertebrate Zoology: A Functional Evolutionary Approach*, 504–516. Brooks/Cole.

Russel-Hunter, W. D. 1988. The gills of chitons (Polyplacophora) and their significance in molluscan phylogeny. *Am. Malacol. Bull.* 6: 69–78.

Rychel, A. L., and B. J. Swalla. 2007. Development and evolution of chordate cartilage. *J. Exp. Zool. B Mol. Dev. Evol.* 308: 325–335.

Sanders, H. L., and R. R. Hessler. 1962. *Priapulus atlantisi* and *Priapulus profundus.* Two new species of priapulids from bathyal and abyssal depths of the North Atlantic. *Deep Sea Res. Oceanogr. Abstr.* 9: 125–130.

Sanger, T. J., and R. Rajakumar. 2018. How a growing organismal perspective is adding new depth to integrative studies of morphological evolution: Organismal evo-devo. *Biol. Rev.* 24: 107.

Santagata, S. 2004. Larval development of *Phoronis pallida* (Phoronida): Implications for morphological convergence and divergence among larval body plans. *J. Morphol.* 259: 347–358.

Santagata, S., and R. L. Zimmer. 2002. Comparison of the neuromuscular systems among actinotroch larvae: Systematic and evolutionary implications. *Evol. Dev.* 4: 43–54.

Scherholz, M., E. Redl, T. Wollesen, A. L. de Oliveira, C. Todt, and A. Wanninger. 2017. Ancestral and novel roles of Pax family genes in mollusks. *BMC Evol. Biol.* 17: 81.

Scherholz, M., E. Redl, T. Wollesen, C. Todt, and A. Wanninger. 2013. Aplacophoran mollusks evolved from ancestors with polyplacophoran-like features. *Curr. Biol.* 23: 2130–2134.

Schmidt-Rhaesa, A. 2007. *The Evolution of Organ systems*. Oxford University Press.

Schmidt-Rhaesa, A., and J. Kulessa. 2007. Muscular architecture of *Milnesium tardigradum* and *Hypsibius* sp. (Eutardigrada, Tardigrada) with some data on *Ramazottius oberhaeuseri. Zoomorphology* 126: 265–281.

Schmidt-Rhaesa, A., and B. H. Rothe. 2006. Postembryonic development of dorsoventral and longitudinal musculature in *Pycnophyes kielensis* (Kinorhyncha, Homalorhagida). *Integr. Comp. Biol.* 46: 144–150.

Schock, F., and N. Perrimon. 2002. Molecular mechanisms of epithelial morphogenesis. *Annu. Rev. Cell Dev. Biol.* 18: 463–493.

Scholtz, G. 2002. The Articulata hypothesis – or what is a segment? *Org. Divers. Evol.* 2: 197–215.

Scholtz, G. 2010. Deconstructing morphology. *Acta Zool.* 91: 44–63.

Schulze, C., R. C. Neves, and A. Schmidt-Rhaesa. 2014. Comparative immunohisto-chemical investigation on the nervous system of two species of Arthrotardigrada (Heterotardigrada, Tardigrada). *Zool. Anz.* 253: 225–235.

Schulze, C., and A. Schmidt-Rhaesa. 2013. The architecture of the nervous system of *Echiniscus testudo* (Echiniscoidea, Heterotardigrada). *J. Limnol.* 72: 44–53.

Schuster, J., S. Atherton, M. A. Todaro, A. Schmidt-Rhaesa, and R. Hochberg. 2018. Redescription of *Xenodasys riedli* (Gastrotricha: Macrodasyida) based on SEM analysis, with first report of population density data. *Mar. Biodivers.* 48: 259–271.

Seaver, E. C. 2003. Segmentation: Mono- or polyphyletic? *Int. J. Dev. Biol.* 47: 583–595.

Sedgwick, A. 1884. On the origin metameric segmentation and some other morphological question. *J. Cell Sci.* s2–24: 43–82.

Segers, H. 2004. Rotifera: Monogononta. In *Freshwater Invertebrates of the Malaysia region*, Eds. C. M. Yule, and H.-S. Yong, 106–120. Kuala Lumpur: Academy of Sciences; Monash University.

Semmler, H., X. Bailly, and A. Wanninger. 2008. Myogenesis in the basal bilaterian *Symsagittifera roscoffensis* (Acoela). *Front. Zool.* 5: 14.

Semmler, H., M. Chiodin, X. Bailly, P. Martínez, and A. Wanninger. 2010. Steps towards a centralized nervous system in basal bilaterians: Insights from neurogenesis of the acoel *Symsagittifera roscoffensis*. *Dev. Growth Differ.* 52: 701–713.

Semmler, H., and A. Wanninger. 2010. Myogenesis in two polyclad platyhelminths with indirect development, *Pseudoceros canadensis* and *Stylostomum sanjuania*. *Evol. Dev.* 12: 210–221.

Shimazaki, A., A. Sakai, and M. Ogasawara. 2006. Gene expression profiles in *Ciona intestinalis* stigmatal cells: Insight into formation of the ascidian branchial fissures. *Dev. Dyn.* 235: 562–569.

Shipley, A. E. 1883. On the structure and development of *Argiope* (with two plates). *Mitt. Zool. Stn. Neapel* 4: 494–520.

Shirley, T. C., and V. Storch. 1999. *Halicryptus higginsi* n. sp. (Priapulida): A giant new species from Barrow, Alaska. *Invertebr. Biol.* 118: 404–413.

Siewing, R. 1973. Morphologische Untersuchungen zum Archicoelomatenproblem. *Z. Morphol. Tiere* 74: 17–36.

Smith, F. W., T. C. Boothby, I. Giovannini, L. Rebecchi, E. L. Jockusch, and B. Goldstein. 2016. The compact body plan of tardigrades evolved by the loss of a large body region. *Curr. Biol.* 26: 224–229.

Smith, F. W., and B. Goldstein. 2017. Segmentation in Tardigrada and diversification of segmental patterns in Panarthropoda. *Arthropod Struct. Dev.* 46: 328–340.

Smith, F. W., and E. L. Jockusch. 2014. The metameric pattern of *Hypsibius dujardini* (Eutardigrada) and its relationship to that of other panarthropods. *Front. Zool.* 11: 1–16.

Sørensen, M. V. 2005. Musculature in three species of *Proales* (Monogononta, Rotifera) stained with phalloidin-labeled fluorescent dye. *Zoomorphology* 124: 47–55.

Sørensen, M. V., G. Accogli, and J. G. Hansen. 2010. Postembryonic development of *Antygomonas incomitata* (Kinorhyncha: Cyclorhagida). *J. Morphol.* 271: 863–882.

Sørensen, M. V., H. S. Rho, W.-G. Min, and D. Kim. 2012. A new recording of the rare priapulid *Meiopriapulus fijiensis*, with comparative notes on juvenile and adult morphology. *Zool. Anz.* 251: 364–371.

Southwell, T., and J. W. S. Macfie. 1925. On a collection of Acanthocephala in the Liverpool School of Tropical Medicine. *Ann. Trop. Med. Parasitol.* 19: 141–184.

Storch, V., and H. Ruhberg. 1990. Electron microscopic observations on the male genital tract and sperm development in *Peripatus sedgwicki* (Peripatidae, Onychophora). *Invertebr. Reprod. Dev.* 17: 47–56.

Stretton, A. O., R. M. Fishpool, E. Southgate, J. E. Donmoyer, J. P. Walrond, J. E. Moses, and I. S. Kass. 1978. Structure and physiological activity of the motoneurons of the nematode *Ascaris. Proc. Natl. Acad. Sci. U. S. A.* 75: 3493–3497.

Stricker, S. A., T. L. Smythe, L. Miller, and J. L. Norenburg. 2002. Comparative biology of oogenesis in nemertean worms: Oogenesis in nemerteans. *Acta Zool.* 82: 213–230.

Subbotin, S. A. 2013. Order Tylenchida Thorne, 1949. In *Nematoda*, Eds. A. Schmidt-Rhaesa, 613–636. Berlin, Boston: De Gruyter.

Sulston, J. E., and H. R. Horvitz. 1977. Post-embryonic cell lineages of the nematode, *Caenorhabditis elegans. Dev. Biol.* 56: 110–156.

Sulston, J. E., E. Schierenberg, J. G. White, and J. N. Thomson. 1983. The embryonic cell lineage of the nematode *Caenorhabditis elegans. Dev. Biol.* 100: 64–119.

Sundberg, P., and M. Strand. 2007. *Annulonemertes* (phylum Nemertea): When segments do not count. *Biol. Lett.* 3: 570–573.

Szmygiel, C., A. Schmidt-Rhaesa, B. Hanelt, and M. G. Bolek. 2014. Comparative descriptions of non-adult stages of four genera of Gordiids (Phylum: Nematomorpha). *Zootaxa* 3768: 101–118.

Tautz, D. 2004. Segmentation. *Dev. Cell* 7: 301–312.

Temereva, E. 2012. Ventral nerve cord in *Phoronopsis harmeri* larvae. *J. Exp. Zool. B Mol. Dev. Evol.* 318: 26–34.

Temereva, E. N., and V. V. Malakhov. 2006. Trimeric coelom organization in the larvae of *Phoronopsis harmeri* Pixell, 1912 (Phoronida, Lophophorata). *Dokl. Biol. Sci.* 410: 396–399.

Temereva, E. N., and V. V. Malakhov. 2011. The evidence of metamery in adult brachiopods and phoronids. *Invert. Zool.* 8: 91–112.

Temereva, E., and E. B. Tsitrin. 2013. Development, organization, and remodeling of phoronid muscles from embryo to metamorphosis (Lophotrochozoa: Phoronida). *BMC Dev. Biol.* 13: 14.

Temereva, E., and A. Wanninger. 2012. Development of the nervous system in *Phoronopsis harmeri* (Lophotrochozoa, Phoronida) reveals both deuterostome- and trochozoan-like features. *BMC Evol. Biol.* 12: 121.

Teuchert, G. 1967. Zum Protonephridialsystem mariner Gastrotrichen der Ordnung Macrodasyoidea. *Mar. Biol.* 1: 110–112.

Teuchert, G. 1968. Zur Fortpflanzung und Entwicklung der Macrodasyoidea (Gastrotricha). *Z. Morphol. Tiere* 63: 343–418.

Thomson, C. T. W. 1865. On the embryogeny of *Antedon rosaceus*, Linck (*Comatula rosacea* of Lamarck). *Philos. Trans. R. Soc. London* 31(155): 513–544.

Tyler, S., and G. S. Hyra. 1998. Patterns of musculature as taxonomic characters for the Turbellaria Acoela. *Hydrobiologia* 383: 51–59.

Vellutini, B. C., and A. Hejnol. 2016. Expression of segment polarity genes in brachiopods supports a non-segmental ancestral role of *engrailed* for bilaterians. *Sci. Rep.* 6: 32387.

von Döhren, J. 2016. Development of the nervous system of *Carinina ochracea* (Palaeonemertea, Nemertea). *PloS One* 11: e0165649.

Voronezhskaya, E. E., S. A. Tyurin, and L. P. Nezlin. 2002. Neuronal development in larval chiton *Ischnochiton hakodadensis* (Mollusca: Polyplacophora). *J. Comp. Neurol.* 444: 25–38.

Walthall, W. W. 1995. Repeating patterns of motoneurons in nematodes: The origin of segmentation? In *The Nervous Systems of Invertebrates: An Evolutionary and Comparative Approach*, Eds. O. Breidbach, and W. Kutsch, 61–75. Basel: Birkhäuser Basel.

Wanninger, A., and G. Haszprunar. 2001. The expression of an *engrailed* protein during embryonic shell formation of the tusk-shell, *Antalis entalis* (Mollusca, Scaphopoda). *Evol. Dev.* 3: 312–321.

Wanninger, A., and G. Haszprunar. 2002. Chiton myogenesis: Perspectives for the development and evolution of larval and adult muscle systems in molluscs. *J. Morphol.* 251: 103–113.

Wanninger, A., and T. Wollesen. 2015. Mollusca. In *Evolutionary Developmental Biology of Invertebrates 2*, Ed. A. Wanninger, 103–153. Vienna: Springer.

Wennberg, S. A., R. Janssen, and G. E. Budd. 2009. Hatching and earliest larval stages of the priapulid worm *Priapulus caudatus*. *Invertebr. Biol.* 128: 157–171.

Wheeler, W. M. 1894. *Syncælidium pellucidum*, a new marine triclad. *J. Morphol.* 9: 167–194.

White, J. G., E. Southgate, J. N. Thomson, and S. Brenner. 1976. The structure of the ventral nerve cord of *Caenorhabditis elegans*. *Philos. Trans. R. Soc. London B Biol. Sci.* 275: 327–348.

Whitington, P. M., and G. Mayer. 2011. The origins of the arthropod nervous system: Insights from the Onychophora. *Arthropod Struct. Dev.* 40: 193–209.

Wicht, H., and T. C. Lacalli. 2005. The nervous system of amphioxus: Structure, development, and evolutionary significance. *Can. J. Zool.* 83: 122–150.

Wilhelmi, J. 1909. *Tricladen*. R. Friedländer & Sohn.

Willey, A. 1893. Studies on the Protochordata. I. On the origin of the branchial stigmata, praeoral lobe, endostyle, atrial cavities, etc. in *Ciona intestinalis*, Linn., with remarks on *Clavelina lepadiformis*. *J. Cell Sci.* s2-34: 317–360.

Willmer, P. 1990. Body divisions—metamerism and segmentation. In *Invertebrate Relationships: Patterns in Animal Evolution*, 39–45. Cambridge University Press.

Wingstrand, K. G. 1985. On the anatomy and relationships of recent Monoplacophora. *Galathea Rep.* 16: 7–94.

Wollesen, T., S. V. Rodríguez Monje, C. Todt, B. M. Degnan, and A. Wanninger. 2015. Ancestral role of *Pax2/5/8* in molluscan brain and multimodal sensory system development. *BMC Evol. Biol.* 15: 231.

Wollesen, T., M. Scherholz, S. V. Rodríguez Monje, E. Redl, C. Todt, and A. Wanninger. 2017. Brain regionalization genes are co-opted into shell field patterning in Mollusca. *Sci. Rep.* 7: 5486.

Wray, G. A., and C. J. Lowe. 2000. Developmental regulatory genes and echinoderm evolution. *Syst. Biol.* 49: 28–51.

Yasui, K., S. Tabata, T. Ueki, M. Uemura, and S. C. Zhang. 1998. Early development of the peripheral nervous system in a lancelet species. *J. Comp. Neurol.* 393: 415–425.

Younossi-Hartenstein, A., M. Jones, and V. Hartenstein. 2001. Embryonic development of the nervous system of the temnocephalid flatworm *Craspedella pedum*. *J. Comp. Neurol.* 434: 56–68.

Zakrzewski, A.-C., A. Suh, and C. Lüter. 2012. New insights into the larval development of *Macandrevia cranium* (Müller, 1776) (Brachiopoda: Rhynchonelliformea). *Zool. Anz.* 251: 263–269.

Zaniolo, G., N. J. Lane, P. Burighel, and L. Manni. 2002. Development of the motor nervous system in ascidians. *J. Comp. Neurol.* 443: 124–135.

Zantke, J., C. Wolff, and G. Scholtz. 2008. Three-dimensional reconstruction of the central nervous system of *Macrobiotus hufelandi* (Eutardigrada, Parachela): Implications for the phylogenetic position of Tardigrada. *Zoomorphology* 127: 21–36.

Zapotosky, J. E. 1975. Fine structure of the larval stage of *Paragordius varius* (Leidy, 1851) (Gordioidea: Paragordidae). II. The postseptum. *Proc. Helm. Soc. Wash.* 42: 103–111.

Zelinka, C. 1908. Zur Anatomie der Echinoderen. *Zool. Anz.* 33: 629–647.

10 Axial Regeneration in Segmented Animals
A Post-Embryonic Reboot of the Segmentation Process

Eduardo E. Zattara

CONTENTS

10.1 INTRODUCTION

Post-traumatic regeneration, or the ability of organisms to grow back body parts lost to injury, is a widespread trait that has historically drawn great interest from researchers and the general public. Such interest stems primarily from the fact that in contrast with many animal groups, humans along with most other mammals are largely incapable of regrowing large or complex structures like heads or limbs. Thus, much of regeneration research is fueled by its potential to inform development of medical therapies, with aims ranging from scar-free healing to amputated limb recovery. But besides its biomedical promises, the study of regeneration also offers a window into how developmental systems can resort to different trajectories to achieve and maintain the same stable morphologies. Similarities and differences in how structures are patterned and built both during embryogenesis and regeneration can be used to better understand how developmental mechanisms and gene networks driving them have evolved. The evolution of regenerative ability is a topic of interest in itself: regeneration is a complex trait potentially subject to adaptive trade-offs, given that it brings an obvious survival advantage against accidents and sublethal predation but can also become a resource sink large enough to be selected against under certain scenarios. All these aspects of regeneration biology can be informed by comparative studies of the distribution and mechanisms of regeneration across organisms.

Regeneration is a process that can occur at a wide range of biological levels, from cells to whole organisms (Bely and Nyberg 2010). This chapter will focus on whole-structure regeneration, and thus regeneration at lower levels, such as tissues or cells, will not be covered.

Structure regeneration abilities can be classified according to the position of the structure relative to the main body axis (usually the anteroposterior axis in bilaterian animals) in three main types: appendage regeneration, involving the loss and regrowth of structures located outside the main body axis (e.g., limbs, barbels, palps, fins, and other outgrowths); axial regeneration, which involves the loss and regrowth of structures located along the main body axis (e.g., heads, tails, and other structures removed by transverse amputations); and whole-body regeneration, in which organisms can reorganize and regrow missing structures after extensive tissue loss along any amputation plane. Axial regeneration can be further categorized according to whether regeneration can restore any length of tissue lost or is limited to restoring terminal regions only. This later distinction is particularly relevant in segmented animals that contain terminal non-segmental regions.

Within the overarching theme of this book, this chapter is organized in three parts. The first part describes the distribution of axial regenerative ability across animals. The second part presents some common patterns of regenerative processes across Metazoa. The third part focuses on current knowledge on annelid regeneration.

10.2 THE PHYLOGENETIC DISTRIBUTION OF REGENERATION

The ability to regenerate structures lost to damage varies widely across and within Metazoan phyla (Sánchez Alvarado 2000; Brockes and Kumar 2008; Bely and Nyberg 2010). This variation is far from random, and adequate knowledge about the distribution patterns of regeneration in animals can help understand both the proximal causes (i.e., the cellular and molecular mechanisms of regeneration) and the ultimate causes (i.e., the evolutionary forces acting on regeneration) of this variation.

An overview of regenerative abilities across metazoans suggests that the earliest animals were good at regenerating (Figure 10.1). Phyla radiating near the base of the Metazoan tree, like Porifera (sponges), Cnidaria (jellyfishes, corals, sea anemones,

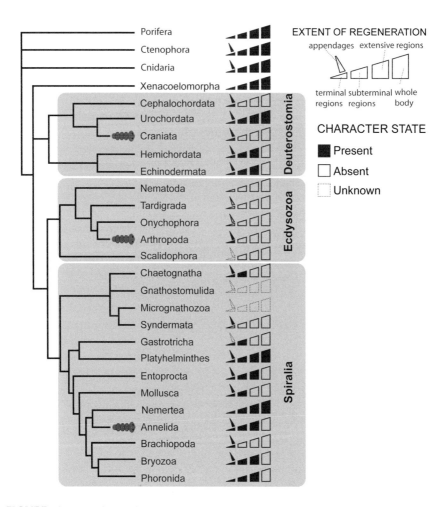

FIGURE 10.1 Phylogenetic distribution of regenerative abilities across metazoans. Ability to regenerate appendages (whenever appendages are part of a body plan) is shown by upward-pointing triangles to the left. Ability to regenerate parts of the main body axis are shown as four increasing levels: (1) only terminal structures regenerate, (2) limited extent of subterminal structures regenerate, (3) large extent of structures regenerate, and (4) whole bodies can regenerate from small fragments. The extent of known regenerative capabilities for each phylum is shown based on the highest level known for species within the phylum and may not reflect the typical abilities for the phylum or its ancestral condition. Phylogenetic relationships after Laumer et al. (2019). Modified after Zattara (2012).

box jellies, and many others), and Ctenophora (comb jellies), are all characterized by extensive regenerative abilities (Bely and Nyberg 2010; Morgan 1901; Slack 2017). Systematic surveys are still lacking, but high regenerative abilities are likely part of the ancestral abilities of these phyla, suggesting they were also present in the common ancestors of all animals. The same goes for groups originating near the base of the Bilateria: species of Placozoa and Xenacoelomorpha all show excellent regeneration (Thiemann and Ruthmann 1991; Martinelli and Spring 2004; Haszprunar 2016), supporting that the developmental toolkit of the Urbilateria, the last common ancestors of all bilaterians, included ample regenerative abilities.

Among the Deuterostomia, extensive regenerative abilities are reported for many species of Echinodermata (sea urchins, starfishes, sea cucumbers, and sea lilies), Hemichordata (acorn worms) and Urochordata (sea squirts and salps), and more modest abilities—appendage and terminal axial regeneration—are known from Cephalochordata (lancelets) and Craniata (the vertebrates) (Tsonis 2000; Bely and Nyberg 2010). This phylogenetic pattern suggests that regenerative abilities inherited from Urbilateria were retained in the deuterostome ancestor.

In contrast, all phyla within Ecdysozoa (which includes nematodes, tardigrades, onychophorans, arthropods, and other groups characterized by a periodically molted external cuticle) show poor axial regeneration (Bely 2010), suggesting this ability was lost during the evolution of the last common ecdysozoan ancestor. Interestingly, poor axial regeneration in ecdysozoans does not imply a lack of regenerative mechanisms: for example, imaginal discs of holometabolous insects can readily regenerate lost parts (Worley, Setiawan, and Hariharan 2012). Furthermore, many arthropods are capable of regenerating their articulated limbs (Maruzzo and Bortolin 2013); since arthropod limbs represent a novel structure relative to the ecdysozoan stem, it is likely that arthropod appendage regeneration represents an evolutionary gain rather than a plesiomorphy inherited from the Urbilateria.

The third main clade of animals, the Spiralia, comprises about 13 phyla including rotifers, flatworms, mollusks, nemerteans, brachiopods, phoronids, bryozoans, and annelids, among others (Edgecombe et al. 2011). Recent works on spiralian phylogeny propose that this clade is comprised of Gnathifera (Rotifera, Acanthocephala, and Gnathostomulida), Rouphozoa (Gastrotricha and Platyhelminthes) and Lophotrochozoa (Cycliophora, Ectoprocta, Mollusca, Nemertea, Annelida Brachiopoda, Phoronida, and Bryozoa) (Laumer et al. 2019). Spiralian regenerative abilities vary widely, from the almost total absence of regeneration in Rotifera to the whole-body regeneration seen in some species of flatworms, annelids, and nemerteans (Bely, Zattara, and Sikes 2014). Within Gnathifera, some rotifers and acanthocephalans can regenerate appendages, but there are no reports of axial regeneration; there is no information about regenerative capabilities of Gnathostomulida. Among Rouphozoa, both axial and appendage regeneration has been reported in species of Gastrotricha (Manylov 2010), but most of the attention has been given to the outstanding regenerative abilities of many species of Platyhelminthes. Several groups of flatworms can regenerate whole animals from small fragments, an ability most intensely studied in species from Tricladida, but also reported from other groups, including the basally branching catenulids. Lophotrochozoan phyla show substantial within-group variation in their regenerative ability, but overall display good to outstanding regeneration. Mollusks lack whole-body regeneration, but several species have been reported to regenerate various axial and appendicular structures, including the head and limbs (Bely, Zattara, and Sikes 2014). Nemerteans show widespread ability for posterior axial regeneration, but the ability to regenerate the head is rare and seems unlikely to represent the ancestral condition for the phylum (Zattara et al. 2019). Brachiopods can regenerate their lophophore, shell, and pedicle (Chuang 1994), but given their body plan, it is difficult to determine if reconstruction of any of these organs represent axial regeneration. In contrast, phoronids and ectoprocts (bryozoans) have several species capable of whole-body regeneration (Emig 1973; Nielsen 1994). Annelids also show great axial and appendage regenerative ability, although many groups seem to have lost the ability to regenerate their anterior ends and even the posterior ends (Bely 2006; Zattara and Bely 2016). Overall, spiralians show abundant examples of good structure regeneration, suggesting that their last common ancestor had good to great regenerative ability. However, systematic surveys and formal phylogenetic analysis and ancestral trait estimation of regeneration have only been conducted in two phyla, Annelida and Nemertea (Zattara and Bely 2016; Zattara et al. 2019); these studies have shown that species showing outstanding regenerative ability might not be representative of all other species of the phylum, or of their last common ancestor.

Of all metazoan phyla mentioned earlier, only three are organized in true segments: the deuterostome Craniata, the ecdysozoan Panarthropoda, and the spiralian Annelids. Segmental organization along the main, anteroposterior body axis in all three groups is achieved by generation of iterated modules at the posterior end during embryonic development (see Chapters 2 to 5) and in many cases during post-embryonic growth. Upon amputation transverse to the main body axis, if axial regeneration occurs, it usually restores the missing ends including the terminal regions and a variable number of intervening segments. This implies that after amputation, axial regeneration is achieved by restoring the terminal regions and rebooting the segmentation process.

10.3   COMMON ASPECTS OF ANIMAL REGENERATION

Embryogenesis is a complex process in which a single cell, the zygote, initiates a developmental trajectory that will lead to a hatching, multicellular organism. Embryogenesis may lead to a final adult morphology in direct developing species, or to an intermediate, larval form in indirect developing species. Indirect development implies post-embryonic developmental trajectories, like metamorphosis, that transform the larvae into the adult form. Regeneration is another post-embryonic developmental trajectory, but differs from embryogenesis and metamorphosis in that it starts from a highly unpredictable situation, as traumatic injury triggering regeneration can happen in many possible ways, each resulting in a different initial condition. Thus, although converging to the same final morphology as embryogenesis and/or metamorphosis, regeneration's developmental trajectory requires additional degrees of freedom to adapt to unpredictable initial conditions.

For successful regeneration, an organism needs to sequentially overcome three main challenges: (1) survive the injury and close the wound to avoid uncontrolled flow between the internal and external environments of the body; (2) gather the resources it will need to reconstruct the structures lost to injury; and (3) deploy and convert these resources into structures functionally equivalent to those lost. Since these three challenges are common to all injured organisms, it is not surprising that regeneration studies in very distantly related organisms usually report the same three main phases for this developmental trajectory: (1) wound healing; (2) cellular reorganization from stem cells, dedifferentiated cells or transdifferentiated cells; and (3) cellular differentiation, morphogenesis, growth, and rescaling.

10.3.1 WOUND HEALING

Organisms keep a strict separation between their internal and external environments. Separation is usually achieved by a layer of epithelial tissue that controls passage between these domains. The integrity of this epithelial barrier is fundamental to keep the proper balance of substances between the inside and the outside, and to prevent infection by pathogens or entry of toxic molecules. Traumatic injury compromises the barrier by generating an opening, the wound. Most if not all organisms deploy some form of wound healing after enduring (and surviving) a wound-opening injury.

The most urgent aim of wound healing is generating a quick response to stop direct contact between the internal and external environment. This can be achieved through different strategies, including covering of the wound by mucus secretions or protein coagulates, blockading the wound by agglomeration of cells into a wound plug, and/or closing of the wound by contraction of nearby muscle fibers. In many cases, wounding also activates a response from the immune system, usually involving the migration of several cell types to the injured area to fight potential infection.

The initial quick wound response is followed by growth and/or expansion of the epithelial layers adjacent to the wound, followed by fusion, until the opening is closed. During or after this second step, wound cleanup takes place to remove internal leftover structures like wound plugs, broken cells, and other debris. Immune cells, especially phagocytes, will often take part in this effort. At the end of this step, tissues adjacent to the wound site, often collectively referred to as the *stump*, no longer suffer uncontrolled exchange with the external environment. In many cases, an epidermal layer known as the *wound epithelium* forms over the wound that is histologically and transcriptionally different from the original epidermis. The wound epithelium has been shown to play critical roles in the earlier phases of regenerative responses (Carlson 2007; Campbell and Crews 2008; Eming, Martin, and Tomic-Canic 2014).

Wound healing is a critical phase for all regenerative processes. Improper wound healing can delay or stall further regeneration. This might be due to several reasons: on the one hand, the wound-healing process generates a cleaner starting point for the next phase of regeneration; on the other, tissues involved in wound healing are often responsible for releasing molecular signals that trigger regeneration (Brockes and Kumar 2008). Wound-healing strategies might also interfere with regenerative potential: in mammals, for example, poor regeneration is thought to be at least partially caused by the inflammatory nature of the immune response and the formation of a non-cellular scab of clotted proteins (Goss 1987; Harty et al. 2003). Faulty signaling during and after wound healing can be a cause for a loss of regenerative ability, as the amputated organism fails to transition from this initial phase to the next, cell reorganization, and blastema formation.

10.3.2 CELL REORGANIZATION AND BLASTEMA FORMATION

Once the challenge of an open wound has been overcome, the regeneration process requires assembling the resources necessary to rebuild the missing structures. The most common strategy in this phase is the buildup of a regeneration *blastema*, a mass of morphologically undifferentiated cells, from which the new structures will develop. Blastema formation is arguably the process that gathers the most interest and debate among regeneration researchers, as this is where regeneration and embryogenesis differ the most. During embryonic development, structure formation goes through a trajectory in which cells move from pluripotent to differentiated states, while during regeneration, structures must be derived from cells found in adult tissue. Since most adult tissues are composed of differentiated cells, regenerating organisms need to either have a reserve supply of pluripotent, undifferentiated stem cells, or else have the ability to dedifferentiate or transdifferentiate already differentiated cells. Evidence for sourcing from reserve stem cells, differentiated cells, or both are found in many lineages showing extensive regenerative abilities. Conversely, lack of stem cells or irreversibility of cell differentiation is often hypothesized to be a cause for diminished regenerative abilities.

A related question is where do cells contributing materials for regeneration come from. In general, most cells contributing to regenerated tissues originate from regions close to the wound site, as in newt limb regeneration (Kragl et al. 2009). However, there are several cases where cells migrate from more distant regions, sometimes being the main source of regenerating materials, as seen in planarians and hydrozoans (Aboobaker 2011; Bosch 2007). Irrespective of their origin, cells contributing to a regeneration blastema usually undergo extensive proliferation; however, some species are known to be able to complete regeneration in the absence of cell proliferation. This latter observation led early regeneration researchers to classify regeneration processes on two general categories: epimorphic processes characterized by active cell proliferation and generation of a visible blastema, and morphallactic processes driven by transdifferentiation of existing cells without cell proliferation (Morgan 1901). Although Morgan explicitly stated that these process categories "are not sharply separated, and may even appear combined in the same form," later authors tried (often forcefully) to fit regeneration of different structures and organisms exclusively into either of these categories, a habit that persists to the present. While some cases do exist where regeneration proceeds exclusively by epimorphosis or morphallaxis, in most situations it can be shown that both cell proliferation and tissue remodeling play fundamental and complementary roles. Thus, if these terms are to remain useful, they should be used to refer to specific processes occurring during regeneration, rather than labels applied to the whole developmental trajectory (Agata, Saito, and Nakajima 2007).

In summary, during the phase of cell reorganization, regenerating organisms deploy epimorphic and/or morphallactic processes to gather cellular resources that will be used for reconstructing the lost structures at the wound site.

10.3.3 CELL DIFFERENTIATION AND MORPHOGENESIS

Once cellular resources are available, the last phase of regeneration initiates: rebuilding the lost structure. During this phase, morphogenetic mechanisms are deployed to pattern the new structure, driving cells to multiply, sort out, and differentiate into the regenerated tissues. Once patterning is complete, the regenerate becomes evident, usually as a smaller, more rudimentary version of the original structure. This rudiment keeps growing and developing, eventually reaching a size and morphology closely resembling the replaced structure.

In many cases, regeneration reconstructs less than the original amount of tissues lost: for example, squamate lizards often regenerate tails that are smaller and less flexible than the original (Alibardi 2014); many annelids show a maximum number of anterior segments rebuilt during head regeneration (Berrill 1952). This can result in a discontinuity of tissue identity between the regenerate and the stump. For example, after anterior regeneration in an annelid that lost ten anterior segments to amputation but can regenerate only four, segment 4 will form adjacent to segment 11, and the worm will be missing all structures specific to segments 5 to 10. Such discontinuities usually trigger a transformation process of stump tissues to adjust to their new positional identities. In our example, stump segments transform so that segment 11 adopts the morphology of segments 6, and so on. Other body-wide adjustments besides positional identity are possible: for example, in many worms, both growth of the regenerate and thinning plus elongation of the stump combine to restore the original body proportions (Coe 1929). End results may vary; while in some species regenerated structures eventually become indistinguishable from the original, in others the regenerate might differ morphologically or even functionally. In some cases, the morphogenetic process results in the regeneration of a different structure than the one that was lost (e.g., a posterior end regenerating after amputation of an anterior end). Such cases of heteromorphic regeneration tend to occur at a low frequency and are thought to be due to developmental errors caused by faulty signaling.

Since in most cases injuries remove terminal structures from appendages or the main body axis, the regenerate develops along a proximodistal axis. Interestingly, in most cases, morphogenesis proceeds by first rebuilding the distalmost, terminal tissues, and then intercalating the remaining structures. When regeneration is incomplete, tissues not restored are usually those that were proximal to the wound site; correspondingly, species that show poor structure regeneration are often still capable of restoring the terminal tips of many structures. This pattern is particularly evident in axial regeneration of segmented animals, where both anterior and posterior terminal regions share a non-segmental origin that differentiates them from the intervening segmental tissues.

In this final phase of regeneration, developmental trajectories converge the most with those found during embryogenesis. While detailed mechanisms are still far from understood, many studies have uncovered a large degree of similarity between regeneration and embryonic development in the genetic toolkit deployed during tissue patterning and cell differentiation. For this reason, regeneration is often used as a proxy of embryogenesis to study development of specific tissues and organs.

10.4 REGENERATION IN ANNELIDS

The ability to make new segments during axial regeneration varies across the truly segmented phyla. Among vertebrates, examples of axial regeneration involving formation of new segments are rather rare (Mochii, Taniguchi, and Shikata 2007; Hutchins et al. 2014), and the regenerated segments are often simpler and more disorganized than those formed during embryonic development. Consequently, studies comparing segmentation mechanisms between embryogenesis and regeneration are lacking. Among arthropods, axial regeneration is exceedingly infrequent, and only involves reconstruction of the terminal region, with no known cases of segment reconstruction. In contrast, most annelids show excellent axial regeneration and can reconstruct segments similar to those made during embryonic development and post-embryonic posterior growth (Bely 2006). Since segments produced by regeneration are not only indistinguishable from those made during embryogenesis and growth, but also develop faster, regeneration (in particular, posterior regeneration) offers a convenient way to study the developmental genetic mechanisms of annelid segmentation.

Annelida is one of the few phyla where phylogenetic distribution of axial regenerative abilities has been formally surveyed and analyzed (Zattara and Bely 2016). This comparative work has shown that the stem group annelids likely had very good axial regenerative abilities, being capable of restoring both posterior and anterior ends. The basic annelid architecture is relatively well conserved across the phylum, minimizing the problem of homology and facilitating comparisons of the regenerative process. Furthermore, annelid regeneration has been a focus of research for a long time, resulting in a rich literature that is now being actively updated and expanded thanks to a recent surge in interest that is applying powerful new tools to dissect the developmental and evolutionary mechanisms underlying this process.

10.4.1 STAGES OF AXIAL REGENERATION

In most annelids that have been studied, axial regeneration proceeds through a similar series of steps that begin with an injury response and end in either a complete anterior or posterior end (in species capable of regeneration) or an incomplete stump stalling at some point in the process (in species with diminished or absent regeneration). The non-segmental, terminal structures—the prostomium anteriorly and the pygidium posteriorly—differentiate first, followed by a variable number of segments. The complete regeneration process is usually divided in five stages: (1) wound healing; (2) blastema formation; (3) blastema differentiation; (4) resegmentation; and (5) growth (Figure 10.2).

Stage 1, wound healing, begins immediately after amputation. After closure of the wound by muscle contraction, cells migrate into the wounded segment to provide an immune response and generate a wound plug by clotting. This stage is characterized

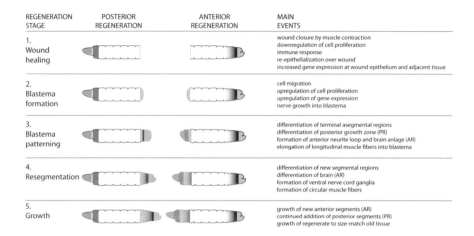

FIGURE 10.2 Generic stages of annelid posterior and anterior regeneration, along with their main events. The wound/blastemal epithelium is shown by a dotted red line; brown anterior regions represent non-segmental prostomium and peristomium; ocher posterior region represents the pygidium; dark red posterior subterminal band and graded dark red region represent the segment addition zone and remaining posterior growth zone, respectively; green regions represent undifferentiated blastemal tissues.

by local to body-wide downregulation of cell proliferation. The adjacent epidermis then extends over the wound forming a wound epithelium; transition to the next stage is marked by a sharp upregulation of several genes known to be key developmental regulators in other systems.

Stage 2, blastema formation, begins with invasion of the wound site by neurites originating in nerves from the ventral nerve cord and peripheral nerves, and the upregulation of local cell proliferation, which results in the formation of a blastema composed of undifferentiated cells. Blastema formation is characterized by expression of a large suite of genes and intense mitotic activity. This stage eventually grades into the next stage.

Stage 3, blastema differentiation, is characterized by morphological differentiation of the non-segmental terminal caps (prostomium and peristomium at the anterior end, pygidium at the posterior end) and the intercalation of additional tissues that form segment addition zones and segmental tissues. Muscular fibers originating from the longitudinal muscle bands of the stump extend until they reach the developing terminal caps, while the brain starts differentiating at the anterior end.

Stage 4, resegmentation, is when segmental units start to be distinguishable at the segmental region of the blastema, and primordia of segmental organs and structures appear. At this stage, the brain completes differentiating, and new ventral nerve cord ganglia form. Fibers of circular muscle form fine rings between the epidermis and the longitudinal muscle.

Stage 5, growth, involves finishing differentiation of all regenerated structures, and growth of the regenerated tissues to reach the adult proportions relative to the stump.

Key processes occurring along these five stages are a wound-healing response, cell migration, cell proliferation, formation of a regeneration blastema, neural regeneration, muscle regeneration, and segmentation. These processes are common to most species able to regenerate. However, details and relative timing vary across species. Furthermore, the aforementioned staging system is based on convenient median landmarks rather than clear developmental boundaries. As a result, many of these processes span two or more stages. Since processes are more comparable across species than timing, further discussion focuses on describing processes rather than stages.

10.4.2 Wound Healing

In annelids, wound healing is elicited by nonlethal amputation across the main body axis. It consists of an initial fast wound-sealing response followed by a slower reepithelialization process.

Wound sealing is achieved by rapid contraction of the circular muscles proximal to the amputation plane, which brings together the edge of the wound while closing off the exchange between the external and internal environments (Bely 2014). Sealing of the wound is sometimes aided by extrusion of the gut tissues. This initial muscular response is complemented by the migration of coelomic cells to the wound site to form a mass known as a wound plug (Stein and Cooper 1983).

Wound sealing is followed by reepithelialization: the severed edges of the epidermis fuse together over the wound, forming a continuous epithelial sheet. Internally, the severed gut edges also fuse, forming a blind end. In some cases, the edges of the gut and body wall might fuse instead, sealing the wound and restoring the gut outlet in the same step—as seen during posterior "open regeneration" in nereids (Kostyuchenko et al. 2019). The newly formed outer epithelial layer is usually called wound epithelium and is different in histological and ultrastructural details, including a lack of basement membrane and a modified extracellular matrix that may play key roles in triggering cell dedifferentiation and proliferation in adjacent tissues (Bely 2014). Although more studies specifically addressing the role of the wound-healing process in enabling the regenerative response in annelids are needed, it is likely that the wound epithelium and wound extracellular matrix play a pivotal function in regeneration, as found in many other phyla (Brockes and Kumar 2008).

10.4.3 Cell Migration

Wound healing and regeneration are achieved through changes in cell behavior, as cells are the main effectors of all developmental processes required to rebuild the lost structures. While many such processes are studied at the tissue or organ levels, two aspects of regeneration are best approached by looking at the cell level: cell migration (this section) and cell proliferation (next section).

Traditionally, cell migration during annelid regeneration has been studied through indirect observation of individuals fixed at different time points along the regenerative process. These provided static snapshots upon which cell dynamics were statistically inferred based on changes in spatial distribution of specific cell types; direct observation and quantification of cell migration through live imaging have only recently been reported (Zattara, Turlington, and Bely 2016).

Extensive cell migration can be seen within minutes to hours after injury in many annelids; this increased cell activity lasts for about 1 day after amputation and then subsides (Bely 2014). *In vivo* observations of cell migration during the first day of regeneration using time-lapse photomicrography in the clitellate *Pristina leidyi* found a suite of cell morphotypes showing distinctive sets of directional behaviors during the first day of regeneration, and in noninjured control animals (Zattara, Turlington, and Bely 2016). These morphotypes match reports from histological observations of other species, but despite several attempts to classify them, terminology and homology are still unclear, both between and within species (Herlant-Meewis 1964; Stein and Cooper 1983; Baskin 1974; Cornec et al. 1987; Vetvicka and Sima 2009; Bely 2014). A simple classification based on morphology and behavior is given in Table 10.1. These morphotypes may however only reflect transient behavioral states rather than differentiated cell types, as *in vivo* observations often show the presence of intermediate types, or a dynamic switch among morphologies.

Amoebocytes (Figure 10.3A–B) are amoeboid cells that move by extending pseudopodia along and between the surface of tissues like the gut, inner lining of the coelomic cavity, or between muscle and connective layers. They range in size from small (5–8 μm) to quite large (up to 30 μm). They may contain a variable number of cytoplasmic granules and are sometimes classified into granular (or granulocyte; Figure 10.3A) and hyaline (Figure 10.3B) amoebocytes. Amoebocytes most likely play a role in immune responses and in phagocytosis of pathogen invaders, foreign particles, and damaged cells.

Eleocytes (Figure 10.3C) are large (up to 40–60 μm), round cells containing vesicles or granules, similar to those found in the chloragogen cells that line the peritoneal surface of the gut in many annelids. They are present in nearly all annelids (Stein and Cooper 1983). They are believed to play roles in nutrition, excretion, and osmotic balance (Vetvicka and Sima 2009), although functions in immunity and regeneration have also been proposed (Liebmann 1942; Stein and Cooper 1983). Though typically found freely in the peritoneal cavity or clinging to septal or peritoneal tissues, eleocytes can attach to the epithelial linings, adopt a more amoeboid shape, and crawl over that substrate (Figure 10.3C, lower panel).

Sarcolytes (Figure 10.3D) are small- to medium-sized (5–15 μm) hyaline cells that vary from spherical to rod-shaped and are free floating or attached to the peritoneal lining of the body wall. They derive from autolysis of muscle fibers near the wound site (Cornec et al. 1987; Zattara, Turlington, and Bely 2016), and are eventually phagocytized by amoebocytes.

Neoblasts or dissepimentary cells (Figure 10.3E) are small- to medium-sized (5-15 μm) spindle-shaped cells usually found sliding over the ventral nerve cord or along the peritoneal lining. Their role has been long debated (Randolph 1891; Krecker 1923; Stone 1932; Stephan-Dubois 1954; Bilello and Potswald 1974; Cornec et al. 1987; Tadokoro et al. 2006; Myohara 2012; Sugio et al. 2012; Zattara, Turlington, and Bely 2016). These cells are normally found on segmental septa and are characterized as spherical in shape with a large nucleus-to-cytoplasm ratio, suggesting they might not be fully differentiated. After amputation, they are thought to enlarge, adopt a spindle

TABLE 10.1

Cell Morphotypes Observed during Wound Healing and Regeneration

Cell Type	Size Range	Features	Putative Function	Found in	Reference
Amoebocytes, granular/granulocyte	5–30 μm	Small to large, irregular but often amoeboid in shape; few to many granular inclusions	Cellular immunity, phagocytosis	Polychaetes, Clitellates	Baskin 1974; Vetvicka and Sima 2009; Zattara, Turlington, and Bely 2016
Amoebocytes, hyaline	5–30 μm	Small to large, irregular but often amoeboid in shape; no granules	Cellular immunity, phagocytosis	Polychaetes, Clitellates	Baskin 1974; Stein and Cooper 1983; Vetvicka and Sima 2009; Zattara, Turlington, and Bely 2016
Eleocytes	10–60 μm	Large, round, with numerous granular inclusions	Trophic, recycling of phagocytized sarcolytes	Polychaetes, clitellates	Baskin 1974; Vetvicka and Sima 2009; Zattara, Turlington, and Bely 2016
Sarcolytes	5–15 μm	Small to medium, round or ovoid, hyaline; free or clinging from body wall	Free muscle cells, originated from body wall, usually phagocytized by amoebocytes	Polychaetes, clitellates	Baskin 1974; Zattara, Turlington, and Bely 2016
Dissepimentary cells/"neoblasts"	5–15 μm	Small to medium, spindle shaped, no granules; move by sliding over substrate	Progenitors of certain mesodermal structures (which ones is contested)	Clitellates	Randolph 1892; Krecker 1923; Cornec et al. 1987; Zattara, Turlington, and Bely 2016

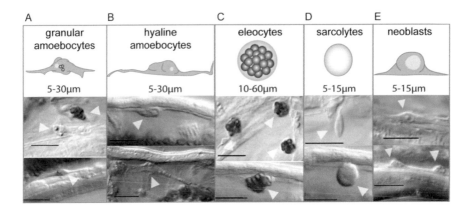

FIGURE 10.3 Free coelomic cell types observed during wound healing and regeneration. The top row shows schematic depictions; middle and bottom row show examples from the clitellate *Pristina leidyi*. Scale bars: 20 μm. Modified after Zattara, Turlington, and Bely 2016.

shape, and move to the dorsal surface of the ventral nerve cord, migrating toward the wound site, and accumulating there. Proposed functions vary from a relatively minor role as the source cells for new peritoneal lining to as far as claiming that they proliferate and differentiate in all kinds of mesodermal, ectodermal, and even endodermal tissue in the regenerate. These inferences are based mostly on static snapshots from histological sections; to date, *in vivo* imaging studies in a single clitellate species have supported presence and directional migration of neoblast-like cells, but gave no information on their origin or eventual fate. In non-clitellate annelids, migration of these cell types has been reported only for short distances, or not at all, indicating that any potential major role of neoblasts in regeneration would be an evolutionary innovation of clitellates, and not an ancestral annelid feature (Hill 1970; Potswald 1972; Paulus and Müller 2006; Bely 2014). Such a scenario is supported by the recent finding that cell migration to the wound site contributes materials to the regenerating tissues on the capitellid *Capitella teleta*, member of a lineage close to the base of the clitellates (Jong and Seaver 2017).

Current data strongly suggest that cell morphotypes are driven by their current function rather than by their cell lineage. Additional *in vivo* studies that combine time-lapse microscopy, cell ablation, and use of drugs inhibiting cell migration, performed across a wider range of species, are needed to better characterize cell types, their behavior, and their roles in wound healing, immunity, and regeneration in annelids.

10.4.4 CELL PROLIFERATION

Cell proliferation patterns after amputation have been characterized by observation of mitotic nuclei in histological sections and more recently by incorporation of thymidine analogs during S-phase and immunodetection of phosphorylated histones and proliferating cell nuclear antigen (PCNA) as markers for cell replication. Post-amputation cell proliferation patterns vary throughout the process of regeneration, also showing differences across species. In most cases, however, cell proliferation is a key step in regeneration, and experimental blockage of proliferation usually abrogates or strongly delays the regenerative process (Bely 2014; Planques et al. 2018).

Studies on several annelid species show that axial amputation triggers a fast and sharp decrease in cell proliferation; this effect is more marked after anterior amputation, and can be weak or absent after posterior amputation (Paulus and Müller 2006; Zattara and Bely 2011, 2013; Jong and Seaver 2016; Planques et al. 2018). When present, this shutdown becomes evident as soon as 30 minutes after amputation and can involve exclusively segments close to the wound site or extend to the whole body. Proliferation remains low until the onset of blastema formation, usually within 1 day after amputation. This proliferation shutdown and delay might be part of a resource allocation strategy aimed to stall current resource sinks (i.e., growth) while early post-amputation preparatory steps, like wound-healing, immune response, and cell dedifferentiation, take place (Zattara and Bely 2013; Bely 2014).

After this shutdown, initiation of the next phase of regeneration, blastema formation, is evidenced by a marked increase in proliferation near the wound site. The earliest proliferative activity is often observed near the severed end of the ventral nerve cord, suggesting that this structure has an important role in initiating blastema formation (Bely 2014). Cell proliferation soon spreads to all tissues, including endodermal, mesodermal, and ectodermal derivatives, adjacent to the wound site. Proliferation usually extends away from the cut site to a variable number of proximal segments, but is more intense in the blastema and developing regenerated structures, supporting that most of the regenerate is derived from proliferation of proximal cells and not from cells that migrate from more distal regions.

10.4.5 CELLULAR SOURCES AND DEVELOPMENT OF THE BLASTEMA

A sharp increase in cell proliferation in tissues adjacent to the wound site indicates the onset of blastema development. Another early indicator of blastema formation is the upregulated expression of several sets of genes strongly associated with the regenerative process. These sets include *Hox* genes; members of the germline multipotency program (GMP) that include annelid orthologues of *piwi*, *vasa*, *nanos*, *PL10*, and *myc*; signaling genes including WNT pathway members; other transcription factors like *even-skipped*, *engrailed*, and paired domain proteins; and RNA-binding proteins like *elav* (Kozin and Kostyuchenko 2015; Özpolat and Bely 2016; Planques et al. 2018). Similar to cell proliferation, earliest gene expression is usually found within the wound epithelium and/or underlying mesoderm, and is first seen at the ventral side at or near to the end of the ventral nerve cord, supporting the hypothesis that the ventral nerve cord plays a key role in blastema initiation and patterning during annelid regeneration (Özpolat and Bely 2016; Planques et al. 2018; Boilly, Boilly-Marer, and Bely 2017).

As cell proliferation continues, a mass of morphologically undifferentiated cells continues to accrue beneath the highly proliferative wound epithelium, forming a clear regeneration blastema. Both direct observation and experiments with thymidine analog markers show that this blastema forms by contribution of ectoderm, mesoderm, and endoderm (Bely 2014; Jong and Seaver 2017; Planques et al. 2018).

The blastema shows *de novo* expression of several GMP genes, strongly suggesting that blastema formation requires dedifferentiation of previously differentiated cells (Özpolat and Bely 2016; Planques et al. 2018). Alternatively, regeneration might be driven by a pool of reserve stem cells present in each tissue that become activated during wound healing. This later hypothesis has weaker support, as it predicts the presence at most tissues of cells showing stem cell–like expression patterns scattered through the body. Such patterns have so far been observed only for *piwi*-expressing cells in a few species and are most likely related to regeneration of the germline (Tadokoro 2018). Thus, current data suggests that the regeneration blastema forms from proliferation of differentiated cells from tissues proximal to the wound site.

A related question that follows is what is the fate of blastemal cells, and whether these undifferentiated cells regain pluripotency or are still restricted to differentiate back into the same tissues from which they arose. Although several authors initially proposed various scenarios in which a few cell sources were able to reconstruct most of the missing tissues, current knowledge does not support extensive transdifferentiation across tissue types and origins (Stephan-Dubois 1954; Cornec et al. 1987; Bely 2014). The epidermis of the regenerate seems to derive from proliferation of cells forming the wound epidermis. The proliferative epidermis also produces cells that become internalized and participate in the regenerate's brain and ventral nerve cord ganglia; at the anteroventral surface, this tissue invaginates to form the mouth opening. The origin of new mesodermal tissues is less well understood, but most data supports that they derive from cells coming from old mesodermal tissues. Finally, gut tissues also appear to derive from proliferation of endodermal cells (Tweeten and Reiner 2012; Planques et al. 2018). Thus, and despite the extensive cell movements reported during annelid regeneration, current evidence paints a scenario where most tissues from the regenerate are derived from nearby stump tissues through dedifferentiation into blastemal cells followed by redifferentiation. In other words, the dedifferentiation process does not seem to confer enough cell pluripotency to allow crossing of embryonic germ layer boundaries.

An interesting observation is the fact that the anteriormost non-segmental region, the prostomium, derives in its entirety from anterodorsal blastemal cells (Zattara 2012). Experiments using iontophoretic labeling of cells with the carbocyanine dye DiI in anteriorly amputated individuals of the clitellate *Pristina leidyi* show that labeled cells located at an anterodorsal region of the early blastema always end up forming part of the prostomium (Figure 10.4A–E). In turn, cells labeled at the anteroventral region of the early blastema are displaced posteriorly and end up in a ventral region posterior to the mouth (Figure 10.4A, F–H). A dorsal origin of the prostomium can also be seen during paratomic reproduction in this species (Bely and Wray 2001; Zattara and Bely 2011). In annelids with indirect development, the prostomial region is derived from the larval episphere, which is also is characterized by expression of the anterior patterning gene *optix/six3* (Steinmetz et al. 2010). Expression of *six3* is also seen early during paratomic reproduction in *Pristina longiseta* in dorsolateral stripes located immediately behind the fission plane (Steinmetz et al. 2010). Thus, both regeneration and fission seem to recapitulate prostomial development by upregulating the same genes that generate this region during embryogenesis; the mouth opening, which forms *de novo* by epidermal

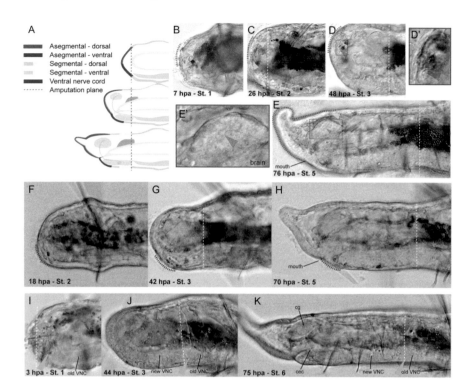

FIGURE 10.4 Cell tracing experiments showing the dorsal origin of the regenerated peristomium in the clitellate *Pristina leidyi* and the participation of neurites from the old VNC in forming the circumenteric connectives and part of the brain. (A) Fate map diagram of the wound epidermis. (B–K) Confocal images of representative experiments. Merge of reflected (red, peak emission of DiI) and transmitted (gray scale, morphology) light channels. Anterior is to the left in all panels. (B–E) DiI labeled cells (yellow arrowheads) at an anterodorsal epidermal patch (green dashed line); an internal labeled cell (red arrowhead) is found at later stages (inset of boxed areas). (B) Stage 1, 7 hpa; (C) Stage 2, 26 hpa; (D) Stage 3, 48 hpa; (D') Detail of boxed area in D; (E) Stage 5, 76 hpa. (E') Detail of boxed area in E. (F–H) DiI labeling of epidermal anteroventral patch (blue dashed line) showing posteroventral displacement. (F) Stage 2, 18 hpa; (G) Stage 3, 42 hpa; (H) Stage 5, 70 hpa. (I–K) DiI labeling of neurons within the ventral nerve cord shows incorporation of neuronal elements from the old cord into the regenerated new CNS. (I) Stage 1, 3 hpa; (J) Stage 3, 44 hpa; (K) Stage 6, 75 hpa. Green arrowhead shows position of the dye injection. VNC, ventral nerve cord; cec, circumenteric connectives; cg, cerebral ganglion. B–K modified after Zattara (2012).

invagination, takes place at the same point where dorsal and ventral tissues had come into contact as a result of wound closure. Another related observation is that in annelids, a dorsal fate seems to be the default state for the body wall, and ventral fate is imposed by interaction with the ventral nerve cord (Boilly, Boilly-Marer, and Bely 2017). This suggests that the timing and spatial extent of neural regeneration (see next section) must be precisely timed to avoid ventralization of the anterodorsal region of the early blastema.

10.4.6 NEURAL REGENERATION

Annelids have a relatively well-conserved nervous system morphology (Figure 10.5A) (Bullock and Horridge 1965; Purschke 2015; Helm et al. 2018). The central nervous system (CNS) is characterized by an anterodorsal cerebral ganglion or brain, and a ventral nerve cord (VNC) that runs along the whole body down to the posterior end. In many groups, the nerve cord has a clearly segmented organization reminiscent of a rope ladder, with iterated sets of transverse connectives linking two longitudinal

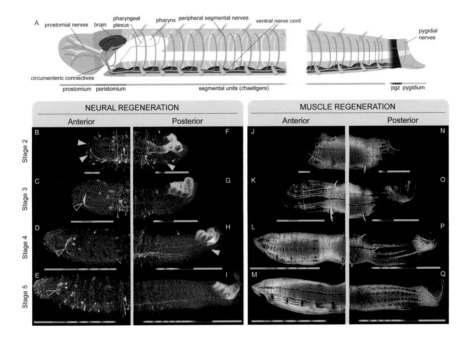

FIGURE 10.5 Annelid nervous system ground pattern, and regeneration of neuronal and muscular components. (A) Diagrammatic representation of a typical annelid nervous system. Brain and ventral nerve cord ganglia are shown in blue, and simplified nerves and neural tracts shown in green. Non-segmental terminal regions shown in shades of brown, and posterior growth zone (pgz) shown in graded dark red. (B–I) Neural development during anterior and posterior regeneration in the naidid clitellate *Dero (Aulophorus) furcata*. Maximum intensity projection of confocal laser scanning microscopy (CLSM) Z-stacks of acetylated alpha-tubulin immunoreactive structures (axons of the nervous system, gut ciliation, and nephridia; green), serotonin immunoreactive structures (central nervous system axons and perykaria; yellow), and DAPI as a nuclear counterstain (DNA; blue). Filled arrowheads show neurites extended from the old ventral nerve cord; empty arrowheads show the anterior neural loop that becomes the new circumenteric connectives; arrows point at axonal extensions of the peripheral nerves over the blastema. (J–Q) Muscle development during anterior and posterior regeneration in the naidid clitellate *Dero (Aulophorus) furcata*. Color-coded projections of CLSM Z-stacks of Alexa-Fluor 488 phalloidin showing muscle development. In B–K, color-coded bars indicate new tissues: brown, prostomium and peristomium; orange, pygidium; graded dark red, new posterior growth zone; solid green, undifferentiated tissue; striped green, developing segments. A modified after Zattara and Bely (2015); B–Q modified after Zattara (2012).

sets of neurites. The number and arrangement of longitudinal connectives that form the neuropil of the VNC varies across annelid groups, but it might include an unpaired median connective running along the ventral midline, inner paired paramedian connectives, and outer paired main connectives (Müller 2006). The dorsal brain and ventral cord are connected by paired sets of circumenteric connectives that circumvent the foregut. Connectives are usually seen as single bundles of nerves, but sometimes they split before reaching the brain in two roots, one ventral and one dorsal. At the anterior end, a variable number of prostomial nerves branch off the circumenteric connectives and arborize to innervate the sensory organs at the front of the worm. Similarly, peripheral nerves branch off at the posterior end of the worm to innervate the pygidial structures. Within the segmental region, bilateral pairs of peripheral segmental nerves branch off perpendicularly from the VNC, go through the body wall, and then extend dorsally, sometimes reaching the dorsal midline and forming transverse rings. These peripheral nerves usually innervate the parapodia and numerous epidermal sensory structures. Peripheral nerves are segmentally iterated, although some groups show a reduction in the number of nerves per segment at the anteriormost segments; furthermore, these patterns are conserved within taxonomic groups but vary across the phylum (Zattara and Bely 2015; Purschke 2015).

Early studies on the regeneration of the annelid nervous system were based on microscope observation of histological sections labeled using traditional staining techniques; newer studies have taken advantage of immunohistochemical detection of specific molecular components of neurons, such as acetylated α-tubulin, serotonin, and FMRF-amide-like peptides, using fluorescent tags and laser scanning confocal microscopy (Purschke 2015).

After amputation, the process of regeneration needs to restore the nervous structures at the new terminal regions and new intercalated segments, including a new stretch of ventral nerve cord and terminal and segmental peripheral nerves. Anterior regeneration additionally implies growing a new brain. In most species where the process has been studied, neural regeneration is first evidenced by early invasion of the wound site by neurites originating from the old ventral nerve cord (Figure 10.5B and F, filled arrowheads). At least some of these neurites are axonal extensions of existing neurons: labeling experiments using the cell membrane–diffusible tracer DiI in the clitellate *Pristina leidyi* show that axons from neurons labeled at the nerve cord ganglion proximal to the wound site invade the blastema and even become part of the circumenteric connectives linking to the regenerated brain (Figure 10.4I–K) (Zattara 2012).

During anterior regeneration, neurites extending toward the anterior end from the amputated nerve cord fuse in a series of loops (Figure 10.5C, empty arrowhead). In most species, paired sets of neurites extend from the paramedian and main connectives after wound healing (Yoshida-Noro et al. 2000; Müller, Berenzen, and Westheide 2003; Müller and Henning 2004; Müller 2004; Zattara 2012; Weidhase, Bleidorn, and Helm 2014; Weidhase, Helm, and Bleidorn 2015; Weidhase et al. 2017). As the anterior blastema forms, each set merges at the anterior tip forming a loop; the medial loop formed by paramedian neurites usually forms first, while the lateral loop formed by main neurites closes later. In the only two species of cirratulids studied so far, *Cirratulus cf. cirratus* and *Timarete cf. punctata*, each main neurite extension instead forms a separate lateral loop, resulting in a trefoil-like structure; lateral loops later fuse with the medial loop. Some species also extend neurites from an unpaired median connective (Müller and Henning 2004; Weidhase et al. 2017; Kozin, Filippova, and Kostyuchenko 2017), but they do not participate in the distal loops, perhaps due to the suggested role of the median nerve in innervating the longitudinal musculature (Müller 2006). At this stage, anterodorsal epidermal cells from the prospective prosto-mium become internalized and congregate around the distal end of the loops, forming the brain anlagen (Figure 10.5D and E, asterisk), while the lateral sides of the loops fuse forming the circumenteric connectives (Figure 10.5D and E, empty arrowhead). In groups where the circumenteric connectives split in ventral and dorsal roots, neurites in the ventral root derive from the medial loop, while neurites in the dorsal root derive from the lateral side of the loop (or lateral loops, in cirratulids). The nerve plexus that innervates foregut structures of many annelids also develops around this time. Development of the prostomial/pygidial CNS precedes development of the segmental features of the ventral nerve cord; these features, which include nerve cord ganglia and peripheral nerves, form as the blastemal tissues that are intercalated between the regenerated peristomium and the anterior boundary of the old tissue differentiate into segments (Özpolat and Bely 2016; Weidhase et al. 2017). The most likely source of blastemal cells fated to form the ganglia is ingression of cells from the lateroventral epidermis, but this cell origin still needs to be experimentally verified through cell tracing or live time-lapse microscopy.

During posterior regeneration, neurites initially invade the distal end branch and develop into the characteristic innervation of the rudiments of pygidial structures (Figure 10.5H, filled arrowheads); thus, differences across species reflect differences in pygidial morphology. In general, regeneration of pygidial neural structures precedes development of species-specific segmental features of the nervous system, which occurs at later stages, after the formation of a new segment addition zone between the regenerated pygidium and the old tissue. These segmental features, which include segmental ganglia of the nerve cord and associated peripheral nerves, develop in the same way as during normal posterior growth; the rate of segment addition, however, is usually much higher during regeneration than during growth.

In many annelids, peripheral nerves from the proximal segment have been reported to extend numerous subepidermal axonal extensions toward the wound site and form a neural plexus over the developing blastema (Figure 10.5B–D, F–H, filled arrowheads). In contrast to neurites extending from the ventral cord, these axons are seen only during the regeneration process and disappear during its later stages. While neurites extending from the VNC have a clear role in building the new CNS, the function of the temporary neural plexus formed by neurite outgrowth from peripheral nerves is unknown. Interestingly, earlier studies found this plexus only in clitellate annelids and concluded that its presence was a clitellate-specific regeneration trait (Müller 2004; Zattara and Bely 2011). However, more recent studies have since described a similar phenomenon for a non-clitellate Sedentaria, *Capitella teleta* (Jong and Seaver 2016, 2017) and two Errantia, *Platynereis dumerilii* and *Alitta virens* (Kozin, Filippova, and Kostyuchenko 2017; Planques et al. 2018). Although variations at a fine taxonomic scale cannot be discarded, it is also possible for this trait to have been overlooked by earlier studies. Indeed, most reports that do not find formation of a peripheral plexus also fail to explicitly state its absence (Özpolat and Bely 2016). Furthermore, peripheral nerve neuronal extensions are usually much finer and harder to detect than outgrowths from the VNC. Thus, it is quite possible that variation in the reports of the presence of a transient peripheral plexus during annelid regeneration results more from observation bias than actual phylogenetic variability of this trait (Kozin, Filippova, and Kostyuchenko 2017).

Two alternative, non-mutually exclusive hypotheses have been proposed for the role of the transient peripheral nerve plexus (Jong and Seaver 2017): one states that the plexus promotes cell proliferation and patterning of the blastema (Herlant-Meewis 1964; Müller, Berenzen, and Westheide 2003; Varhalmi et al. 2008); the other proposes that the axons of the plexus serve as cell migration tracks (Stephan-Dubois 1954; Cornec et al. 1987). Although data are still inconclusive, the relative timing of plexus formation and cell proliferation during regeneration supports a role in promoting proliferation (Jong and Seaver 2017). In contrast, the only available study tracking cell migration *in vivo* shows most cell migration taking place in the coelomic cavity but rarely near the surface of the body wall, which is where most of the neurites forming the plexus are located (Zattara, Turlington, and Bely 2016). Nonetheless, future experimental studies combining live cell labeling and tracing of both migrating cells and neural components with laser ablation of either cells or neurites might shed light on the function of the peripheral nerve plexus.

Expression of genes related to neural fate determination and neurogenesis has been described only for posterior regeneration in *Platynereis dumerilii* (Planques et al. 2018). The pro-neural gene *elav,* encoding an RNA-binding protein that promotes neuronal fates in metazoans (Pascale, Amadio, and Quattrone 2007), is expressed early on at the ventralmost part of the wound epithelium, next to the cut end of the ventral nerve cord. During blastema formation, its expression domain extends to many ventral cells, and splits into a pygidial domain and a segmental domain, separated by the newly formed segment addition zone. *Neurogenin* (*ngn*), another regulator of neuronal differentiation (Seo et al. 2007), is first expressed during the wound-healing stage at paired sets of cells located on the lateral body wall of the stump, posterior to the parapodia. After blastema formation, expression expands to more ventral cells located at the developing pygidium and segments. Pygidial expression disappears at later stages but remains at the posterior growth zone. Early expression of the axonal guidance gene *slit* has a salt-and-pepper pattern, but once new segments begin to differentiate, it is expressed at the ventral midline and as ventrolateral stripes. Expression of the neurogenic transcription factor *pax6*, which precedes *ngn, slit*, and *elav* during *Platynereis* larval development (Denes et al. 2007), is not observed at early stages of regeneration but is later seen as two longitudinal ventrolateral bands at the posterior growth zone. Neurogenic gene expression patterns thus support that despite some shifts in the order of gene expression, neurogenesis proceeds through the same processes in regeneration and embryonic/larval development.

10.4.7 MUSCLE REGENERATION

The most important and conspicuous muscles in the general annelid body plan are those of the body wall (Tzetlin and Filippova 2005; Purschke and Müller 2006). Body wall musculature comprises an outer layer of circular muscles and an inner layer of longitudinal muscle (Figure 10.2C). The circular muscles are located below the basement membrane of the epidermis and usually form finer bands with fewer fibers than longitudinal muscles. They form complete or incomplete fine rings in planes normal to the main anteroposterior body axis and might be completely absent in many species. The longitudinal muscles usually form discrete bundles or bands that run parallel to the main body axis along the complete body length. The number and arrangement of longitudinal bands varies across groups and can go from as few as four to six separate bands, to several bands forming a continuous layer. In some groups, an additional network of oblique muscle fibers can be seen located between the epidermis and the circular muscle layer (Figure 10.5J–Q).

Traditionally, most descriptions of annelid muscular systems resulted from dissections, histological sectioning, and transmission electron microscopy; more recently, the combination of labeling F-actin with fluorescently labeled phalloidin, laser scanning confocal microscopy, and three-dimensional reconstruction has allowed more sensitive detection of finer muscle fibers and a better understanding of annelid muscular structure (Purschke and Müller 2006).

Regeneration of muscles has been studied in several species of annelids and is similar across the phylum (Zattara and Bely 2011; Zattara 2012; Weidhase, Helm, and Bleidorn 2015; Weidhase, Bleidorn, and Helm 2014; Weidhase, Bleidorn, et al. 2016; Weidhase, Beckers, et al. 2016; Kozin, Filippova, and Kostyuchenko 2017). Immediately after amputation, circular muscles adjacent to the wound site contract to seal the wound. Later, during the initial wound-healing stage of regeneration, these circular muscles are removed, along with the ends of the longitudinal muscle fibers adjacent to the wound site. During this stage, dead or injured myocytes are phagocytized, while many surviving myocytes dedifferentiate and move out of the body wall, resulting in a muscle-free distal end of the stump (Figure 10.5J). At the early blastema stage, muscle fibers originating from the old longitudinal muscle bundles extend distally over the growing blastema (Figure 10.5J–O); in contrast, circular muscle fibers seemingly form *de novo*, presumably from blastemal cells (Figure 10.5K–Q). In all reported cases, development of circular muscles of the newly forming segments takes place at slightly later stages than development of longitudinal muscles (but see later regarding pygidial muscles). Circular muscle shows an anterior-to-posterior gradient that is very evident during posterior regeneration but can also be seen, albeit more subtly, during anterior regeneration. In contrast, longitudinal muscles develop from anterior to posterior during posterior regeneration but show the opposite gradient during anterior regeneration. Thus, whereas circular muscle develops *de novo* following the anteroposterior developmental gradient of new segments (see Section 10.4.7), longitudinal muscle instead develops by extension from fibers from the stump and thus follows a proximodistal gradient.

During posterior regeneration, development of pygidial tissues precedes differentiation of segmental tissues. As a result, in species with prominent or elaborate pygidial structures (like the nereid *Alitta virens* or several naidid clitellates), a variably thick ring of circular muscle forms at the posterior end of the blastema, often before the growing fibers of longitudinal muscle reach that tip (Zattara 2012; Kozin, Filippova, and Kostyuchenko 2017) (Figure 10.5N). Additional circular muscle develops over the prospective pygidium into a dense layer that has a sharp anterior boundary where the segment addition zone develops (Figure 10.5O and P). Anterior to that zone, circular muscle develops following the same anterior-to-posterior gradient seen during normal growth (Figure 10.5P and Q).

During anterior regeneration, in contrast, early development of prostomial or peristomial circular muscle has not been reported. Muscle fibers extending from the old longitudinal muscle bands reach the distal tip of the blastema, often branching and crisscrossing over it (Figure 10.5J and K). Circular muscle develops soon after, completing the prostomial and peristomial musculature, including the mouth and other anterior structures (Figure 10.5L and M). A transient circular muscle ring forming earlier at the prospective mouth has been reported for the amphinomid *Eurythoe* cf. *complanata* (Weidhase, Bleidorn, et al. 2016). This ring might be a feature unique to *Eurythoe* regeneration; alternatively, a homologous but finer, fainter structure might also be present in other annelids but have gone undetected in previous studies. When labeling F-actin with fluorescently tagged phalloidin, it is common to find a large difference in signal intensity between old longitudinal muscle bundles and newly developing circular muscle fibers. This generates a challenging trade-off when setting dynamic range parameters during imaging, as increasing sensitivity enough to detect the finer fibers often results in signal saturation of the thicker bundles from the old tissues. Thus, future studies should scan for the presence of such fine structures and explicitly report their absence if not found.

The expression of genetic markers of muscle development has so far been only studied during posterior regeneration in the nereid *Platynereis dumerilii* (Planques et al. 2018). In this species, weak expression of a *twist* homologue is initially detected after wound healing in scattered cells near the wound site. During early blastema formation, *twist* becomes strongly expressed in a ring of mesodermal cells located where the pygidial circular muscles will form (see earlier). About a day later, *twist* becomes also expressed at bilateral patches of mesodermal cells located anterior to the newly formed segment addition zone. This expression precedes formation of circular muscle in the new segments, and it expands as bands as these segments grow and develop. The fact that *twist* is not expressed in the segment addition zone located between the pygidium and developing segments strongly suggests that cell differentiation does not take place until myocyte precursors leave this zone.

10.4.8 SEGMENTATION

In annelid species with indirect development—likely the ancestral mode for the phylum—embryonic development results in an unsegmented trochophore larva with a subterminal growth zone; segmental units form in front of this subterminal growth zone (see Chapter 4). As the larva metamorphoses into a juvenile, the growth zone keeps adding segments, resulting in a body plan that consists of a variable number of segmental units capped by two terminal regions. The anterior region derives from the larval tissues located anterior to the growth zone and comprises the prostomium and the peristomium. The posterior region derives from larval tissues located posterior to the growth zone and comprises the pygidium. Due to their larval origin, prostomium, peristomium, and pygidium are usually considered to be non-segmental in nature (but see Starunov et al., 2015, for a differing view).

During axial regeneration, tissues from both the terminal cap and segments need to be reconstructed. In general, the terminal caps develop first from the regeneration blastema, and then segments arise and differentiate between the cap and the stump (Figure 10.2). A pattern of initial development of the distal portions of a regenerating structure, followed by intercalation of remaining structures, is seen during regeneration in many other systems outside annelids, and might reflect a common pattern of regenerative processes throughout animals (Agata, Saito, and Nakajima 2007).

Studies of intercalary regeneration of segments suggest this process is developmentally very similar to the addition of segments during normal growth (Gazave et al. 2013; Balavoine 2015; Özpolat and Bely 2016). However, the rate of segment addition is usually much faster during regeneration, at least initially (de Rosa, Prud'homme, and Balavoine 2005; Niwa et al. 2013). The rate and extent of segmental regeneration varies widely across species (Berrill 1952). Marked differences in rate are also seen between posterior and anterior regeneration. In general, segments built during posterior regeneration show a clear developmental gradient where older segments are more anterior and farther away from the growth zone. During anterior regeneration, a similar gradient is seen clearly in some species (Allen 1923; Zattara and Bely 2011; Aguado et al. 2014; Weidhase, Helm, and Bleidorn 2015; Ribeiro, Bleidorn, and Aguado 2018); however, in other species, simultaneous development of all segments has been reported, leading to the suggestion that segment formation during posterior and anterior regeneration are fundamentally different processes (Balavoine 2015). Although near-simultaneous segment development can be explained by fast segment formation rates and relatively low numbers of anteriorly regenerated segments, a more accurate test of whether the process of segment addition differs between posterior and anterior regeneration will come from comparative studies of gene expression. Unfortunately, most species in which gene expression has been well characterized during regeneration are not capable of anterior regeneration (Kozin and Kostyuchenko 2015; Niwa et al. 2013; Zattara and Bely 2016; Jong and Seaver 2016; Planques et al. 2018). Thus, the answer will have to wait until gene expression studies are conducted on species capable of both anterior and posterior regeneration.

10.4.8.1 Segmentation during Posterior Regeneration

Segmentation during posterior regeneration has been studied in many species and almost always follows the same general pattern: after the posterior end of the blastema differentiates into the pygidium, morphogenesis of segmentally iterated structures (septa, parapodia, chaetoblasts, ventral nerve cord ganglia, nephridia, etc.) becomes evident at the proximal regions of the blastema, showing a clear anteroposterior developmental gradient. Cell proliferation, which is initially high throughout the entirety of the early blastema, is greatly diminished at the proximal region of the newly formed pygidium but continues at high levels in the newly formed growth zone immediately anterior to the pygidium. Developing segments also show high levels of proliferation until they reach a size and developmental stage similar to that of the segment adjacent to the regenerate. Both cell proliferation and rate of segment formation gradually slow down, until the process can no longer be distinguished from normal posterior growth.

Gene expression patterns during posterior regeneration have been studied in several species. The more complete descriptions so far belong to the errant nereids *Platynereis dumerilii* and *Alitta virens*, and the sedentary capitellid *Capitella teleta*. More circumscribed data are available from other polychaete and clitellate species. With the sole exception of a novel gene discovered in the clitellate *Enchytraeus*, shown through RNA interference experiments to be necessary for regeneration (Takeo, Yoshida-Noro, and Tochinai 2010), empirical verification of gene function is still lacking. Thus, in most cases, putative gene functions have usually been inferred from temporal and spatial patterns of expression assessed by *in situ* mRNA hybridization assays.

While expression of any given gene might vary across species, several general patterns hold for most species studied so far. Soon after the initial fast wound-healing response is complete, genes associated with patterning the posterior end of bilaterians (*caudal/cdx*, *wingless/wnt1*, *brachyury/bra*) are expressed at the wound epithelium (Figure 10.6) (Özpolat and Bely 2016; Planques et al. 2018). Other genes expressed at this tissue include *even-skipped/evx*, *engrailed/en*, *elav*, *hox2*, and *hox3* (Figure 10.7). Soon after, GMP genes associated with maintenance of stem cells and germ cell are upregulated, preceding (and likely promoting) blastema formation (Figure 10.6) (Özpolat and Bely 2016; Planques et al. 2018). These genes include homologues of *piwi*, *vasa*, *nanos*, *PL10*, and *myc*, which are expressed *de novo* in the wound epithelium and/or the underlying mesoderm, and are thought to promote and sustain local cell dedifferentiation during blastema development. At this point, the blastema shows a tripartite pattern of gene expression: a posterior region that corresponds to non-segmental tissues, an anterior region expressing *evx*, *en*, *hox2*, and *hox3,* and a narrow ring in between where *evx/hox2/hox3* expression overlaps with *cdx*. The former region gives rise to the pygidium. The latter two regions form the posterior growth zone (PGZ).

GMP genes remain active at the blastema throughout regeneration but become excluded from the pygidium once this terminal structure begins its differentiation (Figure 10.6). Indeed, differentiation of the pygidium is marked by downregulation of most patterning genes found elsewhere in the blastema, except for the posterior markers *cdx*, *bra*, and *wnt1*, and genes directing development of pygidial muscles (*twist*) and innervation (*neurogenin/ngn*, *elav*, *hox1*, *hox4*, *post2*, *post1*). In species with pygidial appendages, like caudal cirri, their development is marked by expression of *distalless/dlx*, which promotes appendage outgrowth (Figure 10.6). In general, pygidial gene expression patterns during and after regeneration are markedly different from the region anterior to it, supporting the existence of a fundamental difference between the non-segmental pygidium and the adjacent segmental tissues.

As the blastema grows, the expression domains of *hox2*, *hox3*, and *evx* become restricted to a subterminal narrow ring of endodermal and mesodermal cells adjacent to the anterior boundary of the developing pygidium (Figure 10.6). This ring

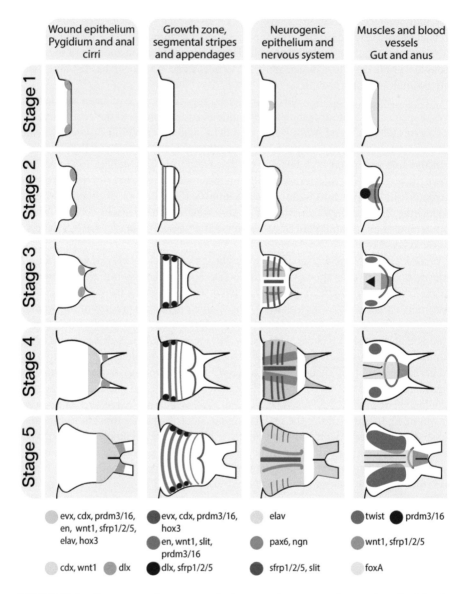

FIGURE 10.6 Gene expression patterns during posterior regeneration, as described for the errant nereid *Platynereis dumerilii*. Modified after Planques et al. (2018) and Kostyuchenko et al. (2019).

becomes the segment addition zone (SAZ), a region showing intense gene expression and cell proliferation that generates cells destined to form segmental tissues. The SAZ behaves as a stem cell niche, and its constituent cells have been proposed to act like the conspicuous stem cells (known as teloblasts) that give rise to segmental tissues during embryonic development in direct developing clitellates like *Tubifex tubifex* and *Helobdella robusta* (see Chapters 4 and 7) (Gazave et al. 2013;

Balavoine 2015). While most studies have failed to find sets of clearly larger cells showing asymmetric division within the SAZ in any adult or regenerating annelid, several lines of evidence (including cyclic, coordinated, and oriented division; lower mitotic rate; cyclic gene expression; differential methylation pattern; large nucleus-to-cytoplasm ratio) suggest that cells within the SAZ act like stem cells (Gazave et al. 2013; Niwa et al. 2013; Jong and Seaver 2017; Planques et al. 2018). The hypothesis that the subterminal region adjacent to the pygidium represents a stem cell niche is also supported by data on GMP genes, which shows a strong expression gradient with peak intensity at the SAZ, tapering anteriorly and presenting a sharp posterior boundary adjacent to the pygidial tissues.

As the SAZ proliferates and intercalates new cells, older cells are displaced anteriorly. These latter cells continue proliferating and begin to differentiate into young segmental tissues. This results in the characteristic developmental gradient seen in the PGZ of both posterior regenerates and growing adults. Exactly how segmentation of the nascent tissues is achieved is still unresolved, but experiments on the nereid *Perinereis nuntia* suggest a boundary-driven segmentation process (see Chapter 4) in which segment boundaries form through a combination of cell cycle synchronization and cell row addition, mediated by the interaction between *wnt1* and *hedgehog/hh* (Niwa et al. 2013). According to this model, cells at the posteriormost row of the SAZ are located in a zone of cell-cycle synchronization (ZCS; Figure 10.7). Cells at the ZCS enter cyclic mitosis with their mitotic spindles always oriented parallel to the body axis. As a nascent row leaves the ZCS, *wnt1* is expressed and a WNT1 gradient forms, inducing expression of *hh* in the younger row behind it and setting the future segmental boundary (Figure 10.7A–D). This WNT1 gradient also coordinates cell-cycle entry of the next rows and inhibits them from expressing *wnt1*. As the ZCS keeps adding rows, WNT1 levels fall below a threshold (Figure 10.7E–M), and *wnt1* is expressed in the newest row, which becomes the posterior boundary of the prospective segment and resets the cycle (Figure 10.7N–P). Thus, the number of cell rows incorporated into each nascent segment depends on the strength of *wnt1* expression, a finding supported by the fact that treatment of regenerating *P. nuntia* with lithium chloride results in an abnormally large number of rows being incorporated into each segment (Niwa et al. 2013). Lithium chloride inhibits GSK3beta, an enzyme that phosphorylates β-catenin, the key intracellular transducer of Wnt signaling; inhibition of GSK3beta thus mimics elevated canonical Wnt signaling. Available data on gene expression and cell proliferation patterns during posterior regeneration in *Platynereis dumerilii*, *Alitta virens*, and *Capitella teleta* are also compatible with this model of segment addition (Novikova et al. 2013; Jong and Seaver 2016; Planques et al. 2018).

Besides *wnt1* and *hh*, several other genes are expressed in stripes at the PGZ (Figure 10.6), including *engrailed*, members of the Wnt signaling pathways (*frizzled*, *sfrp1/2/5*, *tcf*), the axonal guidance protein *slit*, and the *prdm3/16* transcription factors (Niwa et al. 2013; Planques et al. 2018). Their roles have not yet been assessed experimentally, but they are likely to participate in the anteroposterior patterning of each segment. Segmentally iterated expression of other genes (including *hox* cluster members) associated with neurogenesis and myogenesis have also been described

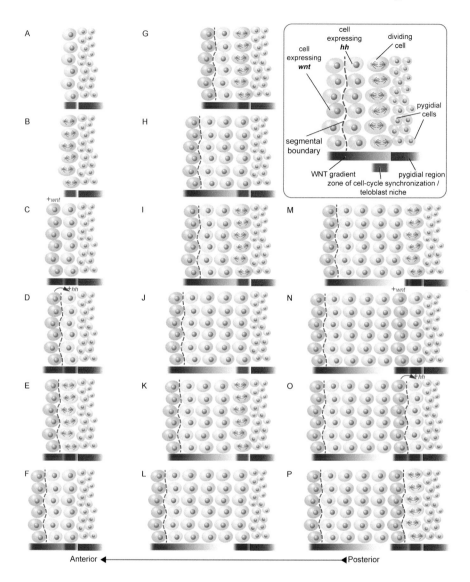

FIGURE 10.7   Cellular model of segment addition and intersegmental boundary specification. Schematic representation of cell organization at the segment addition zone and neighboring pygidial cells. Anterior is to the left. The purple and gray bars represent fixed regions specifying the non-segmental pygidial region and a subterminal zone of cell-cycle coordination. The graded red bar represents a posteriorly oriented gradient of WNT protein generated from stripes of cells expressing *wnt* (red cells). The red arrows represent initial induction of *hedgehog* expression by WNT (green cells), and dashed lines represent the intersegmental boundary. Cells with mitotic spindles represent rounds of polarized cell division. See text for details.

(see earlier). There are several genes whose expression has been reported during larval growth and normal posterior growth but their expression during regeneration remains unassessed (Prud'homme et al. 2003; de Rosa, Prud'homme, and Balavoine 2005; Seaver and Kaneshige 2006; Dray et al. 2010; Gazave, Guillou, and Balavoine 2014; Balavoine 2015). Considering the similarities seen in the segment addition process between normal posterior growth and posterior regeneration, it is highly likely that those genes also show similar expression patterns.

10.4.8.2 Segmentation during Anterior Regeneration

In contrast to posterior regeneration, segment addition during anterior regeneration is more limited. All anteriorly regenerating species regenerate the anterior non-segmental regions (prostomium and peristomium). Most species also regenerate one or more segments: exactly how many varies according to each species, the amount of anterior segments amputated, and often also with the position of the cut along the main body axis (Berrill 1952). In many cases, fewer anterior segments regenerate than the number removed by amputation, generating a positional mismatch at the boundary between the regenerate and the stump and triggering a morphallactic shift of the segmental identities of the stump segments to restore the missing axial identities (see later). Development of regenerated anterior segments is either simultaneous or shows an anteroposterior gradient, suggesting the transient formation of a growth zone at the posterior margin of the blastema, where it joins the stump (Allen 1923; Berrill 1952).

Reports of gene expression patterns associated with formation of segmental tissues during anterior regeneration are much sparser than for posterior regeneration (Özpolat and Bely 2016). Segmental tissue precursor cells share the blastema-wide expression of GMP genes and Wnt pathway genes (Bely and Wray 2001; Tadokoro et al. 2006; Nyberg et al. 2012; Özpolat et al. 2016). Interestingly, in the clitellate *Pristina leidyi*, overactivation of the Wnt pathway during anterior regeneration by treatment with the GSK3beta inhibitor azakenpaullone results in formation of fewer, larger, and less developed anterior segments (Balarezo, Zattara, and Bely 2011). This result is similar to that reported during posterior regeneration in the nereid *Perinereis nuntia* (Niwa et al. 2013) and hints that the role of Wnt signaling in the segmentation clock might be common to most if not all annelids.

Out of the several genes putatively associated with the segmentation process in nereids, only *engrailed* expression has been characterized during anterior regeneration, in the clitellate *Pristina leidyi*, where it shows a salt-and-pepper pattern in likely ventral nerve cord cells of the blastema (Bely and Wray 2001). However, this same pattern was found for posterior regeneration, suggesting this gene might not play the same role in *Pristina* and nereids. Given the scarcity of molecular data of anteriorly regenerating species, it is currently not possible to adequately assess the commonalities and differences in gene expression between posterior and anterior regeneration. This knowledge gap will have to be addressed if we are to gain a better understanding of segment formation during anterior regeneration, and gain insight into why anterior regeneration has shown to be much more labile than posterior regeneration throughout annelid evolutionary history (Bely 2010; Zattara and Bely 2016).

10.4.9 MORPHALLACTIC PROCESSES

Although the generic annelid body plan is built upon relatively similar segmental units, many groups show regional differentiation along the main body axis. Some tube-building species like chaetopterids or sabellids have strongly marked differences between different sets of segments, forming regions termed "tagmata," a term borrowed from arthropod morphology. But more subtle differences in gut regionalization, nervous system organization and gonadal distribution can be found even in groups showing much more segmental homogeneity (Takeo, Yoshida-Noro, and Tochinai 2008; Zattara and Bely 2015; Jong and Seaver 2016). The process of regeneration restores the non-segmental terminal regions and a limited number of segments. When the regenerated segments are fewer than those removed by amputation, a positional mismatch between the proximal new segment and the stump results. This mismatch is usually solved by remodeling the old tissue into the morphology that corresponds to its new positional identity, a process known as morphallaxis.

Morphallaxis ranges from striking changes in overall morphology seen during the transformation of abdominal segments to thoracic segments in sabellids (Berrill 1978), to quite subtle changes associated with remodeling the gut in the clitellates *Pristina leidyi* and *Enchytraeus japonensis* (Takeo, Yoshida-Noro, and Tochinai 2008; Zattara and Bely 2011) or the ventral nerve cord in *Lumbriculus variegatus* (Drewes and Fourtner 1991; Martinez, Menger III, and Zoran 2005; Martinez, Reddy, and Zoran 2006).

In the sabellid *Sabella pavonina*, the anterior 5–11 thoracic segments have dorsal bristles and ventral hooks, while the remaining abdominal segments have ventral bristles and dorsal hooks. Upon amputation at the abdominal region and formation of an anterior blastema, segments closest to the wound site lose their bristles and hooks and redevelop them in reversed position 2 or 3 days later. This morphallactic process proceeds in anteroposterior steps and finishes in the last affected segment about the time that anterior regeneration is complete (Berrill and Mees 1936).

In the clitellate *Pristina leidyi*, the gut forms a stomach-like bulge at segment 7. After amputation at a position posterior to that position, anterior regeneration restores only four segments; thus, during anterior blastema formation and development, the gut within the three segments adjacent to the wound site loses cilia and adopts a stomach-like morphology at the third segment from the original amputation plane (Zattara and Bely 2011). A similar process occurs in another clitellate, *Enchytraeus japonicus*; in this species, expression of three genes (α-*tubulin*, *mino*, and *horu*) localizes to specific regions of the gut, and amputation causes initial downregulation of all three genes, followed by upregulation that reestablishes the normal expression pattern in each regenerating fragment, and consequent morphallactic changes in the gut morphology (Takeo, Yoshida-Noro, and Tochinai 2008).

Morphallactic changes in the gut, blood vessels, and nephridia have also been reported for the clitellate *Lumbriculus variegatus* (Berrill 1952). In this species, morphallaxis of the ventral nerve cord at both functional and molecular levels has been reported. Electrophysiological measurements following amputation have characterized rapid shifts in the anteroposterior range of the sensory fields mediating fast escape responses (Drewes and Fourtner 1991). Regeneration also induces changes in the expression of neural proteins associated with shifts in positional identities and neurobehavioral plasticity (Martinez, Menger III, and Zoran 2005).

Morphallactic adjustments of the ventral nerve cord have also been evidenced by reorganization of the expression domains of *Hox* genes during posterior regeneration in *Alitta virens* and *Capitella teleta* (Novikova et al. 2013; Jong and Seaver 2016). In both species, amputation of posterior fragments causes a forward shift in the anterior expression boundary of *lox4*, *lox2*, and *Post2*. While in *Alitta* several other *hox* genes also adjust their expression domains, in *Capitella* most *Hox* gene expression remains stable after amputation. This might be explained by different deployment strategies for these genes among the two species: in nereids (*Alitta* and *Platynereis*) *Hox* genes are expressed in overlapping domains with variable anterior boundaries and a common posterior boundary located at the posterior growth zone, whereas in *Capitella* most of these genes have overlapping expression domains comprising the anterior, thoracic segments, with staggered anterior and posterior boundaries (Jong and Seaver 2016). Such differences in how *Hox* genes are used to specify body regions within the same body plan highlight the concept that developmental genes belong to a toolkit, and strategies for their deployment, while usually conserved, are not fundamentally constrained and can evolve in different directions.

10.5 CONCLUDING REMARKS

Regenerative ability in annelids is highly variable, but many groups are capable of reconstructing extensive regions of the body lost to injury, and mapping these traits to molecular phylogenies indicate that such regenerative powers are ancestral to the phylum (Zattara and Bely 2016). The process of axial regeneration in annelids presents many similarities to regeneration in other phyla: after wounding and wound healing, the wound epithelium initiates a reconstructive process that begins by reestablishing the terminal regions of the body, and then intercalates the remaining missing structures. This order of events is also seen during regeneration of flatworms, nemerteans, echinoderms, vertebrates, and many other groups, and combines epimorphic and morphallactic processes to restore the original morphology of the injured individual (Agata, Saito, and Nakajima 2007).

Regenerative trajectories (and related fission trajectories) have many fundamental differences with embryogenesis. On the one hand, while embryogenesis always begins from a highly stereotyped and predictable point, the fertilized egg, the starting point for regeneration is variable and unpredictable. Such unpredictability is likely an important challenge, as evidenced by the fact that many groups, including vertebrates, arthropods, and annelids, have evolved specific breakpoints, and might undergo corrective autotomy (self-amputation) at those points if amputated elsewhere (Bely and Nyberg 2010; Bely 2014). On the other hand, even similar morphogenetic processes often occur with different timing between embryogenesis and regeneration. For example, muscles and neurons develop simultaneously in annelid embryogenesis, while neurons develop earlier than muscle cells during regeneration (Kozin, Filippova, and Kostyuchenko 2017). Thus, regeneration cannot be assumed to represent a simple recapitulation of embryonic development.

Prior to morphogenetic events, the regenerative process needs to amass adequate resources: cells capable of moving, proliferating, reorganizing, and differentiating into new tissues. In annelids, such cells are sourced from existing tissues. What fraction of those cells are reserve stem cells and what fraction represent dedifferentiated cells is not yet fully understood. Furthermore, such fraction is likely to vary from group to group. However, current evidence does not support a fully stem cell–based regeneration, as seen in turbellarian flatworms. Evidence for transdifferentiation of cells across germ layers is also scant: most observational and experimental evidence supports that each germ layer furnishes material for its own derivatives.

However, after the stages of wound healing and blastema formation, the morphogenetic processes of regeneration converge with those of embryogenesis, both developmentally and at the gene expression level. This makes perfect sense: millions of years of evolution of embryonic developmental pathways and processes have already generated complex and robust developmental genetic networks that can be readily deployed by activation of relatively few hub genes. After forming a blastema, the regenerative process reaches a more stereotypical, predictable point and is ready to reconstruct the missing structures by launching the developmental pathways that are also used during embryogenesis. Thus, the developmental trajectory of regeneration can be seen as a process that starts as a wound-healing response and leads to the reboot of embryonic development.

The ability of annelid regeneration to rebuild the segment formation zone and reboot the segment formation process is not only amazing on its own, but also highly fortunate, as it offers a unique window to study both conservation and evolution of the segmentation process. Breeding most species of annelids is not an easy task, and thus only a handful of species currently make good systems to study embryonic development. In contrast, many annelids can regenerate at least their posterior end (Bely 2006; Zattara and Bely 2016), providing a widely diverse set of systems for comparative studies of the developmental and genetic mechanisms underlying the formation of repeated body segments.

ACKNOWLEDGMENTS

The author is thankful to the reviewers of this manuscript for helpful comments and feedback, and to Ariel Chipman for the invitation to contribute to this book. Several pieces of data presented here were obtained during the author's graduate research at the University of Maryland, College Park, under the mentorship of Alexa Bely. Images shown in Figures 10.4 and 10.5 were acquired using a Leica SP5X laser scanning confocal microscope at the UMCP-CMNS-CBMG Imaging Core Facility. All figures are original compositions based on or modified from the sources cited in each figure legend.

REFERENCES

Aboobaker, A. A. 2011. Planarian stem cells: A simple paradigm for regeneration. *Trends Cell Biol.* 21(5): 304–311.

Agata, K., Y. Saito, and E. Nakajima. 2007. Unifying principles of regeneration I: Epimorphosis versus morphallaxis. *Dev. Growth Differ.* 49(2): 73–78.

Aguado, M. T., C. Helm, M. Weidhase, and C. Bleidorn. 2014. Description of a new syllid species as a model for evolutionary research of reproduction and regeneration in annelids. *Org. Divers. Evol.* (October): 1–21.

Alibardi, L. 2014. Histochemical, biochemical and cell biological aspects of tail regeneration in lizard, an amniote model for studies on tissue regeneration. *Prog. Histochem. Cytochem.* 48(4): 143–244.

Allen, E. J. 1923. Regeneration and reproduction of the syllid *Procerastea. Philos. Trans. R. Soc. London B Biol. Sci.* 211(382–390): 131–177.

Balarezo, M. G., E. E. Zattara, and A. E. Bely. 2011. Role of Wnt/Beta-catenin signaling pathway in post-amputation cephalic regeneration in the annelid worm *Pristina leidyi*. Paper presented at the 2011 Society for Advancement of Hispanics/Chicanos and Native Americans in Science National Conference in San Jose, CA. https://dx.doi.org/10.13140/RG.2.1.2705.0481.

Balavoine, G. 2015. Segment formation in Annelids: Patterns, processes and evolution. *Int. J. Dev. Biol.* 58(6–8): 469–483.

Baskin, D. G. 1974. The coelomocytes of nereid polychaetes. In *Contemporary Topics in Immunobiology: Volume 4 Invertebrate Immunology*, Ed. M. G. Hanna, and E. L. Cooper, 55–64. Boston, MA: Springer US.

Bely, A. E. 2006. Distribution of segment regeneration ability in the Annelida. *Integr. Comp. Biol.* 46(4): 508–518.

Bely, A. E. 2010. Evolutionary loss of animal regeneration: Pattern and process. *Integr. Comp. Biol.* 50(4): 515–527.

Bely, A. E. 2014. Early events in annelid regeneration: A cellular perspective. *Integr. Comp. Biol.* 54(4): 688–699.

Bely, A. E., and K. G. Nyberg. 2010. Evolution of animal regeneration: Re-emergence of a field. *Trends Ecol. Evol.* 25(3): 161–170.

Bely, A. E., and G. A. Wray. 2001. Evolution of regeneration and fission in annelids: Insights from *engrailed*- and *orthodenticle*-class gene expression. *Dev. Camb. Engl.* 128(14): 2781–2791.

Bely, A. E., E. E. Zattara, and J. M. Sikes. 2014. Regeneration in spiralians: Evolutionary patterns and developmental processes. *Int. J. Dev. Biol.* 58(6–8): 623–634.

Berrill, N. J. 1952. Regeneration and budding in worms. *Biol. Rev.* 27(4): 401–438.

Berrill, N. J. 1978. Induced segmental reorganization in sabellid worms. *J. Embryol. Exp. Morphol.* 47(1): 85–96.

Berrill, N. J., and D. Mees. 1936. Reorganization and regeneration in *Sabella*. I. Nature of gradient, summation, and posterior reorganization. *J. Exp. Zool.* 73(1): 67–83.

Bilello, A. A., and H. E. Potswald. 1974. A cytological and quantitative study of neoblasts in the naid *Ophidonais serpentina* (Oligochaeta). *Wilhelm Roux Arch. Entwicklungsmechanik Org.* 174(3): 234–249.

Boilly, B., Y. Boilly-Marer, and A. E. Bely. 2017. Regulation of dorso-ventral polarity by the nerve cord during annelid regeneration: A review of experimental evidence. *Regeneration* 4(2): 54–68.

Bosch, T. C. G. 2007. Why polyps regenerate and we don't: Towards a cellular and molecular framework for *Hydra* regeneration. *Dev. Biol.* 303(2): 421–433.

Brockes, J. P., and A. Kumar. 2008. Comparative aspects of animal regeneration. *Annu. Rev. Cell Dev. Biol.* 24(1): 525–549.

Bullock, T. H., and G. A. Horridge. 1965. *Structure and Function in the Nervous System of Invertebrates*. London: W.H. Freeman and Company.

Campbell, L. J., and C. M. Crews. 2008. Wound epidermis formation and function in urodele amphibian limb regeneration. *Cell Mol. Life Sci. CMLS* 65(1): 73–79.

Carlson, B. M. 2007. *Principles of Regenerative Biology*. Burlington, MA: Academic Press.

Chuang, S. H. 1994. Brachiopoda. In *Asexual Propagation and Reproductive Strategies. Vol. VI Part B*, Adiyodi K. G. and Adiyodi R. G. (eds.), 315–328. Reproductive Biology of Invertebrates. Chichester, UK: John Wiley & Sons.

Coe, W. R. 1929. Regeneration in nemerteans. *J. Exp. Zool.* 54(3): 411–459.

Cornec, J.-P., J. Cresp, P. Delye, F. Hoarau, and G. Reynaud. 1987. Tissue responses and organogenesis during regeneration in the oligochete *Limnodrilus hoffmeisteri* (Clap.). *Can. J. Zool.* 65(2): 403–414.

Denes, A. S., G. Jekely, P. R. H. Steinmetz, F. Raible, H. Snyman, B. Prud'homme, …, D. Arendt. 2007. Molecular architecture of annelid nerve cord supports common origin of nervous system centralization in Bilateria. *Cell* 129(2): 277–288.

Dray, N., K. Tessmar-Raible, M. Le Gouar, L. Vibert, F. Christodoulou, K. Schipany, …, G. Balavoine. 2010. Hedgehog signaling regulates segment formation in the annelid *Platynereis*. *Science* 329(5989): 339–342.

Drewes, C. D., and C. R. Fourtner. 1991. Reorganization of escape reflexes during asexual fission in an aquatic oligochaete, *Dero digitata*. *J. Exp. Zool.* 260(2): 170–180.

Edgecombe, G. D., G. Giribet, C. W. Dunn, A. Hejnol, R. M. Kristensen, R. C. Neves, …, M. V. Sørensen. 2011. Higher-level metazoan relationships: Recent progress and remaining questions. *Org. Divers. Evol.* 11(2): 151–172.

Emig, C. C. 1973. L'histogenèse régénératrice chez les phoronidiens. *Wilhelm Roux Arch. Für Entwicklungsmechanik Org.* 173(3): 235–248.

Eming, S. A., P. Martin, and M. Tomic-Canic. 2014. Wound repair and regeneration: Mechanisms, signaling, and translation. *Sci. Transl. Med.* 6(265): 265sr6.

Gazave, E., J. Béhague, L. Laplane, A. Guillou, L. Préau, A. Demilly, …, M. Vervoort. 2013. Posterior elongation in the annelid *Platynereis dumerilii* involves stem cells molecularly related to primordial germ cells. *Dev. Biol.* 382(1): 246–267.

Gazave, E., A. Guillou, and G. Balavoine. 2014. History of a prolific family: The Hes/Hey-related genes of the annelid *Platynereis*. *EvoDevo* 5(1): 29.

Goss, R. J. 1987. Why mammals don't regenerate—or do they? *Physiology* 2(3): 112–115.

Harty, M., A. W. Neff, M. W. King, and A. L. Mescher. 2003. Regeneration or scarring: An immunologic perspective. *Dev. Dyn.* 226(2): 268–279.

Haszprunar, G. 2016. Review of data for a morphological look on Xenacoelomorpha (Bilateria incertae sedis). *Org. Divers. Evol.* 16(2): 363–389.

Helm, C., P. Beckers, T. Bartolomaeus, S. H. Drukewitz, I. Kourtesis, A. Weigert, …, C. Bleidorn. 2018. Convergent evolution of the ladder-like ventral nerve cord in Annelida. *Front. Zool.* 15(1): 36.

Herlant-Meewis, H. 1964. Regeneration in annelids. *Adv. Morphog.* 4: 155–215.

Hill, S. D. 1970. Origin of the regeneration blastema in polychaete annelids. *Am. Zool.* 10(2): 101–112.

Hutchins, E. D., G. J. Markov, W. L. Eckalbar, R. M. George, J. M. King, M. A. Tokuyama, …, K. Kusumi. 2014. Transcriptomic analysis of tail regeneration in the lizard *Anolis carolinensis* reveals activation of conserved vertebrate developmental and repair mechanisms. *PLoS One* 9(8): e105004.

Jong, D. M. de, and E. C. Seaver. 2016. A stable thoracic Hox code and epimorphosis characterize posterior regeneration in *Capitella teleta*. *PLoS One* 11(2): e0149724.

Jong, D. M. de, and E. C. Seaver. 2017. Investigation into the cellular origins of posterior regeneration in the annelid *Capitella teleta*. *Regeneration* 5(1): 61–77.

Kostyuchenko, R. P., V. V. Kozin, N. A. Filippova, and E. V. Sorokina. 2019. FoxA expression pattern in two polychaete species, *Alitta virens* and *Platynereis dumerilii*: Examination of the conserved key regulator of the gut development from cleavage through larval life, post-larval growth and regeneration. *Dev. Dyn.* 248(8): 728–743.

Kozin, V. V., N. A. Filippova, and R. P. Kostyuchenko. 2017. Regeneration of the nervous and muscular system after caudal amputation in the polychaete *Alitta virens* (Annelida: Nereididae). *Russ. J. Dev. Biol.* 48(3): 198–210.

Kozin, V. V., and R. P. Kostyuchenko. 2015. Vasa, PL10, and Piwi gene expression during caudal regeneration of the polychaete annelid *Alitta virens*. *Dev. Genes Evol.* 225(3): 129–138.

Kragl, M., D. Knapp, E. Nacu, S. Khattak, M. Maden, H. H. Epperlein, and E. M. Tanaka. 2009. Cells keep a memory of their tissue origin during axolotl limb regeneration. *Nature* 460(7251): 60–65.

Krecker, F. H. 1923. Origin and activities of the neoblasts in the regeneration of microdrilous annelida. *J. Exp. Zool.* 37(1): 26–46.

Laumer, C. E., R. Fernández, S. Lemer, D. Combosch, K. M. Kocot, A. Riesgo, …, G. Giribet. 2019. Revisiting metazoan phylogeny with genomic sampling of all phyla. *Proc. R. Soc. B Biol. Sci.* 286(1906): 20190831.

Liebmann, E. 1942. The correlation between sexual reproduction and regeneration in a series of Oligochaeta. *J. Exp. Zool.* 91(3): 373–389.

Manylov, O. G. 2010. Regeneration in Gastrotricha—I. Light microscopical observations on the regeneration in *Turbanella sp. Acta Zool.* 76(1): 1–6.

Martinelli, C., and J. Spring. 2004. Expression pattern of the homeobox gene *Not* in the basal metazoan *Trichoplax adhaerens. Gene. Expr. Patterns* 4(4): 443–447.

Martinez, V. G., G. J. Menger III, and M. J. Zoran. 2005. Regeneration and asexual reproduction share common molecular changes: Upregulation of a neural glycoepitope during morphallaxis in *Lumbriculus. Mech. Dev.* 122(5): 721–732.

Martinez, V. G., P. Reddy, and M. J. Zoran. 2006. Asexual reproduction and segmental regeneration, but not morphallaxis, are inhibited by boric acid in *Lumbriculus variegatus* (Annelida: Clitellata: Lumbriculidae). *Hydrobiologia* 564(1): 73–86.

Maruzzo, D., and F. Bortolin. 2013. Arthropod regeneration. In *Arthropod Biology and Evolution*, Minelli A., Boxshall G. and G. Fusco (eds.), 149–169. Berlin, Heidelberg: Springer.

Mochii, M., Y. Taniguchi, and I. Shikata. 2007. Tail regeneration in the *Xenopus* tadpole. *Dev. Growth Differ.* 49(2): 155–161.

Morgan, T. H. 1901. *Regeneration.* Columbia University Biological Series. Norwood, MA: Macmillan.

Müller, M. C. M. 2004. Nerve development, growth and differentiation during regeneration in *Enchytraeus fragmentosus* and *Stylaria lacustris* (Oligochaeta). *Dev. Growth Differ.* 46(5): 471–478.

Müller, M. C. M. 2006. Polychaete nervous systems: Ground pattern and variations—cLS microscopy and the importance of novel characteristics in phylogenetic analysis. *Integr. Comp. Biol.* 46(2): 125–133.

Müller, M. C. M., A. Berenzen, and W. Westheide. 2003. Experiments on anterior regeneration in *Eurythoe complanata* ("Polychaeta", Amphinomidae): Reconfiguration of the nervous system and its function for regeneration. *Zoomorphology* 122(2): 95–103.

Müller, M. C. M., and L. Henning. 2004. Ground plan of the polychaete brain—I. Patterns of nerve development during regeneration in *Dorvillea bermudensis* (Dorvilleidae). *J. Comp. Neurol.* 471(1): 49–58.

Myohara, M. 2012. What role do annelid neoblasts play? A comparison of the regeneration patterns in a neoblast-bearing and a neoblast-lacking enchytraeid oligochaete. *PLoS One* 7(5): e37319.

Nielsen, C. 1994. Bryozoa entoprocta. In *Asexual Propagation and Reproductive Strategies. Vol. VI Part B*, Adiyodi K. G. and Adiyodi R. G. (eds.), 432. Reproductive Biology of Invertebrates. Chichester, UK: John Wiley & Sons.

Niwa, N., A. Akimoto-Kato, M. Sakuma, S. Kuraku, and S. Hayashi. 2013. Homeogenetic inductive mechanism of segmentation in polychaete tail regeneration. *Dev. Biol.* 381(2): 460–470.

Novikova, E. L., N. I. Bakalenko, A. Y. Nesterenko, and M. A. Kulakova. 2013. Expression of Hox genes during regeneration of nereid polychaete *Alitta (Nereis) virens* (Annelida, Lophotrochozoa). *EvoDevo* 4(1): 14.

Nyberg, K. G., M. A. Conte, J. L. Kostyun, A. Forde, and A. E. Bely. 2012. Transcriptome characterization via 454 pyrosequencing of the annelid *Pristina leidyi*, an emerging model for studying the evolution of regeneration. *BMC Genomics* 13(1): 287.

Özpolat, B. D., and A. E. Bely. 2016. Developmental and molecular biology of annelid regeneration: A comparative review of recent studies. *Curr. Opin. Genet. Dev.* 40: 144–153. Cell Reprogramming, Regeneration and Repair.

Özpolat, B. D., E. S. Sloane, E. E. Zattara, and A. E. Bely. 2016. Plasticity and regeneration of gonads in the annelid *Pristina leidyi*. *EvoDevo* 7: 22.

Pascale, A., M. Amadio, and A. Quattrone. 2007. Defining a neuron: Neuronal ELAV proteins. *Cell Mol. Life Sci.* 65(1): 128.

Paulus, T., and M. C. M. Müller. 2006. Cell proliferation dynamics and morphological differentiation during regeneration in *Dorvillea bermudensis* (Polychaeta, Dorvilleidae). *J. Morphol.* 267(4): 393–403.

Planques, A., J. Malem, J. Parapar, M. Vervoort, and E. Gazave. 2019. Morphological, cellular and molecular characterization of posterior regeneration in the marine annelid *Platynereis dumerilii*. *Dev. Biol.* 445(2): 189–210.

Potswald, H. E. 1972. The relationship of early oocytes to putative neoblasts in the serpulid *Spirorbis borealis*. *J. Morphol.* 137(2): 215–227.

Prud'homme, B., R. de Rosa, D. Arendt, J. F. Julien, R. Pajaziti, A. W. Dorresteijn, …, G. Balavoine. 2003. Arthropod-like expression patterns of engrailed and wingless in the annelid *Platynereis dumerilii* suggest a role in segment formation. *Curr. Biol.* 13(21): 1876–1881.

Purschke, G. 2015. Annelida: Basal groups and Pleistoannelida. In *Structure and Evolution of Invertebrate Nervous Systems*, Ed. A. Schmidt-Rhaesa, S. Harzsch, and G. Purschke, 254–312. Oxford: Oxford University Press.

Purschke, G., and M. C. M. Müller. 2006. Evolution of body wall musculature. *Integr. Comp. Biol.* 46(4): 497–507.

Randolph, H. 1891. The regeneration of the tail in *Lumbriculus*. *Zool. Anz.* 14(362): 154–156.

Randolph, H. 1892. The regeneration of the tail in *Lumbriculus*. *J. Morphol.* 7(3): 317–344.

Ribeiro, R. P., C. Bleidorn, and M. T. Aguado. 2018. Regeneration mechanisms in Syllidae (Annelida). *Regeneration* 5(1): 26–42.

de Rosa, R., B. Prud'homme, and G. Balavoine. 2005. Caudal and even-skipped in the annelid *Platynereis dumerilii* and the ancestry of posterior growth. *Evol. Dev.* 7(6): 574–587.

Sánchez Alvarado, A. 2000. Regeneration in the metazoans: Why does it happen? *Bioessays* 22(6): 578–590.

Seaver, E. C., and L. M. Kaneshige. 2006. Expression of "segmentation" genes during larval and juvenile development in the polychaetes *Capitella* sp. I and *H. elegans*. *Dev. Biol.* 289(1): 179–194.

Seo, S., J.-W. Lim, D. Yellajoshyula, L.-W. Chang, and K. L. Kroll. 2007. Neurogenin and NeuroD direct transcriptional targets and their regulatory enhancers. *EMBO J.* 26(2): 5093–5108.

Slack, J. M. 2017. Animal regeneration: Ancestral character or evolutionary novelty? *EMBO Rep.* 18(9): e201643795.

Starunov, V. V., N. Dray, E. V. Belikova, P. Kerner, M. Vervoort, and G. Balavoine. 2015. A metameric origin for the annelid pygidium? *BMC Evol. Biol.* 15(1): 25.

Stein, E. A., and E. L. Cooper. 1983. Inflammatory responses in annelids. *Am. Zool.* 23(1): 145–156.

Steinmetz, P. R., R. Urbach, N. Posnien, J. Eriksson, R. P. Kostyuchenko, C. Brena, …, D. Arendt. 2010. Six3 demarcates the anterior-most developing brain region in bilaterian animals. *EvoDevo* 1(1): 14.

Stephan-Dubois, F. 1954. Les néoblastes dans la régénération postérieure des Oligochètes microdriles. *Bull. Biol. Fr. Belg.* 88: 181–245.

Stone, R. G. 1932. The effects of x-rays on regeneration in *Tubifex tubifex*. *J. Morphol.* 53(2): 389–431.

Sugio, M., C. Yoshida-Noro, K. Ozawa, and S. Tochinai. 2012. Stem cells in asexual reproduction of *Enchytraeus japonensis* (Oligochaeta, Annelid): Proliferation and migration of neoblasts. *Dev. Growth Differ.* 54(4): 439–450.

Tadokoro, R. 2018. Reproductive strategies in Annelida: Germ cell formation and regeneration. In *Reproductive and Developmental Strategies*, Kobayashi K., Kitano T., Iwao Y. and Kondo M. (eds.), 203–221. Diversity and Commonality in Animals. Tokyo: Springer.

Tadokoro, R., M. Sugio, J. Kutsuna, S. Tochinai, and Y. Takahashi. 2006. Early segregation of germ and somatic lineages during gonadal regeneration in the annelid *Enchytraeus japonensis*. *Curr. Biol.* 16(10): 1012–1017.

Takeo, M., C. Yoshida-Noro, and S. Tochinai. 2008. Morphallactic regeneration as revealed by region-specific gene expression in the digestive tract of *Enchytraeus japonensis* (Oligochaeta, Annelida). *Dev. Dyn.* 237(5): 1284–1294.

Takeo, M., C. Yoshida-Noro, and S. Tochinai. 2010. Functional analysis of *grimp*, a novel gene required for mesodermal cell proliferation at an initial stage of regeneration in *Enchytraeus japonensis* (Enchytraeidae, Oligochaete). *Int. J. Dev. Biol.* 54(1): 151–160.

Thiemann, M., and A. Ruthmann. 1991. Alternative modes of asexual reproduction in *Trichoplax adhaerens* (Placozoa). *Zoomorphology* 110(3): 165–174.

Tsonis, P. A. 2000. Regeneration in vertebrates. *Dev. Biol.* 221(2): 273–284.

Tweeten, K. A., and A. Reiner. 2012. Characterization of serine proteases of *Lumbriculus variegatus* and their role in regeneration. *Invertebr. Biol.* 131(4): 322–332.

Tzetlin, A. B., and A. V. Filippova. 2005. Muscular system in polychaetes (Annelida). *Hydrobiologia* 535/536(1–3): 113–126.

Varhalmi, E., I. Somogyi, G. Kiszler, J. Nemeth, D. Reglodi, A. Lubics, …, L. Molnar. 2008. Expression of PACAP-like compounds during the caudal regeneration of the earthworm *Eisenia fetida*. *J. Mol. Neurosci.* 36(1): 166–174.

Vetvicka, V., and P. Sima. 2009. Origins and functions of annelid immune cells: The concise survey. *Invertebr. Surviv. J.* 6: 138–143.

Weidhase, M., P. Beckers, C. Bleidorn, and M. T. Aguado. 2016. On the role of the proventricle region in reproduction and regeneration in *Typosyllis antoni* (Annelida: Syllidae). *BMC Evol. Biol.* 16: 196.

Weidhase, M., P. Beckers, C. Bleidorn, and M. T. Aguado. 2017. Nervous system regeneration in *Typosyllis antoni* (Annelida: Syllidae). *Zool. Anz.* 269: 57–67.

Weidhase, M., C. Bleidorn, P. Beckers, and C. Helm. 2016. Myoanatomy and anterior muscle regeneration of the fireworm *Eurythoe cf. complanata* (Annelida: Amphinomidae). *J. Morphol.* 277(3): 306–315.

Weidhase, M., C. Bleidorn, and C. Helm. 2014. Structure and anterior regeneration of musculature and nervous system in *Cirratulus cf. cirratus* (Cirratulidae, Annelida). *J. Morphol.* 275(12): 1418–1430.

Weidhase, M., C. Helm, and C. Bleidorn. 2015. Morphological investigations of posttraumatic regeneration in *Timarete cf. punctata* (Annelida: Cirratulidae). *Zool. Lett.* 1: 20.

Worley, M. I., L. Setiawan, and I. K. Hariharan. 2012. Regeneration and transdetermination in *Drosophila* imaginal discs. *Annu. Rev. Genet.* 46(1): 289–310.

Yoshida-Noro, C., M. Myohara, F. Kobari, and S. Tochinai. 2000. Nervous system dynamics during fragmentation and regeneration in *Enchytraeus japonensis* (Oligochaeta, Annelida). *Dev. Genes Evol.* 210(6): 311–319.

Zattara, E. E. 2012. Regeneration, fission and the evolution of developmental novelty in naid annelids. Ph.D. Thesis, College Park, MD: University of Maryland, College Park. http://dx.doi.org/10.13140/2.1.2054.4967.

Zattara, E. E., and A. E. Bely. 2011. Evolution of a novel developmental trajectory: Fission is distinct from regeneration in the annelid *Pristina leidyi*. *Evol. Dev.* 13(1): 80–95.

Zattara, E. E., and A. E. Bely. 2013. Investment choices in post-embryonic development: Quantifying interactions among growth, regeneration, and asexual reproduction in the annelid *Pristina leidyi*. *J. Exp. Zool. B Mol. Dev. Evol.* 320(8): 471–488.

Zattara, E. E., and A. E. Bely. 2015. Fine taxonomic sampling of nervous systems within Naididae (Annelida: Clitellata) reveals evolutionary lability and revised homologies of annelid neural components. *Front. Zool.* 12(1): 8.

Zattara, E. E., and A. E. Bely. 2016. Phylogenetic distribution of regeneration and asexual reproduction in Annelida: Regeneration is ancestral and fission evolves in regenerative clades. *Invertebr. Biol.* 135(4): 400–414.

Zattara, E. E., F. A. Fernandez-Alvarez, T. C. Hiebert, A. E. Bely, and J. L. Norenburg. 2019. A phylum-wide survey reveals multiple independent gains of head regeneration in Nemertea. *Proc. R. Soc. B Biol. Sci.* 286(1898): 20182524.

Zattara, E. E., K. W. Turlington, and A. E. Bely. 2016. Long-term time-lapse live imaging reveals extensive cell migration during annelid regeneration. *BMC Dev. Biol.* 16(1): 6.

Index